A Numerical Approach to Real Algebraic Curves

with the Wolfram Language

A Numerical Approach to Real Algebraic Curves

with the Wolfram Language

BARRY H. DAYTON

Wolfram Media

Wolfram Media, Inc.

www.wolfram-media.com
Champaign, IL 61820-7237, USA
ISBN 978-1-57955-023-3 (ePub)
ISBN 978-1-57955-024-0 (paper)

Library of Congress Cataloging-in-Publication Data
Dayton, Barry H., author.
A numerical approach to real algebraic curves with the Wolfram language / Barry H. Dayton.
Champaign, IL : Wolfram Media, Inc., [2018] | Includes bibliographical references.
LCCN 2018052334 (print) | LCCN 2018053356 (ebook) | ISBN 9781579550233 (ePub) |
ISBN 9781579550240 (pbk. : alk. paper)
LCSH: Curves, Algebraic. | Wolfram language (Computer program language)
LCC QA565 (ebook) | LCC QA565 .D39 2018 (print) | DDC 516.3/52--dc23
LC record available at https://lccn.loc.gov/2018052334

Final typesetting and page production were completed using Wolfram Notebooks.

Table of Contents

Introduction

> [A solution] might be regarded as in some measure complete if it were admitted that the problem is finished with when its solution has been reduced to a finite number of feasible operations. If however the operations are too numerous or too involved to be carried out in practice the solution is only a theoretical one; and its importance then lies not in itself, but in the theorems with which it is associated and to which it leads. Such a theoretical solution must be regarded as preliminary and not the final stage in the consideration of the problem.
>
> — F. S. Macaulay, *The Algebraic Theory of Modular Systems*, 1916

Real algebraic curve theory is a mature subject, but from Macaulay's perspective, the many theoretical results fall short—we still need good ways to compute things. The objective of this book is to show how one can, using the Wolfram Language, start with a bivariate polynomial equation $f(x, y) = 0$ and actually capture useful features of the resulting curve. By "useful" we have in mind shape, size, and the like—the topology and geometry of a planar algebraic curve. Consider, for example, $f(x, y) = (x^2 + y^2)^2 - 4(x\, y)$, a figure eight curve one often encounters in second semester calculus with analytic geometry. How might one analyze such a curve? Now modify it to $g(x, y) = (x^2 + y^2)^2 - 4(x^3\, y^2)$. Do we have tools more powerful/reliable than just attempting to plot the thing? (Spoiler alert: the answer is, "Yes, but you will have to wait for Chapter 10!")

In[1]:=
```
{Labeled[ContourPlot[(x^2+y^2)^2-4xy == 0, {x, -3, 3}, {y, -3, 3}], "(x²+y²)²= 4xy"],
  Labeled[ContourPlot[(x^2+y^2)^2-4x^3y^2 == 0, {x, -3, 3}, {y, -3, 3}],
   "(x²+y²)²= 4x³y²"]}
```

Out[1]=

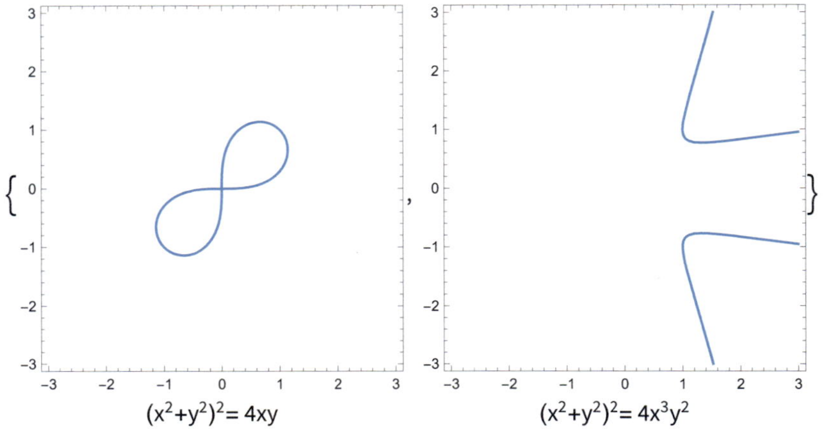

The classical approach to algebraic geometry is to consider explicitly only curves and points described exactly; in particular, only rational numbers are allowed as coefficients and point components. This practice continues in recent texts and journal articles. This has been an efficient way to proceed as it has enabled algebraic number theory and algebraic geometry to develop simultaneously, but for actually describing curves, the problem is that exact curves may have no rational points at all. Perhaps in Macaulay's era there was no good or efficient way for the ordinary mathematician to work with actual real numbers. In this century, we have seen the development of efficient numerical polynomial system solvers. With the Wolfram Language and its capability to work reliably with machine numbers, there is no longer a reason to avoid floating-point numbers. Thus we work in the very practical setting where coefficients are approximate numbers. This is somewhat an oxymoron since the left-hand side of the equation above produces machine numbers while the number zero on the right is of course exact, and machine arithmetic computations rarely give "exact" zeros. Can this be handled with any reliability? (Second spoiler alert: it is the same answer as the first.) En route, we will resolve this in two different ways, an operational description attributed to Newton and a theoretical one attributed to Gauss.

This book is partly motivated by the well-known book by Shreeram Abhyankar *Algebraic Geometry for Scientists and Engineers* (1990), which is basically a theoretical book with proofs replaced by heuristic arguments. However, there are only a few plots and most of them are inaccurate. There is little hope of actually producing visible curves from the theoretical methods described. A better approach, for those who might need to work with actual curves, is to replace the proofs with algorithms that can deliver useful results. This book is thus not written for mathematicians but, rather, for people with some understanding of calculus and partial derivatives and a basic knowledge of working with matrices who would like to develop these curves on their own.

The fact that the focus of this book is on real numbers, numerical coefficients, and algorithms does not mean that I will completely ignore the classical theory of integer coefficient equations and solution curves that live in complex projective space. We will use classical ideas when they are helpful; for example, we will work in the projective plane when convenient using fractional linear transformations (FLTs). In particular, throughout the second half of the book, curves will be considered equivalent if one can be transformed to the other by an FLT. A curve will only be called nonsingular if it has no singularities in the complex projective plane, where Bézout's theorem plays an important role. On the other hand, the Riemann–Roch theorem will only be mentioned in its topological form and then only to compare with this text's *fundamental theorem of real projective curves*. As a companion to this book for the person who wants to know more about curves from the classical point of view, I recommend the excellent book *Plane Algebraic Curves* by Gerd Fischer (2001).

A main feature of this book is that it gives several representations of an algebraic plane curve. From a simply topological point of view, in the projective case, this book's *fundamental theorem* says the topology can be represented by an Euler graph, which is a graph such that each component is an Eulerian cycle. From a geometric point of view, we will plot curves on a representation of the projective plane using a pinched Möbius band to explain the difference between an oval and a pseudo-line. Finally, we will introduce diamond diagrams to try and capture both the global geometry and topology of a curve in a simple plot.

Most of all, this book is computable in the sense that it is printed from an executed Wolfram Notebook that will be made available to readers. The reader can then run and even modify the content to see that it actually works. Unlike other books in mathematics that may add in some code to illustrate the math, in this book, the code is an integral part of the content.

Chapter Overview and Usage Note

For all of the algorithms that are not built in to the Wolfram Language, I give the full code. The goal is to show the actual working of the algorithm, not to produce production-quality code. Ideally, the Wolfram Language is rich enough that one can do most calculations in four lines of code or less. That is not always the case here, but effort has been made to keep it short and understandable to the user without extensive explanation. The reader who is unfamiliar with the Wolfram Language should see Stephen Wolfram's book *An Elementary Introduction to the Wolfram Language* (2017) available at wolfram.com/eiwl.

Even when an algorithm has an obvious extension for more than two variables, the code in this book is given for two variables only for simplicity. Each chapter of this book has a companion notebook for the reader who has access to the Wolfram Language so they can try out the code immediately. Rather than build in all possible options, as would happen with production code, the reader should be able to see how to modify the code to achieve inputs and outputs in a more desirable form. That said, the user of this code needs to be aware that unlike built-in functions, these do not check for correct input or check for all possible special cases that could occur, so one could possibly end up with error messages.

Chapter 1: Basic notation,terminology, and examples of plane algebraic curves are given. In particular, two families of curves, *Gaussian curves* and *Newton hyperbolas*, which will be used as examples throughout the book, are introduced. Parametric curves are also considered.

Chapter 2: For the reader who is not familiar with using machine numbers, we give a short introduction and basic rules of working with these numbers. The operational meaning of a machine-number point on a curve is given. Construction of bivariate polynomials is discussed, leading to an interpolation algorithm.

Chapter 3: Regular points, singular points, critical points, and infinite points are introduced and discussed. Tangent lines are calculated. An algorithm to find the closest point on a curve from a given point in $\mathbb{R}^2$ is given.

Chapter 4: Some concepts of topology are introduced/reviewed, leading to Gauss's perspective on real plane curves. Various algorithms for path tracing are developed. The concept of a graph of a curve is introduced.

Chapter 5: The real and complex projective planes are defined. The intersection multiplicity of two curves is discussed and the important Bézout's theorem is stated. Singular points are counted. Our fundamental theorem of projective plane curves is given and compared with the classical Riemann–Roch theorem.

Chapter 6: Fractional Linear Transformations (FLTs) are introduced. A first application is the calculation of all complex projective singular points of a curve as well as the calculation of all complex projective intersection points of a pair of curves.

Chapter 7: FLTs are applied to well-known curves. Asymptotes and inflection points are calculated. Conics are classified, constructed, and parameterized several ways. Cubics are numerically reduced to Weierstrass standard form and j-invariants are given. Cauchy's theorem on the parameterization of rational curves is given and applied to the hypocycloid.

Chapter 8: Projective real curves are plotted on the hemisphere and Möbius band models of the projective plane. Ovals and pseudo-lines are defined via the Möbius band model.

Chapter 9: Harnack's theorem is given along with a discussion of the first part of Hilbert's 16th problem. Diagrams of numerical curves are given in the spirit of Desargues, Viro, and Gauss.

Chapter 10: Two curves mentioned earlier are analyzed by the methods of this book.

Appendix 1: Basic concepts of real numerical linear algebra are introduced/reviewed using singular value decomposition. Sylvester matrices are defined and used to find the GCD and square-free reduction. Macaulay matrices are introduced and used to calculate intersection multiplicity. Several other applications of numerical linear algebra to curve theory are given.

Appendix 2: All the functions used in more than one chapter are collected in one list.

Usage Note

You can find the notebook version of this book online at wolfram-media.com or barryhdayton.space. These notebooks are fully interactive, and the reader is encouraged to experiment with the code. Many of the functions provided were developed specifically for this book, and definitions for them have been included both within the text (where appropriate) and in Appendix 2, which must be evaluated before each new session in order for the code in the book to work properly.

1 | Introduction to Numerical Plane Curves

The current American mathematics curriculum provides beginning math students with few examples of plane algebraic curves other than lines and simple quadratics. Even advanced math students who are able to take a course in algebraic geometry get few actual examples of plane curves, even if one of the many available books on curves is used. Instead, the emphasis is on developing the big abstract theorem, and most texts have few, if any, illustrations. For this reason, I will start with a number of examples so the reader can have an idea of what this book is about.

First, however, we start with a definition. A *real affine algebraic plane curve* is a real solution set of a bivariate polynomial equation $f[x, y] = 0$.

Historically, the word "real" appeared in mathematics about 100 years after the word "imaginary" was introduced by Cardano describing $\sqrt{-1}$, "real" meaning "not imaginary." Likewise the word "affine" has come to mean "not projective" or "not infinite."

Some bivariate real polynomials are (input and display):

```
In[1]:= f = 3 x^3 - 2 x y + y^2 - 3 x + 1
```

Out[1]= $1 - 3x + 3x^3 - 2xy + y^2$

```
In[2]:= g = 1/9 x^2 - 1/4 y^2 - 1
```

Out[2]= $-1 + \frac{x^2}{9} - \frac{y^2}{4}$

```
In[3]:= h = N[Sqrt[7] x - Sqrt[1/5] y - Sqrt[26]]
```

Out[3]= $-5.09902 + 2.64575 x - 0.447214 y$

The first two are *exact* because the coefficients are integers or rational numbers; the third, although derived from a symbolic polynomial, is numerical because the symbolic square roots have been converted to machine numbers.

Note that by putting the cursor on the output and again hitting enter, we get:

```
In[4]:= -5.0990195135927845` + 2.6457513110645907` x - 0.4472135954999579` y

Out[4]= -5.09902 + 2.64575 x - 0.447214 y
```

Unlike the previous display, which only displays the first few digits, now we can see the full representation of the numbers in the software. We will discuss this fully in Chapter 2, along with what is actually meant when we have an equation of the form $f = 0$ where f is this sort of a polynomial. This book will work mainly with this latter type of bivariate polynomial rather that the first two types seen in classical books on curve theory.

A polynomial may be multiplied by a constant without changing the solution set, but for technical reasons, we may wish to distinguish between f and $-f$ in some cases.

1.1 Linear Curves

The easiest algebraic curves, straight lines, are a basic staple of beginning algebra. Children are taught that an equation of the form $y = b + mx$ or $ax + by = c$ has as a solution set a straight line. For our purposes, we will prefer the more general formulation $f[x, y] = ax + by - c = 0$.

This equation defines an algebraic curve:

```
In[5]:= f = RandomReal[{-1, 1}, 3].{x, y, 1}

Out[5]= -0.118975 + 0.490659 x - 0.358513 y
```

Then generate a contour plot of *f* as a function of *x* and *y*: [f, {x, -10, 10}, {y, -10, 10}]:

```
In[6]:= ContourPlot[f == 0, {x, -10, 10}, {y, -10, 10}]
```

Out[6]=

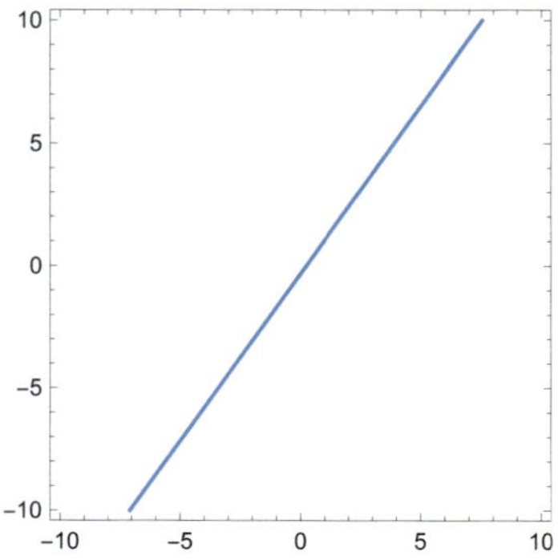

One thing almost everyone knows is that two points determine a straight line. As an example of the spirit of this book, we show how one should find the equation of a line through two points, $p = \{p_1, p_2\}$ and $q = \{q_1, q_2\}$.

We want an equation $ax + by + c = 0$ with at least one of a, b, c not zero. It will become clear why only $c \neq 0$ is acceptable. We need to satisfy the system:

$$\begin{aligned} a\,p_1 + b\,p_2 + c &= 0 \\ a\,q_1 + b\,q_2 + c &= 0 \end{aligned}$$

This is undetermined, so we need another equation. First, we add a dummy variable z for 1 multiplying c, and then a third equation is created, giving us Equation 1:

$$\begin{aligned} a\,p_1 + b\,p_2 + c\,z &= 0 \\ a\,q_1 + b\,q_2 + c\,z &= 0 \\ a\,r_1 + b\,r_2 + c\,r_3 &= 1 \end{aligned}$$

r_1, r_2, r_3 are random real numbers, so the system will give a unique solution. This would be less likely if we used small integers or rational numbers.

Suppose we want to find the line through points {2, 5} and {4, 1}. We could do it in the Wolfram Language:

```
In[7]:= e3 = RandomReal[{-1, 1}, 3].{a, b, c}

Out[7]= 0.0942885 a + 0.763923 b + 0.3196 c
```

```
In[8]:= sol = {a, b, c} /. NSolve[{2 a + 5 b + c, 4 a + b + c, e3 - 1}]

Out[8]= {{-1.03955, -0.519777, 4.67799}}
```

This returns the equation $l = 0$:

```
In[9]:= l = sol[[1]].{x, y, 1}

Out[9]= 4.67799 - 1.03955 x - 0.519777 y
```

However, this is not what the answer section of your standard textbook would show. One may normalize it by:

```
In[10]:= l = Expand[l/Coefficient[l, y]]

Out[10]= -9. + 2. x + 1. y
```

This equation looks neater, but the Wolfram Language would be just as happy with the former in subsequent calculations.

However, our method does give extra dividends. Suppose instead that we want the equation of the line through p with slope m; instead of evaluating $\{x, y, z\}$ with $\{p_1, p_2, 1\}$, we use $\{1, m, 0\}$.

In our example above, instead of the second point, we want slope m = −2 so a = 2 b:

```
In[11]:= e3 = RandomReal[{-1, 1}, 3].{a, b, c}

Out[11]= 0.261515 a + 0.908787 b - 0.268392 c
```

```
In[12]:= sol = {a, b, c} /. NSolve[{2 a + 5 b + c, a - 2 b, e3 - 1}]

Out[12]= {{0.519839, 0.259919, -2.33928}}
```

```
In[13]:= l = sol[[1]].{x, y, 1}

Out[13]= -2.33928 + 0.519839 x + 0.259919 y
```

```
In[14]:= l = Expand[l/Coefficient[l, y]]

Out[14]= -9. + 2. x + 1. y
```

Note that if we have a point far away on the line, say $x = 1000$, then $y = -1991$, so the triple $\{x, y, 1\}$ is approximately $1000\{1, -2, 0\}$.

This suggests that we think of the slope $\{1, m, 0\}$ as being a point on the line but far away, which we call an *infinite point*.

A third way of thinking about a line is parametrically, that is

$$\{x, y\} = \{p_1, p_2\} + t\{v_1, v_2\}.$$

The vector $v = \{v_1, v_2\}$ is called the direction vector or the *tangent vector*, but again, in Equation 1, just use $\{v_1, v_2, 0\}$ as the second point.

In our example above, we want the line through p with tangent vector $\{1, 2\}$, so we use $\{1, 2, 0\}$.

Now the tangent vector also becomes an infinite point on our line, but any nonzero multiple of a tangent vector is also a tangent vector, so in the special case of infinite points, they are homogeneous. However, if we think of a finite point, or as we will call it, an affine point, in the form $\{p_1, p_2, 1\}$, then for the purposes of Equation 1, a constant nonzero multiple by k, $k\{p_1, p_2, 1\} = \{\mathrm{kp}_1, \mathrm{kp}_2, k\}$ works just as well. We will call any one of these equivalent forms the *homogeneous form* of the point. In Chapter 5, we will see that these homogeneous points, affine or infinite, make up the *projective plane*.

With the code for the line function, you can enter points as either pairs or triples (For a complete list of global functions, see Appendix 2):

```
In[15]:= line[p_, q_, x_, y_] := Module[{e1, e2, a, b, c, M},
           e1 = Switch[Length[p], 2, Append[p, 1], 3, p];
           e2 = Switch[Length[q], 2, Append[q, 1], 3, q];
           M = {e1, e2, RandomReal[{-1, 1}, 3]};
           ({a, b, c} /. (Solve[M.{a, b, c} == {0, 0, 1}][[1]])).{x, y, 1}
         ];
```

```
In[16]:= l = line[{2, 5}, {1, -2, 0}, x, y]

Out[16]= 1.70402 - 0.378671 x - 0.189336 y
```

```
In[17]:= Expand[l/Coefficient[l, y]]

Out[17]= -9. + 2. x + 1. y
```

This all seems very complicated, but Equation 1 gives us a general way of constructing lines, and we will see later in Chapter 2 that we can also use it to construct higher-degree curves. In the process, we have discovered the projective plane, which will be important later. The one disadvantage is that even for a simple, exact case, we get the equation in a decimal form, but that is precisely what we are allowed and encouraged to do in this book.

1.2 Conics

Besides lines, high-school algebra students learn about simple quadratic curves, the parabola, hyperbola, and ellipse.

Axes → True simply turns the axes on:

```
In[18]:= Row[
           {ContourPlot[y == 1/4 x^2 - 2, {x, -6, 6}, {y, -4, 8}, Axes → True, ImageSize → Small],
             ContourPlot[y^2 - 1/4 x^2 == 1, {x, -6, 6}, {y, -4, 4},
               Axes → True, ImageSize → Small], ContourPlot[y^2 + 1/4 x^2 == 1,
               {x, -3, 3}, {y, -3, 3}, Axes → True, ImageSize → Small]}]
```

Out[18]=

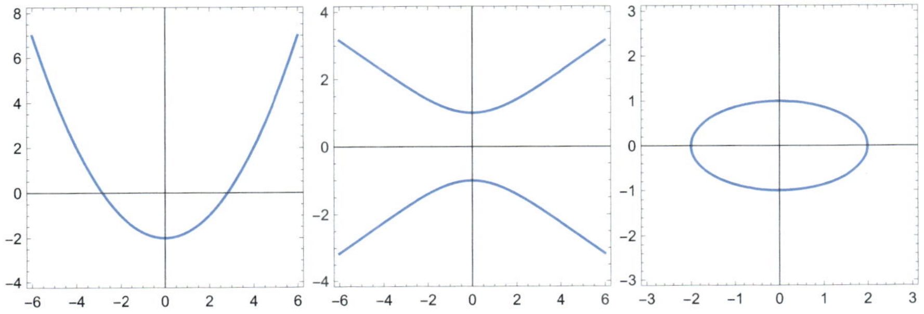

More generally, most curves of the form $a\,x^2 + b\,x\,y + c\,y^2 + d\,x + e\,y + k + x^2 = 0$ will be hyperbolas, parabolas, or ellipses:

In[19]:=
```
ContourPlot[-4-5x+x^2+4y-2xy+3y^2 == 0, {x, -3, 8}, {y, -4, 4}, Axes → True]
```

Out[19]=

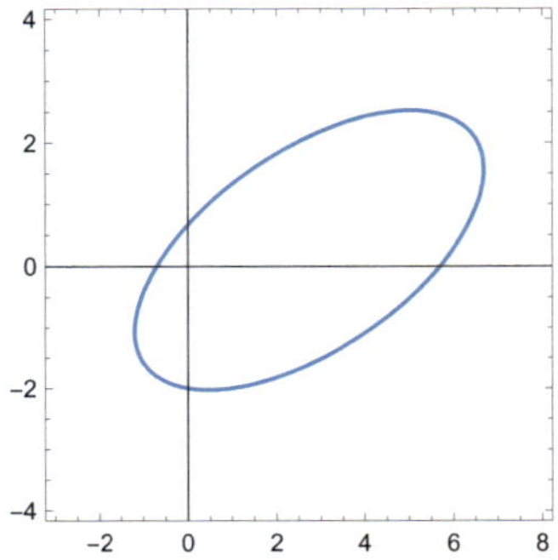

In past years, more advanced students were expected to transform conics, like the one above, into the form of a simple conic. One can rid the equation of the x, y terms by completing the square and then translating it to get the *quadratic form* $a\,b\,x\,y + c\,y^2 + x^2$. This can be rotated to a simple conic form using trigonometry or, better but more advanced, the *principal axis theorem* of linear algebra. In Chapter 7, we will learn some more pleasant ways to transform conics to simple conics or even to circles, depending on the group of transformations you want to allow.

There are some exceptions. For example:

In[20]:=
```
{ContourPlot[y^2-x^2 == 0, {x, -4, 4}, {y, -4, 4}, Axes → True],
  ContourPlot[x^2-2xy+y^2 == 1, {x, -3, 3}, {y, -3, 3}, Axes → True],
  ContourPlot[x^2-2xy+y^2 == 0, {x, -3, 3}, {y, -3, 3}, Axes → True]}
```

Out[20]=

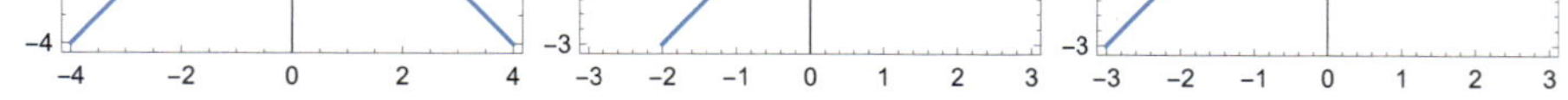

In these cases, we get a pair of lines. In the right-hand plot, the lines are actually the same. In the middle plot, the lines are parallel, that is, they have the same slope $m = 1$, so by the previous section, they meet at the infinite point $\{1, 1, 0\}$. We call these curves *singular*, and in the left picture, the point where they meet is called a *singularity*. We will deal with this concept in Chapter 3.

Make numerical hyperbolas, and use ContourStyle to assign colors to certain lines:

```
In[21]:= f = -1.` - 0.43301270189221946` x^2 + 0.5` x y + 0.43301270189221946` y^2
Out[21]= -1. - 0.433013 x^2 + 0.5 x y + 0.433013 y^2

In[22]:= l1 = 0.` + 0.8583270645009377` x - 1.4866660852270714` y
Out[22]= 0. + 0.858327 x - 1.48667 y

In[23]:= l2 = 0.` - 0.9830458468618561` x - 0.5675617843114374` y
Out[23]= 0. - 0.983046 x - 0.567562 y

In[24]:= ContourPlot[{f == 0, l1 == 0, l2 == 0}, {x, -6, 6},
           {y, -6, 6}, ContourStyle → {Blue, Dashed, Dashed}]
```

Out[24]=

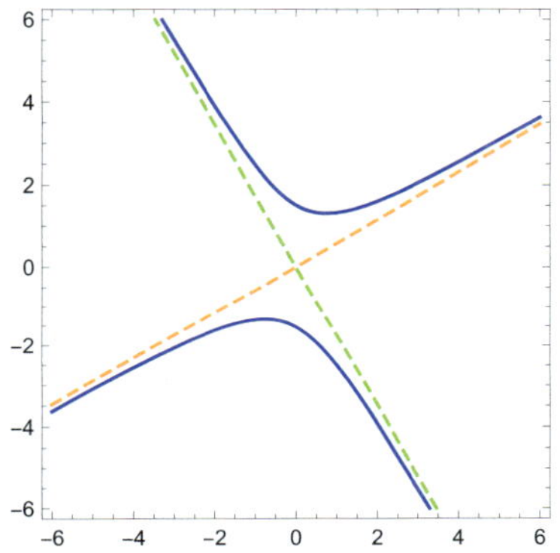

In the plot, we have a numerical hyperbola and two lines that closely approximate the hyperbola when both x, y are large. These are called asymptotes. We will see these again in the next section, but they will reappear in greater detail in Chapter 7.

1.3 Cubics

Many people who have gone further in mathematics, including PhD mathematicians, have never seen algebraic curves of a higher degree, other than function curves like $y = f(x)$, than cubics. Even the cubics they have seen are mostly of the form $f(x, y) = y^2 - (a\,x^2 + b\,x + c + x^3) = 0$. Nonsingular curves of this form are called *elliptic curves*. These form a major topic in mathematics, and especially *number theory*, so I recommend searching the Wolfram Language & System Documentation Center, Wolfram|Alpha, Wolfram.com, or Wikipedia for *elliptic curves* or the related *elliptic integrals*. Among other things, elliptic curves played an important role in Andrew Wiles's 1996 solution of Fermat's conjecture, known by many as Fermat's last theorem.

Here is the Wolfram|Alpha page for elliptic curves:

```
Elliptic Curve »
Input interpretation
EntityClass["PlaneCurve", "Elliptic"]
```

Input interpretation:

```
EntityClass["PlaneCurve", "Elliptic"]
```

elliptic curves

Named curves:

```
{"Burnside curve", "Mordell elliptic curve", "Ochoa elliptic curve",
 "semicubical parabola"}
```

Burnside curve | Mordell elliptic curve | Ochoa elliptic curve | semicubical parabola

Cartesian equation: More

```
Dataset[EntityValue[EntityClass["PlaneCurve", "Elliptic"],
  EntityProperty["PlaneCurve", "CartesianEquation"],
  "EntityAssociation"]]
```

Burnside curve	$y^2 = x(x^4 - 1)$
Mordell elliptic curve	$y^2 = n + x^3$
Ochoa elliptic curve	$y^2 = x^3 - 440\,067\,x + 106\,074\,110$
semicubical parabola	$y^2 = a^2\,x^3$

Related entities:

```
Dataset[EntityValue[EntityClass["PlaneCurve", "Elliptic"],
  {EntityProperty["PlaneCurve", "FilledRegion"],
   EntityProperty["PlaneCurve", "RelatedWolframLanguageSymbols"],
   EntityProperty["PlaneCurve", "AssociatedPeople"]},
  "EntityPropertyAssociation"]]
```

	associated people
Burnside curve	William Burnside
Mordell elliptic curve	Louis Joel Mordell
Ochoa elliptic curve	Juan Ochoa Melida
semicubical parabola	William Neile | John Wallis | Hendrik van Heuraet | Gottfried Leibniz | Christiaan Huygens

Typical examples of elliptic curves are:

```
In[25]:= {ContourPlot[y^2 == x^3+3 x^2-3 x+4, {x, -5, 5}, {y, -15, 15}],
          ContourPlot[y^2 == -12 x+x^2+x^3, {x, -5, 7}, {y, -15, 15}]}
```

Out[25]=

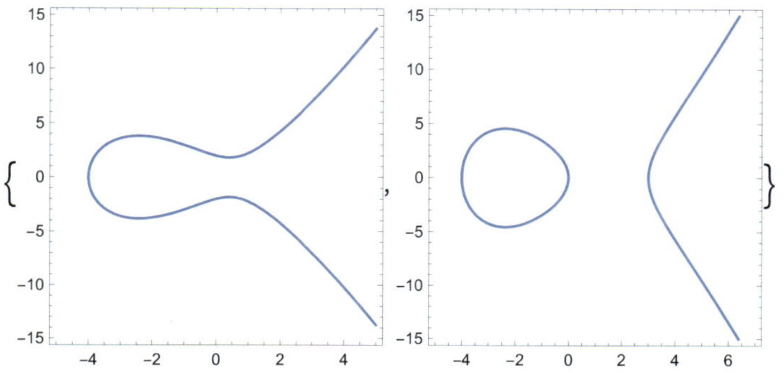

Less well remembered are Issac Newton's explorations with real cubics, see for example Struik (1986). Newton was largely interested in asymptotes, so he called his curves hyperbolas and numbered them from 1 to 72. In honor of Newton, I have established a large catalog of curves with asymptotes using his ideas. I call these *Newton Hyperbolas*, and there are 960 cubics, numbered 64 through 1023, among these curves. There are over 268 million Newton hyperbolas, just 960 are of degree 3.

If you are reading this in a Wolfram Notebook, the code for the global function newtonHyperbola is in the closed cell above, be sure to run it first. Once that is evaluated, the syntax for newtonHyperbola is.

```
In[27]:= newtonHyperbola[836, x, y]
```

Out[27]= $1+x+0.2x^2-0.008x^3-y+xy+0.2x^2y+0.2y^2-0.2xy^2-0.008y^3$

Some examples showing where the Newton hyperbola equation is positive are:

```
In[28]:= Row[{RegionPlot[newtonHyperbola[376, x, y] > 0, {x, -25, 25}, {y, -25, 25},
            ImageSize → Small], RegionPlot[newtonHyperbola[413, x, y] > 0,
            {x, -25, 25}, {y, -25, 25}, ImageSize → Small], RegionPlot[
            newtonHyperbola[836, x, y] > 0, {x, -25, 25}, {y, -25, 25}, ImageSize → Small]}]
```

Out[28]=

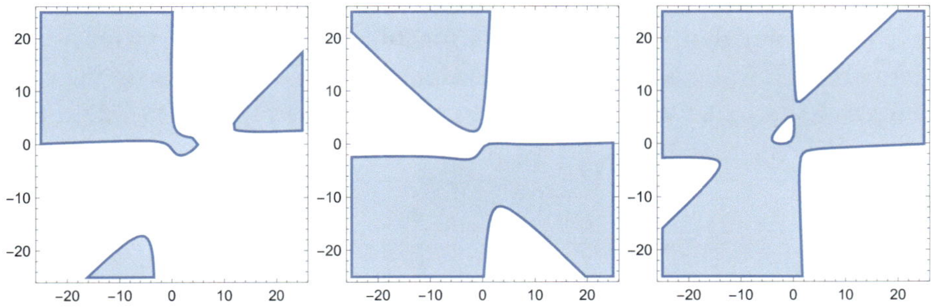

We will connect these Newton hyperbolas to the twentieth-century work by Oleg Viro in Chapter 9 (2006). In particular, we can represent these curves by Gauss diagrams, which represent the topology of both the curve and positive region of the equation (brown dots).

Gauss curves represented as Newton hyperbolas:

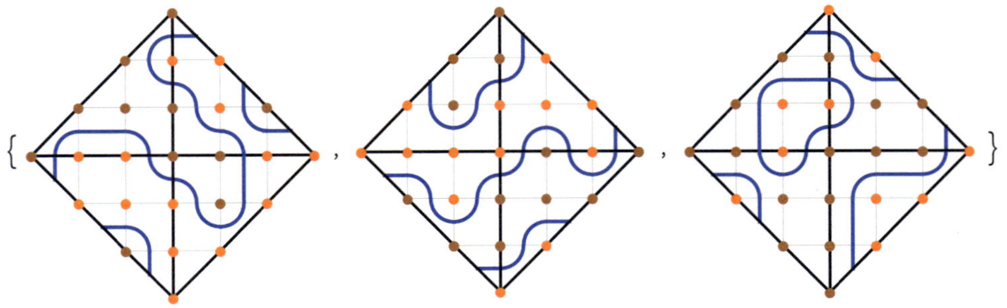

There are also singular cubics. The equation $y^2 = f[x]$, where $f[x]$ is a polynomial of one variable, is called, in general, *hyperelliptic*. If the polynomial $f[x]$ is a cubic with a multiple real zero at $x = r$, then the resulting curve will have a singularity at $\{r, 0\}$.

Here is a look at three singular cubics:

In[29]:=
```
Row[{ContourPlot[y^2 == x^3+2x^2, {x, -3, 3}, {y, -5, 5}, ImageSize → Small],
    ContourPlot[y^2 == x^3, {x, -1, 3}, {y, -5, 5}, ImageSize → Small],
    ContourPlot[x^3+y^3 == 3xy, {x, -2, 2}, {y, -2, 2}, ImageSize → Small]}]
```

Out[29]=

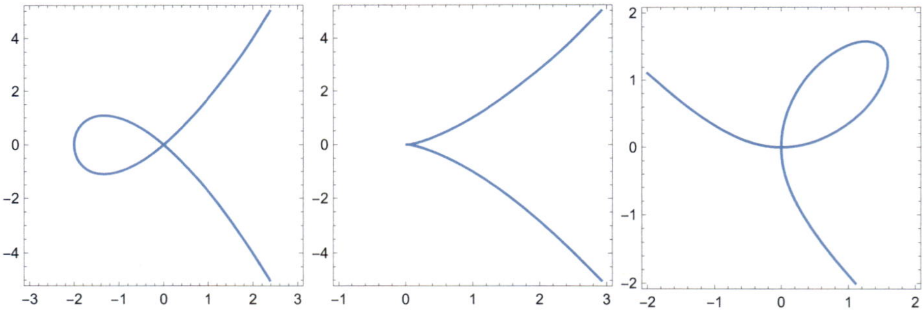

The left-hand singularity is called a *node*; the middle singularity is called a *cusp*. Other cubics may also be singular, such as the folium of Descartes on the right, which also has a node. We will explain in Chapter 5 why an irreducible cubic can have only one singularity.

Here is a look at reducible cubics, which are cubics where the equation factors can have two or three singularities:

```
In[30]:= {ContourPlot[x y (x+y-1) == 0, {x, -2, 2}, {y, -2, 2}],
  ContourPlot[(x^2+y^2-1) (x-y) == 0, {x, -2, 2}, {y, -2, 2}]}
```

Out[30]=

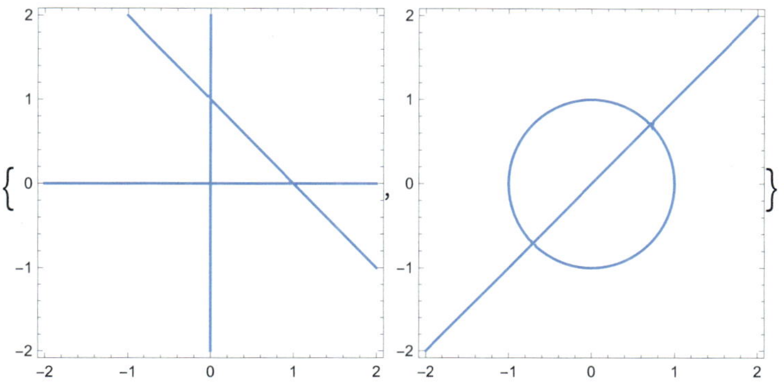

In Chapter 7, I show that each irreducible cubic can be reduced to an equivalent hyperelliptic form. This is well known for nonsingular cubics in theory, but using my method, it actually can be done for almost any cubic, including those with numerical coefficients.

1.4 Higher Degree

Linear curves are of degree 1, conics are of degree 2, and cubics are of degree 3. Some people call curves of degree 4 biquadratics; others call them *quartics*. Otherwise, we simply say that the curve has degree d. Specifically, the *total degree* of a monomial $x^i\, y^j$ is $i + j$. The total degree of a polynomial is the highest total degree of a monomial with a nonzero coefficient.

As you have seen, quite a bit is known about algebraic curves of degree 3 or lower. There is still a lot to discover about higher-degree curves. We will give some well-known examples in this section. Much of the rest of this book will be concerned with methods to represent higher-degree curves, especially when a contour plot is not sufficient.

We start with a few named biquadratics:

```
In[31]:= Row[{Labeled[ContourPlot[x^4+4 x y+y^4 == .25,
      {x, -2, 2}, {y, -2, 2}, ImageSize → Small], "Peanut Curve"],
   Labeled[ContourPlot[x^4+y^4 == 1, {x, -2, 2}, {y, -2, 2}, ImageSize → Small],
    "Fermat Curve"], Labeled[ContourPlot[(x^2+y^2/4-1) (x^2/4+y^2-1) == -.04,
      {x, -3, 3}, {y, -3, 3}, ImageSize → Small], "Klein Curve"]}]
```

Out[31]=

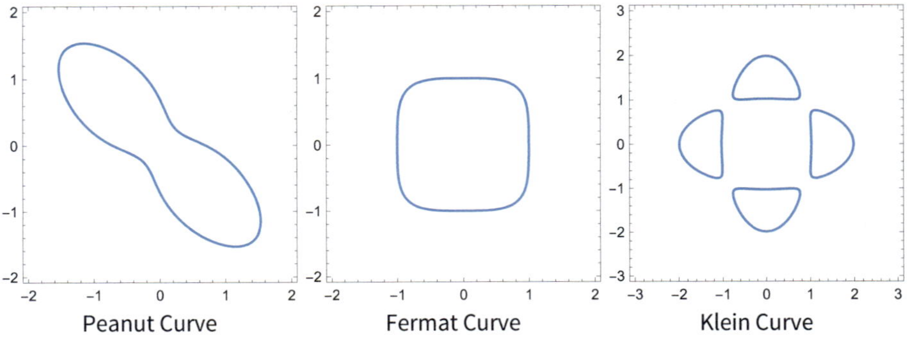

Actually, "peanut curve" is my own name; this curve is similar to certain *Cassini ovals*.

Some singular curves are:

In[32]:=
```
Row[{Labeled[ContourPlot[(x^2-4) (x-2)^2+(y^2-4)^2 == 0,
      {x, -4, 4}, {y, -4, 4}, ImageSize → Small], "Bicuspid Curve"],
   Labeled[ContourPlot[x^4+4 x y+y^4 == 0, {x, -2, 2}, {y, -2, 2},
      ImageSize → Small], "Lemniscate"],
   Labeled[ContourPlot[ 3 (x^2+y^2)^2+8 x (3 y^2-x^2)+6 (x^2+y^2) == 1,
      {x, -2, 2}, {y, -2, 2}, ImageSize → Small], "Hypocycloid"]}]
```

Out[32]=

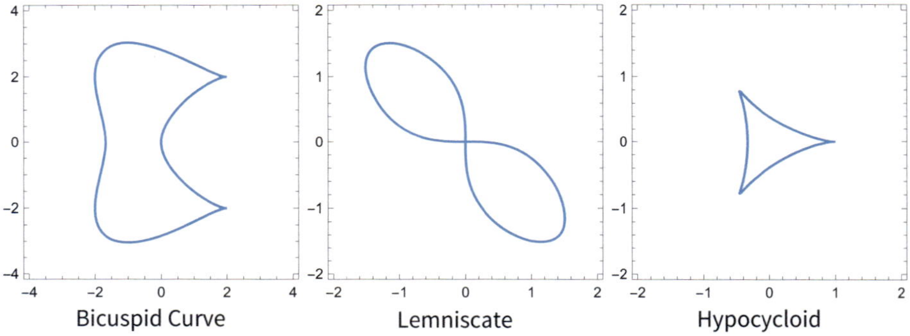

Three more singular curves are plotted below. The first two curves do not look singular but, in fact, are at {0, 0}. The more interesting part of these singularities is hidden in the complex plane. The third curve has an isolated, hence singular, point at {0, 1, 0}. In fact, for technical reasons such as not allowing reducible curves to be nonsingular, a curve will be called *nonsingular* only if there are no infinite or complex singular points as well as no affine real singular points. We will go into more detail in Chapters 3, 5, 6, and 8.

Here are more singular curves:

In[33]:=
```
Row[{Labeled[ContourPlot[x^4+x^2 y^2+y^4 == x (x^2+y^2),
      {x, -.5, 1.5}, {y, -1, 1}, ImageSize → Small], "Bean Curve"],
   Labeled[ContourPlot[y^3 == x^6, {x, -4, 4}, {y, -2, 6}, ImageSize → Small],
     "Smooth Singular Curve"],
   Labeled[ContourPlot[6+9 x^2-8 x^4-7 y+9 x^2 y-3 y^2-8 x^2 y^2 == 0,
      {x , -3, 3}, {y, -4, 2}, ImageSize → Small], "Heart Curve"]}]
```

Out[33]=

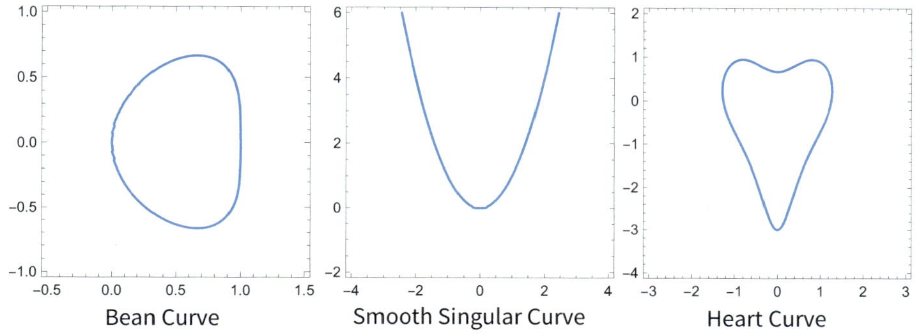

Here are some curves of higher degree (code not given here):

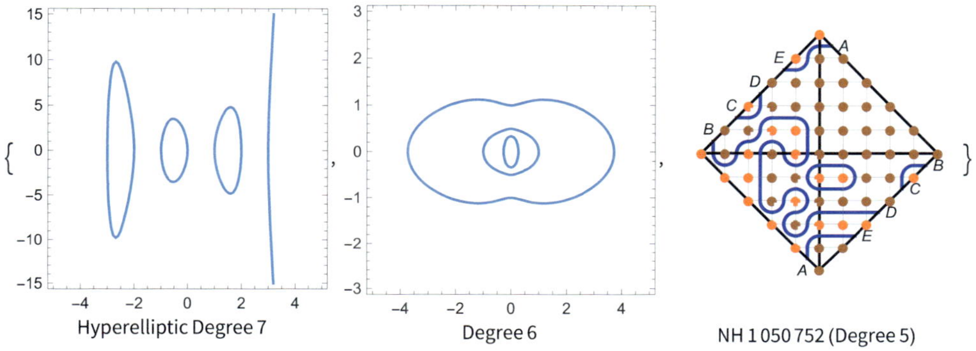

The last curve is given as a Gauss diagram. This curve of degree 5 is discussed in Chapters 8 and 9 and cannot be properly represented by a contour, or similar, plot. The letters label infinite points, and the brown dots indicate where the defining polynomial is positive.

The same can be plotted for a whole family of curves described by Gauss in his 1799 dissertation, now called *Gauss curves*. We start with a univariate complex polynomial in z, possibly with complex coefficients, and use the following procedure to replace z with $x + iy$, expand it, and take the formal real part, which is the monomials with a real coefficient.

For this Gauss curve, *p* is a polynomial in *z* with possibly complex coefficients. ComplexExpand assumes that variables *x*, *y* are real:

```
In[34]:= gaussCurve[p_, x_, y_, z_] := ComplexExpand[Re[p /. {z → x + I * y}]];

In[35]:= gaussCurve[z^2 + 2 z - 1, x, y, z]

Out[35]= -1 + 2 x + x^2 - y^2

In[36]:= gaussCurve[z^2 + I z + (2 - 3 I), x, y, z]

Out[36]= 2 + x^2 - y - y^2
```

Here are some region plots of Gauss curves; note that Gauss explicitly allowed the case where the polynomial $p[z]$ had multiple zeros. These produce singularities in the Gauss curve; however, in the right-hand plot, not all singularities come from multiple zeros of $p[z]$. We show RegionPlot because Gauss was specifically interested in the region where the Gauss curve was positive. Note also that the middle curve is Gauss's original example for his fourth proof of the fundamental theorem of algebra in 1849.

RegionPlot makes and shows the region in which pred is true (pred, {x, −10, 10}, {y, −10, 10}):

```
In[37]:= g1 = gaussCurve[(z^2+I) (z^2-2I) z, x, y, z]
```

Out[37]= $2x + x^5 + 3x^2 y - 10x^3 y^2 - y^3 + 5x y^4$

```
In[38]:= g2 = gaussCurve[z^5-5z^4+9z^3-5z^2, x, y, z]
```

Out[38]= $-5x^2 + 9x^3 - 5x^4 + x^5 + 5y^2 - 27x y^2 + 30x^2 y^2 - 10x^3 y^2 - 5y^4 + 5x y^4$

```
In[39]:= g3 = gaussCurve[(z-I)^3 (z+I)^2, x, y, z]
```

Out[39]= $x + 2x^3 + x^5 + 4x y + 4x^3 y - 6x y^2 - 10x^3 y^2 - 4x y^3 + 5x y^4$

```
In[40]:= Row[{RegionPlot[g1 > 0, {x, -7, 7}, {y, -7, 7}, ImageSize → Small],
           RegionPlot[g2 > 0, {x, -7, 7}, {y, -7, 7}, ImageSize → Small],
           RegionPlot[g3 > 0, {x, -5, 5}, {y, -5, 5}, ImageSize → Small]}]
```

Out[40]=

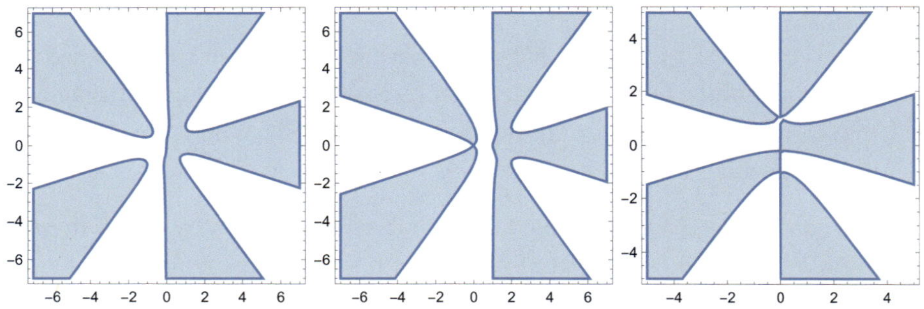

In the right-hand plot, there are three singular points, {0, 1}, {0, − 1}, and {0, −0.2}. The first two come from multiple zeros of $g3$; the last does not. We will discuss Gauss curves more in Chapter 4.

1.5 Rational Functions

Actually, you have seen higher-degree algebraic curves before. For instance, the graph of the polynomial function $y - p[x] = 0$ is an algebraic curve, but the curve properties are simple, and there is no advantage in thinking of them in curve terms. Graphs of rational functions $y = \frac{p[x]}{q[x]}$ can also be thought of as algebraic curves in the form $y\, q[x] - p[x] = 0$.

Some examples showing asymptotes using code developed later in the book are:

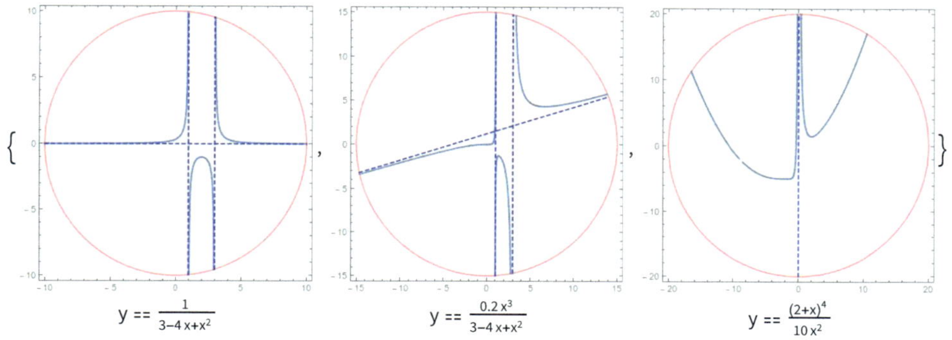

The first has two vertical asymptotes and one horizontal one. As mentioned earlier, the slope of a vertical line can be entered using our line function as {0, 1, 0}, which we interpreted as saying that {0, 1, 0} is a point at infinity, but then both vertical asymptotes pass through that point, so {0, 1, 0} becomes a singular point. The second plot has the same two vertical asymptotes but also a third asymptote called, in high-school algebra, a *slant asymptote*. The right-hand plot shows one vertical asymptote, but here is a case where the curve does not approach the asymptote at both ends, again making the infinite point {0, 1, 0} singular. However, we show in Chapter 7 that this rational function also approaches {0, 1, 0} as $x \to \pm\infty$, making that singularity even worse! We will see later that rational function curves of degree 2 or higher are always singular.

Do not confuse *rational function curves* with *rational curves*; every rational function curve is a rational curve, but the converse is not always true. We will discuss rational curves in the next section.

1.6 Parametric Curves

Aside from algebraic curves, there are also parametric curves. A *parametric curve* is the image of a continuous function $F[t] = \{f_1[t], f_2[t]\}$ of one real variable taking values in the affine plane, possibly augmented by infinite points. This function may have a domain that is a bounded open or closed interval in the line or perhaps the entire real line. The study of parametric curves with differentiable parameterization is part of *differential geometry*, while *algebraic geometry* includes the the study of algebraic curves. Curves in *topology* are required to have only continuous parameters. The connection between these ideas is one of the key ideas in this book:

Every real plane algebraic curve is the union of continuously parameterized curves.

Since we are only requiring the parameterizing functions to be continuous, we will not, in general, be able to explicitly describe the parameterizing functions, but we will be able to approximate these curves well by giving as many points on them as desired.

Consider, as an example, the conic $f = -x^2 + y^2 - 1 = 0$. It is the union of two parametric curves:

```
In[41]:= f = y^2-x^2-1;
         F1 = {t, Sqrt[t^2+1]};
         F2 = {t, -Sqrt[t^2+1]};
```

In both cases, the parameter domain is the whole real line:

```
In[44]:= {ContourPlot[f == 0, {x, -4, 4}, {y, -4, 4}],
          ParametricPlot[{F1, F2}, {t, -4, 4}, PlotRange → 4, Frame → True, Axes → False]}
```

Out[44]=

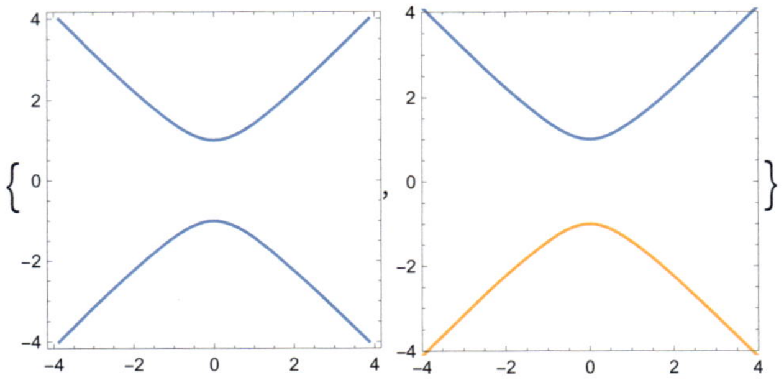

Note above that the Wolfram Language gives the same point sets, but on the right, it treats the two parameterized curves as separate, with different default coloring. In the second case, the two colors represent different *topological components*, while the first picture shows one *algebraic component*. This is because there is no algebraic equation with a full real solution set: just one of the topological components on the right. For that to happen, these equations would have to be factors of the original equation f, but since f is of degree 2, the factors would have to be of degree 1, which are lines, but the two topological curves on the right are clearly not lines.

A different case is the following cubic curve, $f = x^3 - x^2 y + x y^2 - x - y^3 + y$, which we have seen before:

```
In[45]:= {ContourPlot[{(x^2+y^2-1) == 0, (x-y) == 0}, {x, -2, 2}, {y, -2, 2}],
          ParametricPlot[{{t, t}, {Cos[t], Sin[t]}}, {t, -Pi, Pi},
           PlotStyle → Blue, PlotRange → 2, Frame → True, Axes → False]}
```

Out[45]=

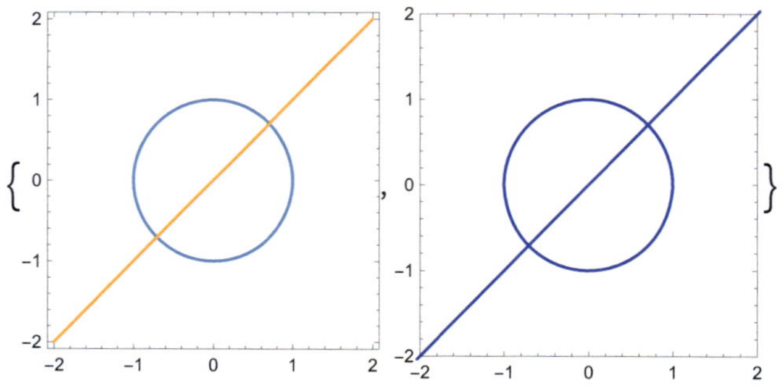

In this case, the curve has two algebraic components because f factors as $f = (x^2 + y^2 - 1)(x - y)$, but for the parametrization on the right, there is only one topological component because, as a point set, it is connected. For those not familiar with topology, see the first section of Chapter 4.

In general, if a curve is defined with an exact (integer or rational number coefficients) polynomial equation, then the Wolfram Language's `Factor` can decide if the curve can be factored into integer factors or not. If a curve can be factored, possibly with non-rational or complex coefficients, we call the curve *reducible*; if not, we call it *irreducible*. The Wolfram Language does not have built-in factoring for numerical polynomials, but we will see that, as in the previous example, reducible curves always have singular points, although possibly complex. Using these singular points, we can often visually guess the decomposition, possibly using the interpolation algorithm, covered in the next section, and `nDivide`, covered in Appendix 1, to verify.

A problem one might run into with reducible curves would be the following degree 5 curve, which should have the same point set as the example above:

```
In[46]:= g = Expand[(x^2+y^2-1)^2 (x-y)]
```

Out[46]= $x - 2x^3 + x^5 - y + 2x^2 y - x^4 y - 2xy^2 + 2x^3 y^2 + 2y^3 - 2x^2 y^3 + xy^4 - y^5$

Attempting a contour plot, we get:

```
In[47]:= ContourPlot[g == 0, {x, -2, 2}, {y, -2, 2}]
```

Out[47]=

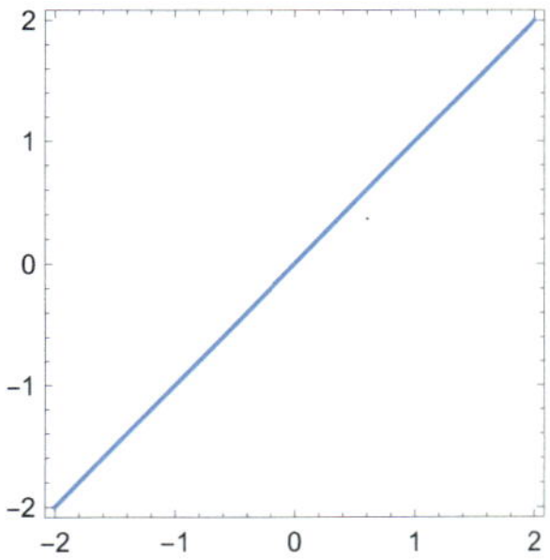

The entire algebraic component defined by $c = x^2 + y^2 - 1 = 0$ is missing. `ContourPlot` works using Gauss's definition of a curve by finding where the sign of $c[x, y]$ changes, but since c occurs squared in the equation of f, the sign of $f[x, y]$ does not change near c. Since we will be using Gauss's idea even in our more sophisticated plotting in Chapter 4, we want to avoid polynomials like g that have a factor that is raised to the second or higher power. Again, if the equation is given exactly, as is g, then often we can use the Wolfram Language's `Factor` to see if there is any repeated factor and discard all repetitions. Unlike factoring, which is hard and theoretically impossible for numerical polynomials, it is easy to find and delete repeating factors numerically to get a square-free polynomial; for that, use `sqFree`, created for this book, which works with both exact and numerical polynomials.

sqFree is explained in Appendix 1, which discusses numerical linear algebra in more detail. However, make sure you evaluate Appendix 2 before experimenting with any of the math in this chapter:

In[46]:=
```
gs = sqFree[g, x, y, 1.*^-12]
```

Out[46]= $-0.108604\,x+0.108604\,x^3+0.108604\,y-0.108604\,x^2\,y+0.108604\,x\,y^2-0.108604\,y^3$

In[48]:=
```
gs = Expand[gs/Coefficient[gs, x^3]]
```

... Power: Infinite expression $\frac{1}{0}$ encountered.

Out[48]= ComplexInfinity

Now, you will notice we have f, except now coefficients are treated as numerical quantities.

For the rest of this book, we will assume all curves are given by square-free polynomials.

If a curve is reducible, then each factor of a square-free defining polynomial defines a sub-curve, which will be called an *algebraic component* of the curve. Topological components, which we will also call *connected components*, will be defined in Chapter 4.

We end this section by looking at several more curves that can be described as parametric or algebraic curves.

The bow curve where t ≠ 0, is given parametrically by:

In[49]:=
```
Fbow = {(-1+t^2)/t^3, (-1+t^2)/t^4}
```

Out[49]= $\left\{\frac{-1+t^2}{t^3}, \frac{-1+t^2}{t^4}\right\}$

We will show in Chapter 2 that this parameterized curve is contained in the algebraic curve fbow = 0, which defines the same affine point set:

In[50]:=
```
fbow = x^4-x^2y+y^3;
```

This curve is quite difficult to plot. Below, we show three plots, ParametricPlot, RegionPlot, and a region plot without boundary, with a plot from Chapter 4:

In[51]:=
```
Row[{Show[ParametricPlot[Fbow, {t, -20, -.1}, PlotRange → .4, Axes -> False,
        Frame → True], ParametricPlot[Fbow, {t, .1, 20}], ImageSize → Small],
    RegionPlot[x^4-x^2y+y^3 < 0, {x, -.4, .4}, {y, -.4, .4}, ImageSize → Small],
```

Show[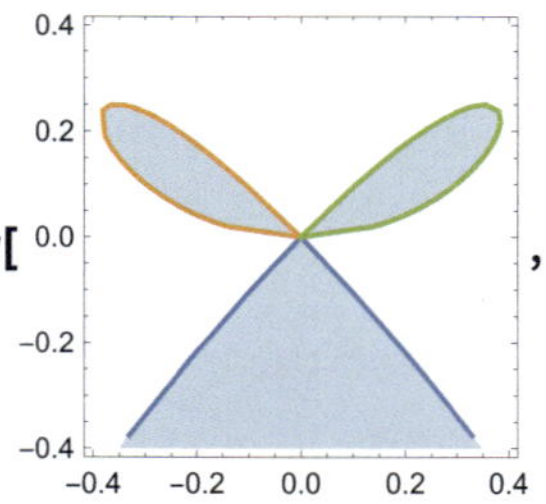 , ImageSize → Small]}]

Out[51]=

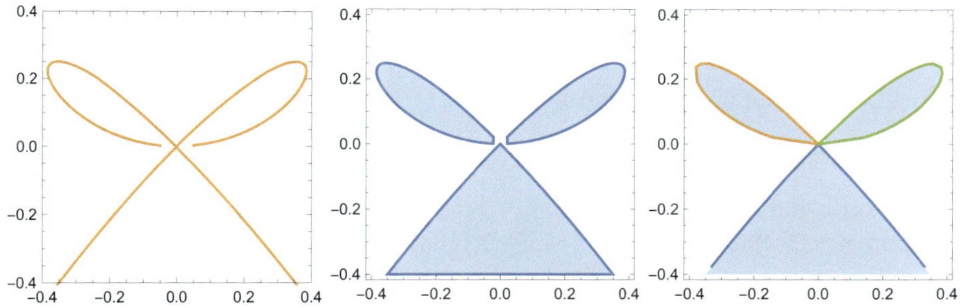

In general, an algebraic curve that can be locally parameterized by a pair of rational functions of one variable is called a *rational curve*. In Chapter 7, we will show that conics are rational curves. Singular cubics are rational curves. In Chapter 5, we will show that an irreducible curve of degree d has at most $\frac{(d-1)(d-2)}{2}$ singular points. A classical theorem of algebraic geometry says that rational curves have many or complicated singularities. Conversely, if a curve has the maximal number of singularities for that degree, as seen above, then it is rational, but there is no simple method to find the rational parametrization.

In Chapter 3 of Fischer (2001), he gives several examples of complicated algebraic curves derived from parametric curves.

The degree *d* Chebyshev curve is given parametrically in the Wolfram Language by:

```
In[52]:= Cheb[d_] := ParametricPlot[{ChebyshevT[d - 1, t], ChebyshevT[d, t]},
           {t, -1, 1}, Axes → False, Frame → True, ImageSize → Small];
```

```
In[53]:= Row[{Cheb[4], Cheb[5], Cheb[6]}]
```

Out[53]=

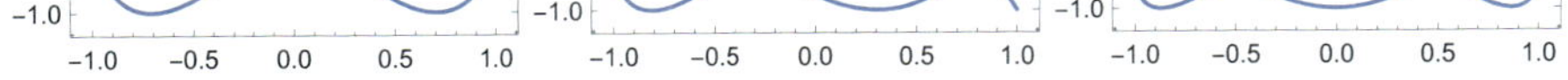

Fischer (2001) shows that **Cheb[d]** is algebraic of degree d with $\frac{(d-1)(d-2)}{2}$ singularities, which, as mentioned, is the maximum number allowed for an irreducible curve of degree d. For example, the equation $8x^4 - 8x^2 - 4y^3 + 3y + 1 = 0$ has three singularities, the maximum for a degree 4 curve.

Another family of interesting parametric curves is the rose curves given by:

```
In[54]:= Rose[k_] := {Cos[t] Cos[k t], Sin[t] Cos[k t]};
```

k can be any constant real number. This gives an algebraic curve for each rational number k, but it is not algebraic for irrational k. For more detail, see mathworld.wolfram.com/Rose.html.

Use ParametricPlot to show these beautiful curves:

In[55]:=
```
{ParametricPlot[Rose[3], {t, −2 Pi, 2 Pi}, Frame → True, Axes → False],
  ParametricPlot[Rose[2/3], {t, −3 Pi, 3 Pi}, Frame → True, Axes → False]}
```

Out[55]=

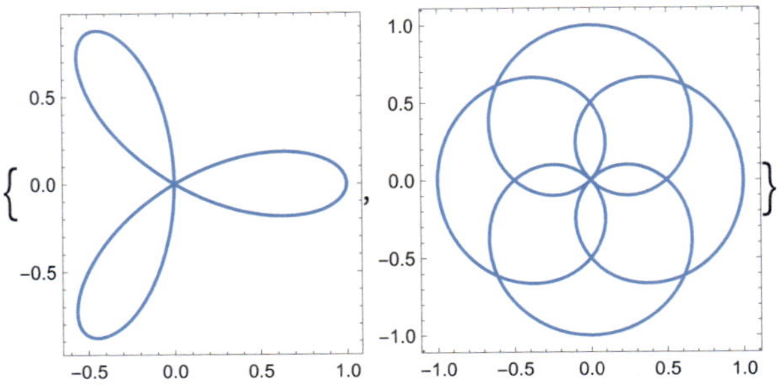

Q&A

Here are two thought questions to be answered later.

In Shreeram Abhyankar's book *Algebraic Geometry for Scientists and Engineers* (1990), he gives the following picture of a quartic C_4 surrounded by a conic Q_2 in the plane:

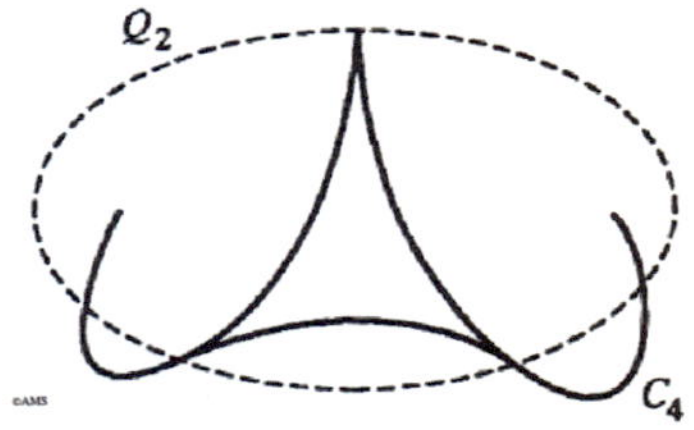

This is meant only to be illustrative, not literal. It is too bad he did not think to use the Chebyshev quartic for C_4. Give at least two reasons why C_4 is not an actual plot of a square-free real plane algebraic curve. This question will be answered in Chapters 4 and 5.

A plot of Newton Hyperbola 336941 is shown below:

In[56]:=
```
ContourPlot[newtonHyperbola[336941, x, y] == 0, {x, −20, 20}, {y, −20, 20}]
```

Out[56]=

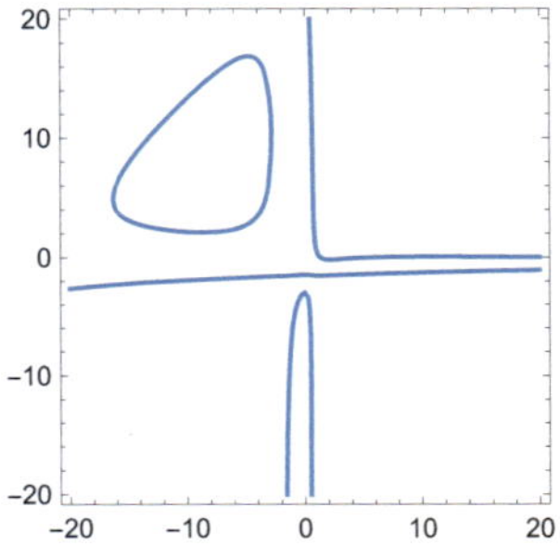

We will see later that there is obviously missing information about the full curve. Without further plotting or calculations, give several possible ways this curve could be completed to the entire affine (or projective) plane. Chapter 10 gives the answer to this question.

More to Explore

For more on Gaussian functions, go to Wolfram MathWorld (wolfr.am/wbbQVTSY)

If you'd like to learn more about Fermat's last theorem, try Wolfram|Alpha (wolfr.am/wbbT5ADJ)

Try R. Dickau's fractal curves Demonstration at the Wolfram Demonstrations Project for some fun with Hilbert and Moore curves you can adjust (wolfr.am/vt3x0r9n)

Per my mention of Cassini ovals, try E. J. O. Martos's Demonstration (wolfr.am/vt3CIAjn)

2 | Classical vs. Numerical Perspective

The split between exact and numerical mathematics dates back 3,500 years. As urbanization began in Egypt and Mesopotamia, there was a need for mass production of essential commodities like bread and beer. Proportional reasoning was needed to get the recipes right. For more in-depth information, see George Gheverghese Joseph's book *The Crest of the Peacock* (1991). The Egyptians solved this problem using very awkward *unit fractions*, which persisted in Europe through the Early Middle Ages. In Mesopotamia, a base-60 number system was easily modified by adding place values for $\frac{1}{60}$, $\frac{1}{3600}$, ... etc. They were aware that a quotient, such as $1 \div 7$, did not have a finite representation, but the Mesopotamians used their numbers for practical things like manufacturing bread and beer and building houses and roads; measurement was not exact anyway, so a sexagesimal place or two was sufficient.

It was Pythagoras, ironically a shaman, who saw in numbers a way to demonstrate nonphysical truths. He took the well-known Mesopotamian right triangle theorem $a^2 + b^2 = c^2$ used in engineering and made it into a Diophantine problem by searching for integer solutions. For an accurate account of the life and work of Pythagoras, read Walter Burkert's book *Lore and Science in Ancient Pythagoreanism* (1972). Pythagoras, who was no doubt aware of Mesopotamian mathematical proficiency, realized that it was the Egyptian numbers that were *true*.

Plato and Aristotle, who had little good to say about Pythagoras, liked this idea that mathematics could be repurposed to illustrate truth by logic and that Mesopotamian numbers were not acceptable for this purpose. Classical mathematics thus began with Euclid, a follower of Plato who eliminated all numbers by using only geometrical concepts and Aristotelian logic.

To be fair, while Mesopotamian mathematics produced many good algorithms, iterative methods for solving quadratic and cubic equations that had noninteger solutions, it is the classical method that allowed for proof and the development of abstract mathematics. The two views of mathematics have persisted to this day; often we call them *pure* and *applied*, but that is not an accurate distinction. It is more a matter of whether we need an actual answer that might not be a rational number or we just want an existence or structural theorem. In the journals of the applied mathematics world, like SIAM or ACM, you will rarely come across a decimal number.

The mathematicians who have been the most influential are often those who are able to work in both worlds, such as Archimedes, Newton, and Gauss. You will see the latter two names often in this book. Keep in mind as you read forward that often numerical methods are available only because we have the classical insight to go on.

As you will see, it is the numerical approach that leads to many simple but effective algorithms. In this book, our goal is to write simple code of a few lines each, and the numerical approach will allow us to do that. However, if your goal is to develop black box algorithms, algorithms that answer difficult situations with no human input, the classical method is preferable, but expect thousands of lines of code rather than a few. The longest code in this book is the non-numerical black box code for diamond diagrams (`dDiagram`), which, with constants and subroutines, takes about four notebook pages, and even then, desired options are omitted.

The reader who is knowledgeable about numerical methods and the Wolfram Language may skim the rest of this chapter, but there may be some things that are slightly non-standard that should be noted. In particular, I explain what is meant by the solution of $f[x, y] = 0$ in the numerical case.

2.1 Numbers in the Wolfram Language

One reason I like the Wolfram Language is that it treats numeric data in a sophisticated way but takes care of the details, so for the most part, the user does not need to pay attention to the details of how it is done. The documentation is quite complete, so I will just give an overview.

Numbers can be exact, symbolic, or numeric. Exact numbers are integers or rational numbers, and arithmetic, addition, subtraction, multiplication, division, and exponentiation with an integer exponent all preserve exactness.

Mathematica can display numbers as far as your computer's CPU can calculate:

```
In[1]:= 34+2/3
```

$$\text{Out[1]=}\ \frac{104}{3}$$

```
In[2]:= (3/17-7/19)^7
```

$$\text{Out[2]=}\ -\frac{3\,521\,614\,606\,208}{366\,790\,143\,213\,462\,347}$$

Symbolic numbers are given by a predefined symbol or letters e or E, π or pi or an expression using exact or symbolic numbers, such as Sqrt[2] or Sqrt[e^4].

The Wolfram Language will try to simplify, but the answer will always return as exact or symbolic:

```
In[3]:= a = Log[e^7]
Out[3]= 7
```

```
In[4]:= Sqrt[e^4]
Out[4]= e^2
```

For example, the function Solve will return exact/symbolic answers to exact/symbolic equations if it can:

```
In[5]:= sol1 = x /. Solve[x^2 - Pi x^2 + E x - 7 == 0, x][[1]]
```

Out[5]= $-\frac{i\left(-i e+\sqrt{-28-e^2+28\pi}\right)}{2-2\pi}$

The output sol1 is now a symbolic number.

The Wolfram Language also has *approximate* decimal numbers of two types, *machine numbers* and *arbitrary-precision numbers*. In this book, we will be satisfied with using machine numbers, which are faster and accurate enough for our needs. However, the reader who wants to extend our results to curves above degree 6 may need to consider using arbitrary-precision numbers and rewriting much of the code.

Any number entered with a decimal point is considered, by default, to be a machine number. The *precision* of a machine number is the constant MachinePrecision, which may be dependent on your hardware.

To find the machine precision for your machine, use $MachinePrecision:

```
In[6]:= $MachinePrecision
Out[6]= 15.9546
```

For technical reasons, a number may be 53 base-2 bits, which is the approximate number of decimal digits available, so what that means is that we can expect about 16-digit precision.

The user must be particularly aware of the difference between the display number and the actual machine number it represents. The display number typically uses only four or five significant digits (leading or trailing zeros are not considered significant), and it is rounded. Internally, Mathematica stores it as a full machine-precision number. If you put your cursor at the end of the numerical output and evaluate the cell, you will get the full precision output.

Here is an example of a display number:

```
In[7]:= a = Sqrt[2.0]

Out[7]= 1.41421
```

When you evaluate the output, you get the machine-precision result ending in `:

```
1.4142135623730951`
```

One of the basic rules for using numerical numbers is *never manually type in a display output that is being used in a calculation, including in other software.* This is part of the more general rule *never round in the middle of a calculation.*

When combining machine numbers with exact or symbolic numbers, the result is returned as a machine number:

```
In[8]:= 5.12*4*Pi

Out[8]= 64.3398
```

One may convert exact or symbolic numbers to machine numbers using the function N:

```
In[9]:= {N[1/7], N[sol1]}

Out[9]= {0.142857, 0.63464 + 1.69288 ⅈ}
```

A machine number should never be converted back to an exact number in the context of this book. In particular, never use the function Rationalize in our context.

Remember that *equals* is not a concept in numerical mathematics, but the relations $<$ and $>$ are allowed. In particular, machine numbers are never equal to zero, even if they display as such. In certain cases, an algorithm is designed to return zero or perhaps a zero coefficient or a zero imaginary part of a complex number. In this case, judicious use of Chop is acceptable.

Using the Wolfram Language function Chop, zero out elements that are very close to 0:

```
In[10]:= InverseFourier[Fourier[{20, 10, 5, .1, 0}]]

Out[10]= {20., 10., 5., 0.1, 7.94411×10^-16}

In[11]:= Chop[%]

Out[11]= {20., 10., 5., 0.1, 0}
```

On the other hand, the central point of this book is understanding the solution of a bivariate polynomial equation, even when the polynomial has numerical coefficients. Exact zero is not a machine number. What we mean by $f[x, y] = 0$ is a central issue in this book and will be discussed in the next two sections.

Certain Wolfram Language built-in functions work fine with numerical data, such as CoefficientRules, Expand, NSolve, and FindRoot. If the bivariate polynomial is exact, like in the previous example, one could use Factor, but it can't be used with numerical polynomials. In general, trying to factor numerical polynomials is tricky, but in principle, numerical polynomials do not factor, so neither the Wolfram Language nor I give a numerical factoring procedure.

Another simple example from the documentation, a multivariate polynomial:

```
In[12]:= Clear[a, x, y];
         Factor[2 x^3 y - 2 a^2 x y - 3 a^2 x^2 + 3 a^4]
```

Out[13]= $(a - x)(a + x)(3a^2 - 2xy)$

Note that only integer factors are found by default:

```
In[14]:= Factor[-2 + x^2 + 2 x y + y^2]
```

Out[14]= $-2 + x^2 + 2xy + y^2$

```
In[15]:= a = (x + y + N[Sqrt[2]])
         b = (x + y - N[Sqrt[2]])
```

Out[15]= $1.41421 + x + y$

Out[16]= $-1.41421 + x + y$

```
In[17]:= Expand[a b]
```

Out[17]= $-2. + x^2 + 2xy + y^2$

In Chapter 1, I do have a reliable procedure to decide if a numerical polynomial is reduced (no repeated factors) or to attempt to find greatest common divisors. Built-in functions such as Simplify or Reduce should not be used on numerical polynomials. In fact, most of the polynomial algebra functions in the Wolfram Language should not be used on numerical polynomials. Mathematica is perhaps the only computer algebra software where computations like GroebnerBasis using numerical polynomials are allowed and do, in fact, work reasonably well, but I avoid these methods in this book. Some numerical methods I like for working with polynomial systems are discussed in Appendix 1.

2.2 Iteration and Newton's Method

We often need to decide if a machine number is a zero of a polynomial, but since zero is not a machine number, evaluating the polynomial at that number may display zero but will not give zero.

For example:

In[18]:=
```
f = x^4+2 x^3-x-.2;
x0 = -0.7813887111276197`;
f /. {x → x0}
```

Out[20]= 2.22045×10^{-16}

The value returned is called the residue. A large residue would indicate that *x*0 is not a solution, but there is no general rule as to when a residue is small enough to be a solution.

A better way to identify a zero of a polynomial equation is to produce that zero by an iterative solution method.

Specifically, we will illustrate using Newton's method, which is both simple and effective at identifying solutions to numerical equations. Actual software may use somewhat different iterative methods. Historically, what we now call Newton's method was actually discovered by Thomas Simpson; Newton himself used a method dating back to Mesopotamia. Simpson named it after his mathematical hero, Newton. The argument as to whether Newton or Raphson is the discoverer of the modern Newton method is moot since neither used this method. Prior to the use of electronic computers, Newton's method was frowned upon partly because it was not exact and thus not suitable for mathematics but also because it works best with complex initial points, and mathematicians did not have the confidence that Newton's method would work to grind through the hand calculations of the complex division in sufficient precision.

In the late 40s, Uspensky wrote about Graeffe's method, which is another numerical but not iterative method using logarithms, which "is the only practical method for [the] computation of imaginary roots" (Uspensky, 1948). With computers, it is now easy to do a lot of experiments to convince ourselves that the complex Newton's method almost always works, but a theoretical analysis is not good since, in theory, there are, for degree 3 and higher, an infinite number of initial points where Newton's method does not work (see Section 4.1).

Newton's method in one variable works as follows. Start with a polynomial $f[z] = a_n z^n + a_{n-1} z^{n-1} + \ldots + a_1 z + a_0$, using the variable z because we may want to work with complex numbers even though the a_j are real. Pick a random initial value p_0, preferably complex, and then iteratively define $p_{j+1} = p_j - \frac{f[p_j]}{f'[p_j]}$ where $f'[p]$ is the derivative of $f[z]$ at p_j.

A simple version of the Newton method using *n* iterations is given below. This will not be used later, so it is not initialized. To use it, execute the code:

```
In[21]:= newton1a[f_, p0_, n_, z_] := Module[{p},
           p = N[p0];
           Do[Print[p]; p = p - (f/D[f, z]) /. {z → p}, {i, n}];
           p]
```

For example:

```
In[22]:= f = -1 - z - 0.2` z^2 + 0.008` z^3 + 0.000064` z^4
```

Out[22]= $-1 - z - 0.2\,z^2 + 0.008\,z^3 + 0.000064\,z^4$

```
In[23]:= p0 = RandomComplex[{-10 - 10 I, 10 + 10 I}]
```

Out[23]= 3.89632 − 8.17423 *i*

```
In[24]:= newton1a[f, p0, 10, z]
```

3.89632 − 8.17423 *i*

1.67568 − 3.4634 *i*

−0.174985 − 1.44594 *i*

−1.09542 − 0.604915 *i*

−1.44255 − 0.181376 *i*

−1.45537 − 0.00314766 *i*

−1.43532 + 0.0000803439 *i*

$-1.43557 + 2.41792 \times 10^{-8}\, i$

$-1.43557 + 9.75833 \times 10^{-16}\, i$

−1.43557 + 0. *i*

Out[24]= −1.43557 + 0. *i*

As noted earlier, in the modern use of Newton's method, an important point is that the stopping criteria is not that the residue $f[p]$ is small but that the change in p is small.

A more sophisticated version of Newton's method would then be (newton1 and newton1a are functions listed in Appendix 2 and must be evaluated first):

```
In[25]:= newton1[f_, p0_, z_] := Module[{p1, p2},
           p1 = N[p0];
           Chop[Catch[Do[p2 = p1 - (f/D[f, z]) /. {z → p1};
                 If[Abs[p1 - p2] < 1.*^-14, Throw[p2]];
                 p1 = p2, {30}]]]
          ]
```

```
In[26]:= p0 = RandomComplex[{-10 - 10 I, 10 + 10 I}]
```

Out[26]= −6.85991 + 2.77851 *i*

In[27]:= **z1 = newton1[f, p0, z]**

Out[27]= −2.97309

In[28]:= **NumberForm[z1, 16]**

Out[28]//NumberForm= −2.973085523660411

The polynomial we used is numerical since at least one coefficient has a decimal point, and even though these decimals are equivalent to rational numbers, they are, by default, treated as machine numbers.

We say the number z1 = −1.435566053338458 is a zero of *f*[*z*] or a solution of *f*[*z*] = 0 even though the residue is not exactly zero:

In[29]:= **f /. {z → z1}**

Out[29]= 1.66533×10^{-16}

If there is a question whether a result of Newton's method with a complex starting point is real, one can take the real part of the answer to one of the Newton programs and run it from that starting point. If the procedure comes close to obtaining the same result, then it is likely a real number since a Newton function with a real function and a real initial point stays real.

Here we use newton1a to test this:

In[30]:= **z1 = newton1a[f, p0, 8, z]**

−6.85991 + 2.77851 *i*

−4.67167 + 1.49594 *i*

−3.54672 + 0.74031 *i*

−3.04594 + 0.289781 *i*

−2.93756 + 0.0383781 *i*

−2.97283 − 0.00192792 *i*

$-2.97308+6.52544\times10^{-7}\,i$

$-2.97309-2.20046\times10^{-12}\,i$

Out[30]= $-2.97309-1.16476\times10^{-23}\,i$

In[31]:= **z2 = Re[z1]**

Out[31]= −2.97309

In[32]:= **z3 = newton1a[f, z2, 5, z]**

```
-2.97309
-2.97309
-2.97309
-2.97309
-2.97309
```

Out[32]= -2.97309

We can conclude this is a real number.

In practice, rather than using either of the Newton functions I created and used perviously, one would use the built-in function FindRoot[f, {z, z0}], which is an implementation of Newton's method.

To get all of the zeros at once, use NSolve[f], or use NSolve[f, z, Reals] to get just the real solutions:

In[33]:= **NSolve[f, z, Reals]**

Out[33]= {{z → −145.715}, {z → −2.97309}, {z → −1.43557}, {z → 25.1238}}

In two variables, which will mostly be our case, one needs two equations and an initial point that has two coordinates. The two-variable version of the Newton method is often called the Gauss–Newton method because sophisticated implementations use ideas of both Gauss and Newton. However, we will use the built-in function FindRoot.

A typical use would be to solve the two simultaneous equations:

In[34]:=
```
f = -1 + x - 0.2 * x ^ 2 + 0.008 * x ^ 3 + y +
      x * y - 0.2 * x ^ 2 * y - 0.2 * y ^ 2 - 0.2 * x * y ^ 2 + 0.008 * y ^ 3;
g = x ^ 2 + y ^ 2 - 64;
```

A good choice of initial point would be a random complex number:

In[36]:= **{p1, p2} = RandomComplex[{−10 − 10 I, 10 + 10 I}, 2]**

Out[36]= {0.326168 + 9.30269 *i*, −8.50218 − 3.55168 *i*}

In[37]:= **q = {x, y} /. FindRoot[{f, g}, {{x, p1}, {y, p2}}]**

Out[37]= $\{-1.20431 + 1.51576 \times 10^{-16}\,i,\ -7.90883 + 2.49741 \times 10^{-17}\,i\}$

This looks like a real number, which we can check:

In[38]:= **q0 = Re[q]**

Out[38]= {−1.20431, −7.90883}

In[39]:= **FindRoot[{f, g}, {{x, q0[[1]]}, {y, q0[[2]]}}]**

Out[39]= {x → −1.20431, y → −7.90883}

Better yet, NSolve finds all real solutions:

In[40]:=
```
NSolve[{f, g}, {x, y}, Reals]
```

Out[40]=
```
{{x → -7.90883, y → -1.20431}, {x → 7.97264, y → -0.661077}, {x → 7.89586, y → -1.28659},
 {x → -1.20431, y → -7.90883}, {x → -0.661077, y → 7.97264}, {x → -1.28659, y → 7.89586}}
```

It should be noted that Newton's method will not work as well if the zero you want is a *multiple zero,* especially in the two-variable case. We will discuss this more complicated situation in Appendix 1. Points that satisfy the criteria as a multiple zero, as defined in Chapter 3 or Appendix 1, are also considered solutions to $f = 0$.

In Chapter 4, we will give Gauss's alternate definition of a point that is on a possibly numerical curve $f = 0$ where p is a topological boundary point of the region $f > 0$. We will then show how to find enough points to approximately plot the curve.

2.3 Randomness in Numerical Algorithms

An important way in which numerical algorithms are simpler than corresponding exact algorithms is the ability to use randomness. In many cases, an algorithm will work most of the time, but there are special cases in which it will not work. In Newton's method, for example, we noted that most initial points gave rise to a sequence of values converging to a zero, but we also mentioned that for univariate polynomials of degree 3 or more, there are many initial points where Newton's method does not, in theory, converge. It is not, however, possible to give a useful description of these points to avoid, especially before finding the solutions, but in most cases, these points are surrounded by points that are good initial points. For this reason, we choose *random* points as initial values and, particularly if we choose a random complex number, the probability that we choose a bad point is essentially zero.

A bad example from a standard calculus book is the following:

In[41]:=
```
f = 3x^4-6x^3-x^2+4x-4
```

Out[41]= $-4+4x-x^2-6x^3+3x^4$

In[42]:=
```
newton1a[f, 1, 6, x]
```

```
1.
0.
1.
0.
1.
0.
```

Out[42]= 1.

Here we see that Newton's method will jump between 0 and 1 indefinitely, so both 0 and 1 are not good initial points.

We use the following testing code using the fact that the zeros of *f* are known:

```
In[43]:= sol = x /. NSolve[f]

Out[43]= {-1., 0.5 - 0.645497 i, 0.5 + 0.645497 i, 2.}

In[44]:= success = 0;
         Do[p = newton1[f, RandomInteger[{0, 4}], x];
           If[Length[Select[sol, Abs[# - p] < .001 &]] > 0, success++], {100}]
         success

Out[46]= 65

In[47]:= success = 0;
         Do[p = newton1[f, RandomInteger[{-9, 9}], x];
           If[Length[Select[sol, Abs[# - p] < .001 &]] > 0, success++], {100}]
         success

Out[49]= 86

In[50]:= success = 0;
         Do[p = newton1[f, RandomComplex[{-9 - 9 I, 9 + 9 I}], x];
           If[Length[Select[sol, Abs[# - p] < .001 &]] > 0, success++], {100}]
         success

Out[52]= 95
```

For this exceptionally tricky polynomial, newton1 works, but in this experiment, it works only 66% of the time for a small integer range of initial points, 91% of the time for a larger range of integer initial points, and 94% of the time for a random complex number. A user-chosen integer point will fail even more often.

With a more generic degree 4 polynomial, our experiment shows success more than 99% of the time:

```
In[53]:= g = randomRealPolynomial[4, x, y] /. {y → 0}
```

Out[53]= $-0.580442 - 0.891308\,x - 0.415966\,x^2 - 0.93218\,x^3 + 0.966202\,x^4$

```
In[54]:= solc = x /. NSolve[g]

Out[54]= {-0.535408, -0.0883733 - 0.813194 i, -0.0883733 + 0.813194 i, 1.67694}

In[55]:= success = 0;
         Do[p = newton1[g, RandomComplex[{-9 - 9 I, 9 + 9 I}], x];
           If[Length[Select[solc, Abs[# - p] < .001 &]] > 0, success++], {1000}]
         success

Out[57]= 1000
```

Another experiment the reader may try in Chapter 3, or better in Chapter 6 where we find complex and infinite singularities, is to compare randomintegerpolynomial[4, 4, x, y] and randomrealpolynomial[4, x, y] to see which produces more singular polynomials. When I do this, I find that about 20% of the time randomintegerpolynomial[4, 4, x, y] produces a singular (including non-square free) polynomial while randomrealpolynomial[4, x, y] has never given me a singular or non-square free polynomial. (See the next section for the code for random polynomials.) In fact, randomintegerpolynomial[d, d + 2, x, y] will be singular most of the time for $d \geq 5$ although the singularities will often be infinite.

In Chapter 1, we used random real numbers in the system for finding the equation of a line so that our system would have a high probability of being of full rank.

In fact, the routine for finding singular points, including infinite and complex, in Chapter 6 uses randomization in an essential way. In Chapter 9, we will also consider an algorithm that starts by making a random transformation on the curve, and the answer depends somewhat on this random choice. This brings up a new question. Is it better to have an algorithm that returns a different random answer each time or a less good algorithm that always returns the same answer to a problem? One can certainly argue the latter, especially when trying to debug other algorithms using this result. One solution to this question is to choose the random item once and hope that no application just happens to conflict with this choice. In the case of the built-in function NSolve, the Wolfram Language makes this second choice when it encounters a system with infinite solutions and wants to return just one. I prefer the compromise of making this choice once per Wolfram Language session.

2.4 Constructing and Working with Bivariate Polynomials

Since an algebraic curve is the "solution set," as defined previously, of a real bivariate polynomial, we discuss in this section some aspects of real bivariate polynomials.

In many Wolfram Language procedures handling multivariate functions, one of the inputs is a list of function variables. However, these procedures often expect each variable to be an atomic quantity, that is, a variable could be x, t, a, or even α or Ω but not X[[2]] where X is a list. So in coding user-defined procedures, as we do often in this book, the variables are treated as independent characters.

Considering the previous code to find real solutions of a pair of bivariate polynomials, we would write a shortcut function where F generally refers to the list {f, g}:

```
In[58]:= rsolve[F_, x_, y_] := NSolve[F, {x, y}, Reals];
```

X replacing {x, y} might be more obvious, but it might not work as well:

```
In[59]:= rsolve[F_, X_] := NSolve[F, X, Reals];
```

The best method in the Wolfram Language to extract coefficients is to use CoefficientRules. Since the coefficients are expected to be Wolfram Language machine numbers in this book, specifying variables is unnecessary unless some of your variables are user-undefined symbolic numbers.

It should be noted that CoefficientRules, by default, will use the function Variables to find variables:

```
In[60]:= Variables[3.2 x y^2 + 5 Pi x^2 y - Sin[.2334] x + c y]

Out[60]= {x, c, y}
```

In this example, *c* was intended as a placeholder. If a polynomial was defined inside a function using a local variable, it is useful to let the Wolfram Language discover the variables:

```
In[61]:= CoefficientRules[0.2500319097983996` - 0.8807521249099297` x +
           0.5727398772490768` x^2 + 0.6298021363965578` x^3 - 0.06969551948469332` x^4 +
           0.12009965854459015` y + 0.3557137203380485` x y + 0.003982230515285234` x^2 y -
           0.03985635320567571` x^3 y - 0.41567653449909425` y^2 -
           0.5547000285700272` x y^2 + 0.6951203004292736` x^2 y^2 +
           0.909106063953359` y^3 - 0.7657856873127402` x y^3 - 0.9546094753841348` y^4]

Out[61]= {{4, 0} → -0.0696955, {3, 1} → -0.0398564, {3, 0} → 0.629802,
           {2, 2} → 0.69512, {2, 1} → 0.00398223, {2, 0} → 0.57274, {1, 3} → -0.765786,
           {1, 2} → -0.5547, {1, 1} → 0.355714, {1, 0} → -0.880752, {0, 4} → -0.954609,
           {0, 3} → 0.909106, {0, 2} → -0.415677, {0, 1} → 0.1201, {0, 0} → 0.250032}
```

Use Association to find the total degree:

```
In[62]:= tDeg[f_] := Max[Table[Total[t], {t, Keys[Association[CoefficientRules[f]]]}]]
```

Do not ever use Round *or* Chop *on coefficients in a numerical polynomial as that could affect the total degree, which is important information encoded in a polynomial!*

For example, the Newton hyperbolas of Chapter 1 could have the highest coefficient:

```
In[63]:= N[.2^15]

Out[63]= 3.2768 × 10^-11
```

It will be useful to construct random curves of degree *d*. To this end, we need a list of ordered pairs with non-negative components totaling *d* or less.

This can be constructed recursively:

```
In[64]:= pExps[d_] := Module[{pex},
           pex[0] = {{0, 0}};
           pex[n_] := pex[n] = Join[pex[n - 1], Table[{n - j, j}, {j, 0, n}]];
           pex[d]]

In[65]:= pExps[3]

Out[65]= {{0, 0}, {1, 0}, {0, 1}, {2, 0}, {1, 1}, {0, 2}, {3, 0}, {2, 1}, {1, 2}, {0, 3}}
```

Then a procedure to construct random numerical polynomials is simply:

```
In[66]:= randomRealPolynomial[d_, x_, y_] :=
           FromCoefficientRules[Table[e → RandomReal[{-1, 1}], {e, pExps[d]}], {x, y}]
```

FromCoefficientRules is the built-in inverse of CoefficientRules:

```
In[67]:= g = randomRealPolynomial[4, x, y]
```

```
Out[67]= 0.124243 + 0.703201 x + 0.337763 x^2 + 0.103069 x^3 - 0.335 x^4 -
           0.582387 y - 0.580046 x y - 0.53027 x^2 y + 0.987303 x^3 y + 0.607222 y^2 -
           0.302423 x y^2 - 0.366698 x^2 y^2 - 0.630866 y^3 - 0.640134 x y^3 + 0.346355 y^4
```

Since zero is not a numerical concept, a numerical polynomial should have a coefficient for each possible monomial. On the other hand, we may want to generate random exact polynomials.

But here, coefficients can be zero, hence the more complicated code (For a complete list of global functions, see Appendix 2):

```
In[68]:= randomIntegerPolynomial[d_, kk_, x_, y_] := Module[{k, S, b, j, m, t},
            b = Binomial[d + 2, 2];
            If[kk < 1 || kk > b, k = b, k = kk];
            S = RandomSample[pExps[d], k];
            m = Max[Table[Total[t], {t, S}]];
            If[m < d, j = RandomInteger[{0, d}]; S = Append[Take[S, k - 1], {j, d - j}]];
            FromCoefficientRules[Table[e → RandomInteger[{-9, 9}], {e, S}], {x, y}]]
```

This next example gives a degree 5 polynomial with six terms:

```
In[69]:= h = randomIntegerPolynomial[5, 6, x, y]
```

```
Out[69]= 8 - 5 x y + 6 x^3 y + 7 x^3 y^2 + 8 y^3
```

At the start of Chapter 1, we mentioned that our, perhaps strange, algorithm for finding the equation of a line could be generalized for finding the equation of a curve through a number of points. As in the line algorithm, we must determine all the coefficients. A counting exercise based on the algorithm pExps shows that there are binomial coefficient $\binom{d+2}{2} = \frac{(d+2)\,(d+1)}{2}$ coefficients of a bivariate polynomial of degree d. We will use the last equation for getting a random solution out of the otherwise homogeneous system. We only need $\binom{d+2}{2} - 1$ conditions. These could be points, tangent directions at points, specification of specific coefficients, etc. We will look at two examples.

For our first example, we look for a conic that has certain properties, including these exponents:

```
In[70]:= ex = pExps[2]
```

```
Out[70]= {{0, 0}, {1, 0}, {0, 1}, {2, 0}, {1, 1}, {0, 2}}
```

We need a 6×6 coefficient matrix with a random last row leaving five conditions. We want the axes to be parallel to the coordinate axes, which means the monomial involving $x\,y$ should be zero. $x\,y$ corresponds to $\{1, 1\}$ in *ex*, which is the fifth coefficient, so one row will have a 1 in column 5 and a zero otherwise. We want this conic to be symmetric about the y axes, so the coefficient of x should be zero. This gives a row with 1 in column 2 and zero elsewhere. Finally, we want the conic to go through points $\{5, 0\}$, $\{2, 3\}$, and $\{2.5, -1\}$. For the first point, the exponents where y has a positive degree have zero coefficients, and the x ones have the 5 to the power, hence the row $\text{r3} = \{1, 5, 0, 25, 0, 0\}$.

Here are the results for the last two rows of this matrix:

```
In[71]:= r4 = Table[2^e[[1]]*3^e[[2]], {e, ex}]

Out[71]= {1, 2, 3, 4, 6, 9}
```

```
In[72]:= r5 = Table[2.5^e[[1]]*(-1)^e[[2]], {e, ex}]

Out[72]= {1., 2.5, -1., 6.25, -2.5, 1.}
```

Here is the complete system matrix:

```
In[73]:= M1 = {{0, 0, 0, 0, 1, 0}, {0, 1, 0, 0, 0, 0}, {1, 5, 0, 25, 0, 0}, r4, r5, RandomReal[{-1, 1}, 6]};
         M1 // MatrixForm
         B1 = {0, 0, 0, 0, 0, 1};
```

Out[74]//MatrixForm=

$$\begin{pmatrix} 0 & 0 & 0 & 0 & 1 & 0 \\ 0 & 1 & 0 & 0 & 0 & 0 \\ 1 & 5 & 0 & 25 & 0 & 0 \\ 1 & 2 & 3 & 4 & 6 & 9 \\ 1. & 2.5 & -1. & 6.25 & -2.5 & 1. \\ 0.413737 & 0.832985 & 0.555725 & 0.28844 & 0.231678 & -0.384001 \end{pmatrix}.$$

Solving M1.X = B1:

```
In[76]:= X = LinearSolve[M1, B1]
```

Out[76]= $\{-1.50738, -5.23415\times10^{-17}, -0.742385, 0.0602952, 1.62148\times10^{-18}, 0.388151\}$

```
In[77]:= XY2 = Table[x^e[[1]]*y^e[[2]], {e, ex}]
```

Out[77]= $\{1, x, y, x^2, x\,y, y^2\}$

Here is our polynomial:

```
In[78]:= f1 = X.XY2
```

Out[78]= $-1.50738 - 5.23415\times10^{-17}\,x + 0.0602952\,x^2 - 0.742385\,y + 1.62148\times10^{-18}\,x\,y + 0.388151\,y^2$

Cleaning up, remember we wanted *x* and *xy* to have coefficient 0:

```
In[79]:= f1 = Chop[Expand[f1/Coefficient[f1, y^2]]]

Out[79]= -3.8835 + 0.15534 x^2 - 1.91262 y + 1. y^2
```

```
In[80]:= ContourPlot[f1 == 0, {x, -6, 6}, {y, -2, 4},
           Epilog → {PointSize[Large], Red, Point[{{5, 0}, {2, 3}, {2.5, -1}}]}]
```

Out[80]=

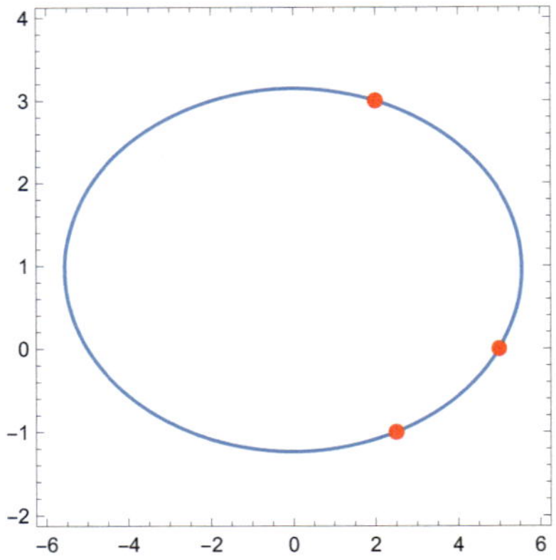

If we just want a curve through $\binom{d+2}{2} - 1$ points, we can automate this with the following function for affine curves (For a complete list of global functions, see Appendix 2):

```
In[81]:= aCurve[pts_, x_, y_] := Module[{d, P, M, B, n, c, pow},
           pow[a_, n_] := If[n == 0, 1, a^n];
           d = Switch[Length[pts], 2, 1, 5, 2, 9, 3, 14, 4, 20, 5,
               27, 6, _, Return["number of points must be 2,5,9,14,20,27"]];
           P = pExps[d];
           n = Length[P];
           M = Table[
               If[Length[p] == 2, pow[p[[1]], e[[1]]] * pow[p[[2]], e[[2]]], pow[p[[1]], e[[1]]] *
                   pow[p[[2]], e[[2]]] * pow[p[[3]], d - e[[1]] - e[[2]]]], {p, pts}, {e, P}];
           AppendTo[M, RandomReal[{-1, 1}, n]];
           B = Append[Table[0, {n - 1}], 1];
           c = LinearSolve[M, B];
           FromCoefficientRules[Table[P[[i]] → c[[i]], {i, n}], {x, y}]
         ];
```

As with our other example, we want an algebraic equation for the bow curve given parametrically:

```
In[82]:= Fbow = {(-1+t^2)/t^3, (-1+t^2)/t^4}
```

Out[82]= $\left\{\frac{-1+t^2}{t^3}, \frac{-1+t^2}{t^4}\right\}$

We see that $t = 0$ is not in the domain; moreover, t near zero points are far away from the interesting part of the curve, so we restrict our parameter interval to $-4 \le t \le -.8$ and $.8 \le t \le 4$. Since we expect a curve of degree 4, we need $\binom{6}{2} - 1 = 14$ points, and by the remarks in the previous section, we should pick points using random t in the interval above.

Here's how that looks:

```
In[83]:= Pts = Table[Fbow /. {t → r}, {r, Join[RandomReal[{-4, -.8}, 7], RandomReal[{.8, 4}, 7]]}]

Out[83]= {{-0.290956, 0.0946819}, {-0.352218, 0.152834}, {-0.289573, 0.0936467},
  {-0.365126, 0.170485}, {-0.378554, 0.195226}, {-0.376192, 0.189945},
  {-0.316144, 0.115272}, {0.246995, 0.0656429}, {0.285977, 0.0909958}, {0.371628, 0.181151},
  {0.234405, 0.0586099}, {0.268706, 0.0790422}, {0.26956, 0.0796053}, {0.236472, 0.0597295}}
```

```
In[84]:= f2 = aCurve[Pts, x, y]
```

Out[84]= $-9.65453\times10^{-15} - 5.87888\times10^{-15}\,x + 1.6435\times10^{-12}\,x^2 - 1.97518\times10^{-11}\,x^3 + 2.76072\,x^4 - 1.24706\times10^{-12}\,y + 2.00332\times10^{-11}\,x\,y - 2.76072\,x^2\,y + 7.16733\times10^{-11}\,x^3\,y - 7.75558\times10^{-12}\,y^2 - 9.59835\times10^{-11}\,x\,y^2 - 3.52011\times10^{-11}\,x^2\,y^2 + 2.76072\,y^3 + 7.88483\times10^{-11}\,x\,y^3 - 1.16926\times10^{-10}\,y^4$

We see there are a few larger coefficients but guess that the little ones are perhaps numerical errors. We clean up our curve using the following:

```
In[85]:= f3 = Expand[Chop[f2, 1.*^-9]/Coefficient[f2, x^4]]
```

Out[85]= $1.\,x^4 - 1.\,x^2\,y + 1.\,y^3$

In case we picked the wrong degree, we would have a curve that went through these 14 points of the bow curve but not the actual bow curve.

Since we have a simple exact parametrization, we can check it:

```
In[86]:= Expand[x^4 - x^2 y + y^3 /. Thread[{x, y} → Fbow]]

Out[86]= 0
```

This says, at least, that the bow curve is contained in the curve defined by $x^4 - x^2 + y^3 = 0$. If we did not have an exact system, we could have randomly picked other random values of s to see if Fbow /. $\{t \to s\}$ still approximately satisfy f3.

Tech Notes

- For style purposes, some of the initialization cells in this chapter have had their background color removed. If you are using the notebook version of this book, you should evaluate the initialization cells when opening a notebook.
- Note the first line *p* = *N*[*p*0];. This is because attempting Newton's method with an exact equation and an initial point may otherwise give iterations as fractions where the denominators will quickly blow up.

Newton's method without the N in the first line is as follows:

```
In[87]:= newton1b[f_, p0_, n_, z_] := Module[{p},
           p = p0;
           Do[Print[p]; p = p - (f / D[f, x]) /. {z → p}, {i, n}];
           p]
```

```
In[88]:= f = x ^ 2 + 3 x - 5;
         newton1b[f, 2 / 3, 5, x]
```

$\frac{2}{3}$

$\frac{49}{39}$

$\frac{10006}{8385}$

$\frac{451661161}{378725295}$

$\frac{921162049720053046}{772409560098046065}$

Out[89]= $\frac{3831622164498734659314458051363799241}{3212878332799132328866938118089080655}$

It would be even worse if we started from an exact complex number and even worse yet if our function contained a radical. We leave it to the reader to do these experiments.

- We use Catch and Throw since the conditional If statement lies inside the Do loop. The Do loop is used because in the unlikely case of no convergence, we don't want an infinite loop. We use Chop since we are expecting a real result and want the imaginary part to be zero; the optional second argument tells Chop to drop all terms where the coefficient is smaller in absolute value than this number.
- Round will not thread over the coefficients of a polynomial, which normally, you should not do anyway.

Use this code to convert a polynomial like the output of f2 below, which appears to be an exact integer equation, to it's theoretically exact form:

```
In[90]:= f2 = 0.9999999996287839` x^4 - 0.999999999431914` x^2 y + 1.` y^3;
         Assoc = Association[CoefficientRules[f2, {x, y}]];
         FromCoefficientRules[Table[k → Round[Assoc[k]], {k, Keys[Assoc]}], {x, y}]
```

Out[92]= $x^4 - x^2 y + y^3$

More to Explore

For some more fun with curves, check out this Wolfram MathWorld page, which includes several links to free downloadable notebooks (wolfr.am/vGS18KUb)

Try this Wolfram Demonstration for Newton's method of root finding by C. Maes (wolfr.am/vGTqmlcJ)

Check out this bit of ancient Egyptian fraction solving on Wolfram MathWorld (wolfr.am/vGRXtbmt)

3 | Special Points and Lines

In this chapter, we learn about special kinds of points and lines associated with a real affine plane algebraic curve.

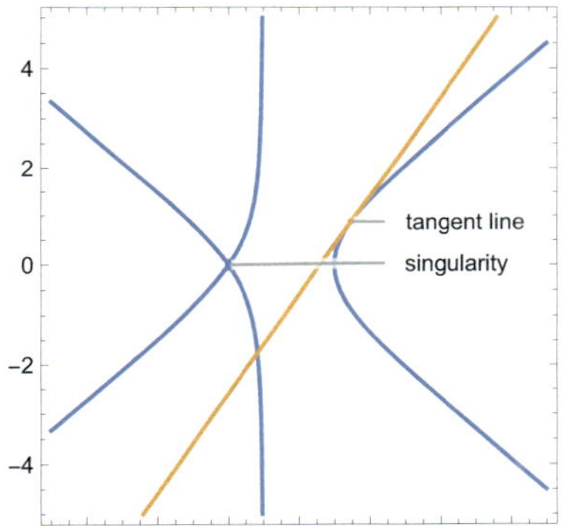

gaussCurve[z^3−3z^2]

3.1 Regular Points and Tangent Lines

Let $f = f[x, y] = 0$ be a bivariate polynomial, that is, the equation of our affine plane curve. Let p be a numerical point on our curve. p is called regular if one of the partial derivatives $D[f, x]|_p$, $D[f, y]|_p$ is nonzero, in the sense of Chapter 2, when evaluated at p. Most points are regular since we are assuming, unless otherwise stated, that curves are square-free.

For a regular point $p = \{x_0, y_0\}$, we will define the *tangent line* to be the line with the equation

$$l = (D[f, x]|_p)\, x + (D[f, y]|_p)\, y - c,$$

where

$$c = (D[f, x]|_p)\, x_0 + (D[f, y]|_p)\, y_0.$$

This choice of c puts p on line l.

Alternatively, we can use the general line function

$$\text{line}\left[\{x_0,\ y_0\},\ \left\{D[f,\ y]\,|_p,\ -D[f,\ x]\,|_p\right\},\ x,\ y\right]$$

since the direction of the tangent line is $\left\{D[f,\ y]\,|_p,\ -D[f,\ x]\,|_p\right\}$.

Find the tangent line of f (x, y) at p for the Gauss curve $z^3 - 3z^2$, as pictured on the previous page:

```
In[1]:= f = x^3 - 3 x^2 + 3 y^2 - 3 x y^2;
        p = {3.4756865177956913`, 0.879614879812377`};
        l = line[p, {D[f, y], -D[f, x], 0} /. Thread[{x, y} → p], x, y]

Out[3]= -1.64203 + 0.632505 x - 0.632505 y
```

In the future we will simply use the following procedure (For a complete list of global functions, see Appendix 2):

```
In[4]:= tLine[f_, p_, x, y] := line[p, {D[f, y], -D[f, x], 0} /. Thread[{x, y} → p], x, y];
```

From the plot at the beginning of this section, we see how closely the tangent line approximates the curve at the point of tangency. In Chapter 5, I note that the *intersection multiplicity* of the curve and its tangent line is greater than 1 or that the point of tangency is a *multiple zero* of the system $\{f,\ l\}$ (See Appendix 1). A key fact we will learn about this situation is that the Jacobian determinant of this system is zero at the point. That is

$$D[\{f,\ l\},\ \{\{x,\ y\}\}] = \text{Det}\left[\begin{pmatrix} D[f,\ x] & D[f,\ y] \\ D[l,\ x] & D[l,\ y] \end{pmatrix}\right] = D[f,\ x]\,D[l,\ y] - D[f,\ y]\,D[l,\ x] = 0$$

when evaluated at p. Conversely, if p is a regular point on the curve and line l and the Jacobian is 0 there, line l is the tangent line. If $l = ax + by + c$, then $D[l,\ x] = a$ and $D[l,\ y] = b$, so the vanishing of the Jacobian determinant is independent of c. It follows that if regular point p of $f = 0$ is a solution to the following nonlinear system

$$\text{System 2}\ \begin{cases} f[x,\ y] & = 0 \\ b\,D[f,\ x] - a\,D[f,\ y] & = 0 \end{cases},$$

then $l = ax + by + c$ is the tangent line for $f = 0$ where c is chosen so that p is a point of $l = 0$.

The following procedure then finds points where the tangent line to $f = 0$ is parallel to the line $ax + by = 0$. Enter the curve f, a line $ax + by$ and the variables.

All real points with tangent lines parallel to a x + b y = 0 will be returned (For a complete list of global functions, see Appendix 2):

```
In[5]:= tangentRealPoints[f_, k_, x_, y_] := Module[{J, p, q, sol},
        J = Det[D[{f, k}, {{x, y}}]];
        sol = NSolve[{f, J}, {x, y}, Reals];
        If[Length[sol] > 0, {x, y} /. sol, {}]];
```

One application is to the case where the curve is bounded and all points lie inside a finite circle. We would like to find the maximal x, y coordinates or the maximal points in certain directions.

Consider the peanut curve of Chapter 1:

```
In[6]:= f = x^4+4 x y +y^4-1/4;
```

Using tangentRealPoints:

```
In[7]:= xrange = tangentRealPoints[f, x, x, y]

Out[7]= {{1.53604, -1.15381}, {-1.53604, 1.15381}}

In[8]:= yrange = tangentRealPoints[f, y, x, y]

Out[8]= {{-1.15381, 1.53604}, {1.15381, -1.53604}}
```

In[9]:= tangentRealPoints[$-\frac{1}{4}+x^4+4\,x\,y+y^4$, y, x, y]

```
Out[9]= {{-1.15381, 1.53604}, {1.15381, -1.53604}}
```

This suggests that the entire curve lies in the bounded region:

```
In[10]:= {xrange[[1, 1]] < x < xrange[[2, 1]], yrange[[1, 2]] < y < yrange[[2, 2]]}
```

The curve appears to be symmetric about the line x + y = 0, so using the perpendicular line x − y = 0:

```
In[11]:= diagrange = tangentRealPoints[f, x-y, x, y]

Out[11]= {{-1.4355, 1.4355}, {1.4355, -1.4355}}
```

We see the maximum norm is:

```
In[12]:= r = Norm[diagrange[[1]]]

Out[12]= 2.0301
```

Thus we can use one of the following plots to see the entire curve:

In[13]:=
```
{ContourPlot[f == 0, {x, xrange[[1, 1]], xrange[[2, 1]]}, {y, yrange[[1, 2]], yrange[[2, 2]]}],
  ContourPlot[{f == 0, x^2+y^2 == r^2}, {x, -r, r}, {y, -r, r}]}
```

Out[13]=

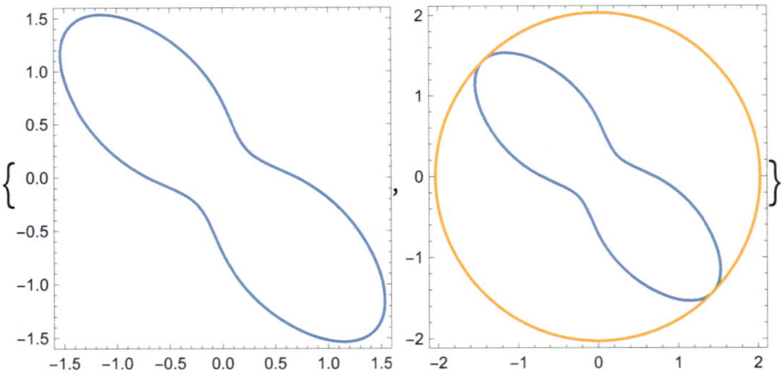

A special case of tangency s is when the tangent line intersects in even higher multiplicity with the curve. These points are called *inflection points.* Inflection points are of most interest regarding cubic curves, so we will save the discussion of inflection points for Chapter 7.

3.2 Singular Points

In the classical case, a singular point is one that is not regular. In particular, the three equations $f[x, y] = 0$, $D[f, x] = 0$ and $D[f, y] = 0$ must hold, but numerically, as we have seen in Chapter 2, the failure of the criterion $D[f, x] = 0$ is not the same as $D[f, x]$ is not zero, which requires $D[f, x]$ to be large enough that it cannot be considered a numerical error. In addition, a system of 3 equations in 2 unknowns is overdetermined, and numerical methods for overdetermined systems require an *a priori* assumption that there is actually a solution, which does not help us find the small number of actual singular points for a square-free curve.

We therefore make the following definition:

A point p on $f = 0$ is singular *if p is a solution to System 2 for all real numbers a, b.*

Numerically, we can, for finding singular points, replace the word *all* with *random.* Thus we can use our function `tangentRealPoints` using two different random lines, and singular points should appear in both cases, whereas regular points should not.

We illustrate this using the Chebyshev quartic:

In[14]:=
```
f = 1 - 8 x^2 + 8 x^4 + 3 y - 4 y^3;
```

In[15]:=
```
l = RandomReal[{-1, 1}, 2].{x, y}
```

Out[15]= 0.701317 x − 0.204217 y

In[16]:=
```
k = RandomReal[{-1, 1}, 2].{x, y}
```

Out[16]=
```
-0.612793 x + 0.770037 y
```

In[17]:=
```
run1 = tangentRealPoints[f, l, x, y]
```

Out[17]=
```
{{8.49708, 21.7562}, {0.707107, 0.5}, {0.707107, 0.5}, {-0.921557, 0.00803236},
 {0., -0.5}, {0., -0.5}, {0.967662, -0.760189}, {-0.707107, 0.5}, {-0.707107, 0.5}}
```

In[18]:=
```
run2 = tangentRealPoints[f, k, x, y]
```

Out[18]=
```
tangentrealpoints[1 - 8 x^2 + 8 x^4 + 3 y - 4 y^3, -0.612793 x + 0.770037 y, x, y]
```

In[19]:=
```
tangentRealPoints[1 - 8 x^2 + 8 x^4 + 3 y - 4 y^3,
 -0.9738578350417568` x + 0.5455043467593894` y, x, y]
```

Out[19]=
```
{{1.01626, 1.02896}, {0.707107, 0.5}, {0.707107, 0.5}, {0., -0.5}, {0., -0.5},
 {0.92585, -0.862551}, {-0.741927, 0.440549}, {-0.707107, 0.5}, {-0.707107, 0.5}}
```

We note that the following points are repeated in both runs:

In[20]:=
```
sing = run1[[{1, 3, -1}]]
```

Out[20]=
```
{{8.49708, 21.7562}, {0.707107, 0.5}, {-0.707107, 0.5}}
```

The following plot gives further confirmation. Note the use of Epilog to get large red points:

In[21]:=
```
ContourPlot[f == 0, {x, -2, 2}, {y, -2, 2}, Epilog → {Red, PointSize[Large], Point[sing]}]
```

Out[21]=

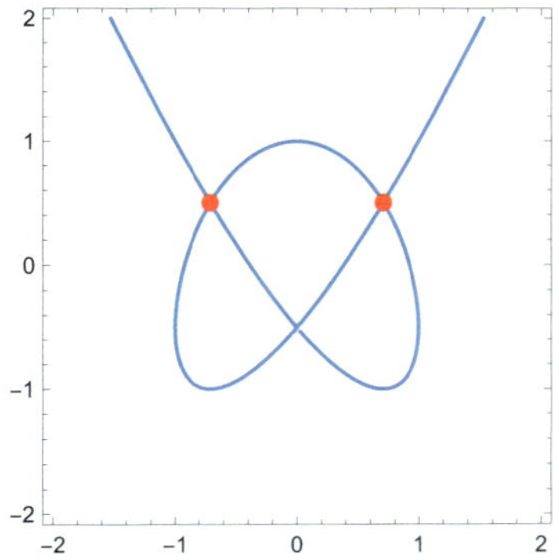

The above example also shows that singular points are actually multiple zeros in System 2, so an alternate strategy for finding singular points is to look for multiple zeros of tangentRealPoints for one random line. The Wolfram Language reports multiple zeros by repetition; however, repeated zeros cause difficulty to numerical solvers and may be reported as zeros that are close but not the same to full precision. See Appendix 1 for more details. Thus we apply a tolerance to check for repeated zeros.

The Wolfram Language function Tally is useful for this:

```
In[22]:= Tally[run1, Norm[#1-#2] < 1.*^-6 &]

Out[22]= {{{8.49708, 21.7562}, 1}, {{0.707107, 0.5}, 2}, {{-0.921557, 0.00803236}, 1},
          {{0., -0.5}, 2}, {{0.967662, -0.760189}, 1}, {{-0.707107, 0.5}, 2}}
```

Therefore, our three singular points are counted twice, whereas the other solutions of System 2 are counted only once. This procedure for counting singular points would then become useful after extracting the distinct singular points from the Tally. The tolerance can be changed by the user by modifying the first line if necessary.

Here we use Flatten to reduce the level of subexpressions and lists, thereby removing the duplicates:

```
In[23]:= realSingularPoints[f_, x_, y_] := Module[{eps},
           eps = 1.*^-6;
           Flatten[
             Take[Select[Tally[tangentRealPoints[f, RandomReal[{-1, 1}, 2].{x, y}, x, y],
                   Norm[#1-#2] < eps &], #[[2]] > 1 &], All, 1], 1]];

In[24]:= realSingularPoints[f, x, y]

Out[24]= {{0.707107, 0.5}, {0., -0.5}, {-0.707107, 0.5}}
```

3.3 Infinite Points

In Chapter 1, we introduced the idea of *infinite points* of a line. We extend this idea to an arbitrary real algebraic plane curve.

But first, given a bivariate polynomial f, we group all the monomials of the same total degree j together and call this the *form* $f_{(j)}$, where the total degree of f is n.

$$f = f_{(0)} + f_{(1)} + \cdots + f_{(n)}$$

We also use the notation $f_{\max} = f_{(n)}$, the *maximal degree form* of f.

For example, if

$$f = 3 + x - 3\,y + 2\,x^2 - 3\,x\,y + y^2,$$

then

$$f_{(0)} = 3$$
$$f_{(1)} = x - 3\,y$$
$$f_{(2)} = 2\,x^2 - 3\,x\,y + y^2.$$

Thus

$$f_{\max} = f_{(2)} = 2\,x\text{^}2 - 3\,x\,y + y^2.$$

Gauss had pointed out in his 1798 dissertation that if $\| \{x, y\} \| = \sqrt{x^2 + y^2}$ is large, the curve $f_{\max} = 0$ approximates the curve $f = 0$. More precisely, for p on the curve $f = 0$, there is a q on the curve $f_{\max} = 0$ so that $\frac{\|p-q\|}{p}$ is as small as you like as long as $\| p \|$ is large enough.

Now $f_{\max}$ is homogeneous, so if $\{x_0, y_0\} \neq \{0, 0\}$ is a real point on $f_{\max}$, then so is $\{r\, x_0, r\, y_0\}$ for any real r; hence, this set of solutions is a line through $\{0, 0\}$ with direction vector $\{x_0, y_0\}$. It is easily seen from the discussion in Chapter 1 that $\{x_0, y_0, 0\}$ is the *infinite point* on this line.

The infinite points described above will be called *infinite points of the curve* f. This definition will be further justified in Chapters 6 and 8.

Examples of infinite points:

```
In[25]:= f = gaussCurve[z^3 - 3 z + 1, x, y, z]

Out[25]= 1 - 3 x + x^3 - 3 x y^2

In[26]:= fmax = x^3 - 3 x y^2;

In[27]:= sol1 = {x, y} /. NSolve[{fmax, x^2 + y^2 - 1}, {x, y}, Reals]

Out[27]= {{-0.866025, -0.5}, {0.866025, 0.5}, {0., 1.}, {0., -1.}, {-0.866025, 0.5}, {0.866025, -0.5}}
```

We can eliminate solutions that differ by the constant −1 and add a 0 as the third coordinate to get the infinite points:

```
In[28]:= infpts = Append[#, 0] &/@sol1[[{1, 4, 5}]]

Out[28]= {{-0.866025, -0.5, 0}, {0., -1., 0}, {-0.866025, 0.5, 0}}

In[29]:= ContourPlot[{f == 0, fmax == 0}, {x, -5, 5}, {y, -5, 5},
           ContourStyle → {Thick, Directive[Orange, Dashed]}]
```

Out[29]=

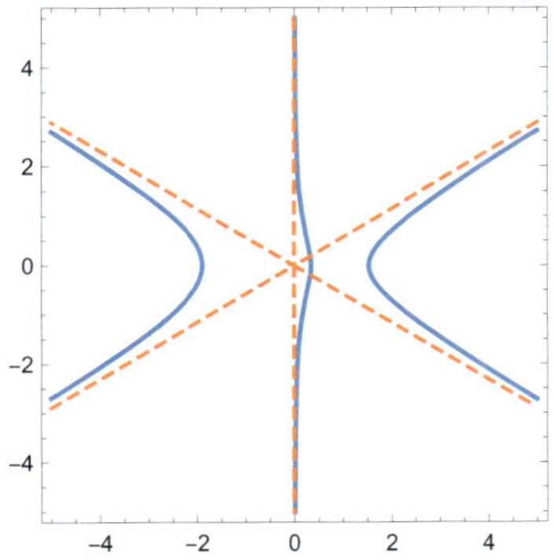

The orange dashed lines here are the lines given by $\{r\, x_0, r\, y_0\}$ for real r. We will call these lines *asymptotes*, which will be precisely defined in Chapter 6. These lines have the infinite points of the curve as infinite points.

Another example is the rational function curve $y = \frac{1}{x^2-4}$:

```
In[30]:= g = Expand[y (x^2-4)-1]

Out[30]= -1-4 y+x^2 y

In[31]:= gmax = x^2 y;
```

The g_{max} is not square free and will not properly plot, so we use the square-free reduction:

```
In[32]:= gmaxsf = x y;
```

We now see the infinite points are {0, 1, 0} and {1, 0, 0}:

```
In[33]:= ContourPlot[{g == 0, gmaxsf == 0, x^2 == 4}, {x, -5, 5}, {y, -5, 5},
           ContourStyle → {Thick, Directive[Orange, Dashed], Directive[Green, Dashed]}]
```

Out[33]=

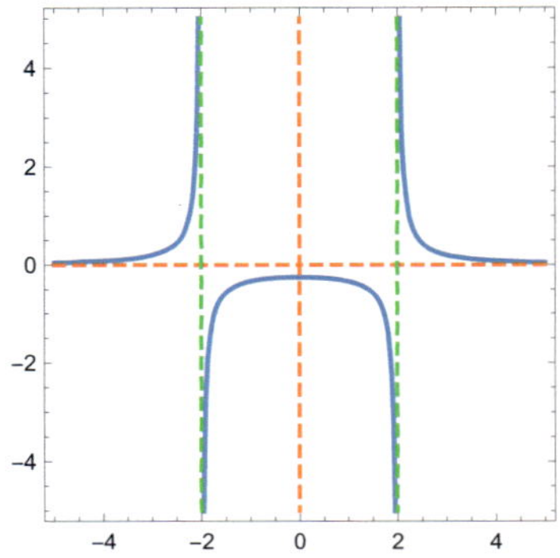

Again, the two orange dashed lines come from g_{max}, but here the asymptotes are given by the green dashed lines and the horizontal orange dashed line. Thus asymptotes are sometimes parallel to the lines in g_{max}. Asymptotes will be discussed in detail in Chapter 7. However, we note the following useful functions in calculating infinite points.

Find the form of degree k:

```
In[34]:= form[h_, k_, x_, y_] :=
           FromCoefficientRules[Select[CoefficientRules[h, {x, y}], Total[#[[1]]] == k &], {x, y}];
```

Find the maximal form of f, f_{max}:

```
In[35]:= maxForm[h_, x_, y_] := form[h, tDeg[h, x, y], x, y];
```

The following function to find distinct infinite points will be given in Chapter 7 (For a complete list of global functions, see Appendix 2):

```
In[36]:= infiniteRealPoints[f_, x_, y_] := Module[{fm, rnd, sol},
             fm = maxForm[f, x, y];
             rnd = RandomReal[{-1, 1}, 2].{x, y} - 1;
             sol = NSolve[{fm, rnd}, {x, y}, Reals];
             If[Length[sol] == 0, Return[{}]];
             sol = DeleteDuplicates[{x, y} /. sol, Norm[#1 - #2] < .001 &];
             Table[Append[sol[[i]], 0], {i, Length[sol]}]
           ];
```

Find the infinite points of Newton Hyperbola 613 (Appendix 2 global functions should be initialized):

```
In[37]:= f = newtonHyperbola[613, x, y]
```

```
Out[37]= -1 + x + 0.2 x^2 - 0.008 x^3 - y + x y + 0.2 x^2 y - 0.2 y^2 + 0.2 x y^2 - 0.008 y^3
```

```
In[38]:= infiniteRealPoints[f, x, y]
```

```
Out[38]= {{-0.0763409, -1.98192, 0}, {1.0436, 0.040198, 0}, {0.657296, -0.657296, 0}}
```

```
In[39]:= Normalize/@%
```

```
Out[39]= {{-0.0384901, -0.999259, 0.}, {0.999259, 0.0384901, 0.}, {0.707107, -0.707107, 0.}}
```

3.4 Ordinary Real Singularities

We start by considering a singularity of $f = 0$ at the origin $\{0, 0\}$. As above, we write $f = f_{(0)} + f_{(1)} + \cdots + f_{(n)}$ where $f_j = \text{Form}[f, j, x, y]$. We will refer to $f_{(j)}$ with the smallest j such that $f_j = 0$ as f_{min}.

Appendix 1 of (Fischer, 2001) devotes several pages to discussing the singularity of $f = y^2 - x^n$ for $n > 2$ and related curves.

In this situation, $f_{\text{min}} = y^2$, but the singularities are quite different:

```
In[40]:= {ContourPlot[y^2 == x^3, {x, -4, 4}, {y, -4, 4}, ImageSize → Tiny],
   ContourPlot[y^2 == x^4, {x, -4, 4}, {y, -4, 4}, ImageSize → Tiny],
   ContourPlot[y^2 == x^8, {x, -4, 4}, {y, -4, 4}, ImageSize → Tiny],
   ContourPlot[(y - x^2) (y - x^7) == 0, {x, -4, 4}, {y, -4, 4}, ImageSize → Tiny]}
```

Out[40]=

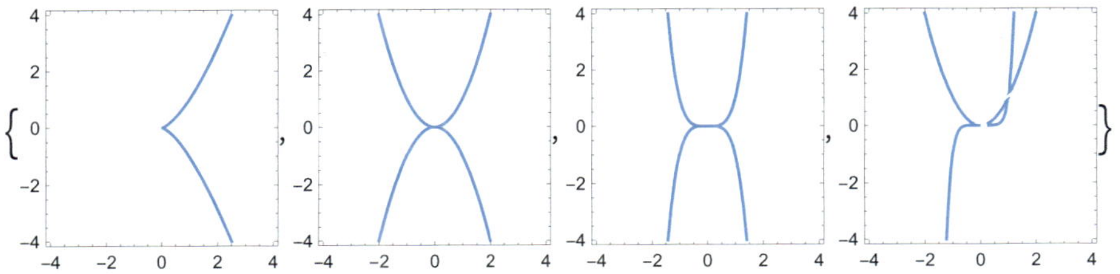

There is one case when f_{min} does tell the whole story.

First, note than any form, in particular f_{min}, factors completely into linear factors. If f_0, f_1 are both 0, then $\{0, 0\}$ is a singular point of *order k* where $f_{(k)} = f_{\text{min}}$. Then we call the singularity a *real ordinary singularity* if there are k distinct real lines in the factorization.

In the discussion of lines in Section 1.1, we noted that an infinite point could be viewed as a direction vector of a line. Thus a quick check of the infinite points of the curve $f_{\text{min}} = f_{(k)} = 0$ will tell us whether this point is an ordinary singularity; mainly, there needs to be k distinct infinite points.

The bow curve of Chapter 1 has an ordinary singularity at the origin because the degree of f_{min} is 3 and there are three distinct infinite points of f_{min}:

```
In[50]:= fbow = x^4-x^2 y+y^3;
         fmin = -x^2 y+y^3;
         InfiniteRealPoints[fmin, x, y]
```

```
Out[50]= InfiniteRealPoints[-x^2 y+y^3, x, y]
```

If the singularity is a point $p = \{x_0, y_0\}$, which is not the origin, we can translate to the origin with the transformation $x \to x + x_0$, $y \to y + y_0$. The new polynomial f will have singularity at $\{0, 0\}$.

Then find f_{min} and check this for the correct number of infinite points (For a complete list of global functions, see Appendix 2):

```
In[51]:= pointMinForm[f_, p_, x_, y_] := Module[{A, h, j},
            h = f /. Thread[{x, y} → {x, y} + p];
            A = <| CoefficientRules[h, {x, y}] |>;
            j = Min[Table[Total[k], {k, Keys[A]}]];
            form[h, j, x, y]];
```

```
In[52]:= pointMinForm[(y-x^2) (y-x^7), {0, 0}, x, y]
```

```
Out[52]= y^2
```

Our first example is the rose curve of three leaves:

```
In[53]:= h = y^3-3 x^2 y+x^4+2 x^2 y^2+y^4;
```

```
In[54]:= hmin = pointMinForm[h, {0, 0}, x, y]
```

```
Out[54]= -3 x^2 y+y^3
```

```
In[55]:= infiniteRealPoints[hmin, x, y]
```

```
Out[55]= {{-1.28632, -2.22798, 0}, {2.20845, 0., 0}, {0.594169, -1.02913, 0}}
```

```
In[56]:= ContourPlot[{h == 0, hmin == 0}, {x, -1.5, 1.5}, {y, -1.5, 1.5}]
```

Out[56]=

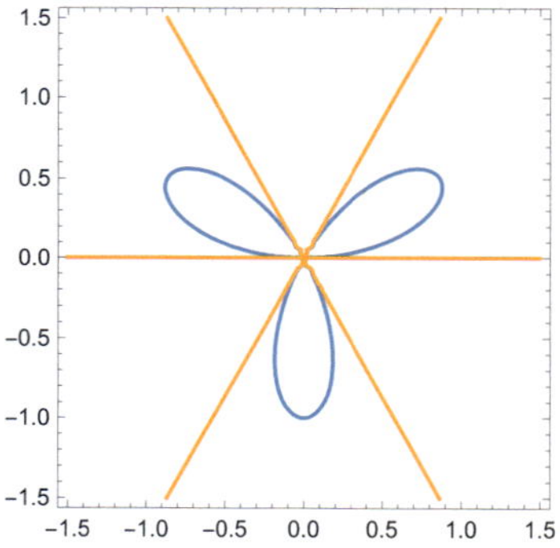

Now let's try another example with the Chebyshev quartic:

In[57]:=
```
f = 1 - 8 x^2 + 8 x^4 + 3 y - 4 y^3;
```

In[58]:=
```
p = N[{-Sqrt[2]/2, 1/2}]
```

Out[58]= {-0.707107, 0.5}

In[59]:=
```
fm = pointMinForm[f, p, x, y]
```

Out[59]= 2.22045×10^{-16}

In[60]:=
```
infiniteRealPoints[fm, x, y]
```

Out[60]= {}

In[61]:=
```
fpmin = fm /. Thread[{x, y} → {x, y} - p]
```

Out[61]= 2.22045×10^{-16}

A ContourPlot is:

In[62]:=
```
ContourPlot[{f == 0, fpmin == 0}, {x, -2, 2}, {y, -2, 2}]
```

Out[62]=

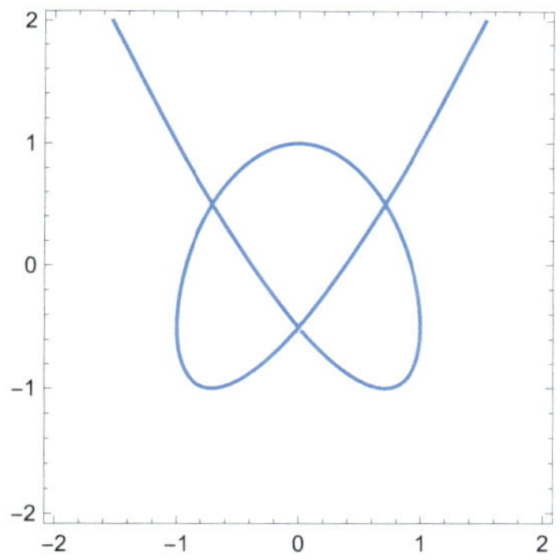

3.5 Critical Points

Most of the ideas in this book derive from classical ideas although some have been refitted to work numerically. The material in this section is based on private correspondence with Daniel Lichtblau, a developer and mathematician at Wolfram Research.

In the currently popular field of numerical algebraic geometry, positive dimensional algebraic sets (called *varieties*) are identified by *witness points.* These form a, possibly somewhat random, set of points on the variety so that every point on the variety can be reached from a witness point by *curve tracing*. Curve tracing will be described and important in the next chapter. In this book, we will mostly use *critical points.* The study of critical points in general is important in the field of optimization.

For plane curves, calculating critical points is especially easy. In this book, we observe that a *Critical Point* of curve $f = 0$ is a solution to the system

$$\text{System 3} \begin{cases} f & = 0 \\ y\,D[f,\,x] - x\,D[f,\,y] & = 0 \end{cases}.$$

In particular, each singular point will be a critical point. In addition, the second equation tells us that at a regular critical point p, the tangent vector will be orthogonal to the vector from the origin to p.

Critical Points are points where the curve achieves a local minimum or maximum distance from the origin. In addition, singular points are critical points.

The following function will give the distinct critical points:

```
In[63]:= criticalPoints[f_, x_, y_] := DeleteDuplicates[
           {x, y} /. NSolve[{f, y D[f, x] - x D[f, y]}, {x, y}, Reals], Norm[#1 - #2] < 1.*^-6 &];
```

Factor is a classic Wolfram Language function that can factor univariate, multivariate and rational functions:

```
In[64]:= f = gaussCurve[(z^2-4)^2, x, y, z]
```

Out[64]= $16 - 8x^2 + x^4 + 8y^2 - 6x^2y^2 + y^4$

```
In[65]:= Factor[f]
```

Out[65]= $(-4 + x^2 - 2xy - y^2)(-4 + x^2 + 2xy - y^2)$

This is the reducible union of two conics:

```
In[66]:= critpts = criticalPoints[f, x, y]
```

```
Out[66]= {{-2., 0.}, {2., 0.}, {-1.55377, -0.643594},
          {1.55377, 0.643594}, {1.55377, -0.643594}, {-1.55377, 0.643594}}
```

```
In[67]:= ContourPlot[f == 0, {x, -4, 4}, {y, -4, 4},
          Epilog → {PointSize[Large], Red, Point[critpts]}, ImageSize → Small, Axes → True]
```

Out[67]=

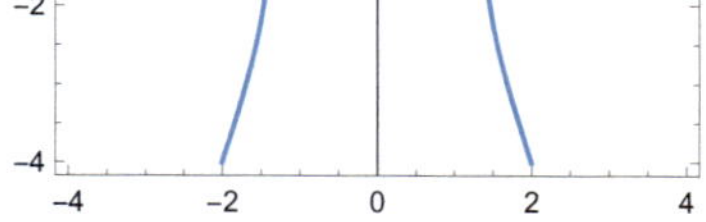

Above, we see that two of the critical points appear to be singular. Running just the defining system for critical points, we can see that $\{-2, 0\}$ and $\{2, 0\}$ are multiple points and hence are probably singular points.

Let's check and find out if they are ordinary points by factoring the point minimal form (something we are able to do since they are integer polynomials):

```
In[68]:= {x, y} /. NSolve[{f, y D[f, x] - x D[f, y]}, {x, y}, Reals]

Out[68]= {{-2., 0.}, {-2., 0.}, {2., 0.}, {2., 0.}, {-1.55377, -0.643594},
          {1.55377, 0.643594}, {1.55377, -0.643594}, {-1.55377, 0.643594}}

In[69]:= fm1 = Factor[PointMinForm[f, {-2, 0}, x, y]]

Out[69]= PointMinForm[16 - 8 x^2 + x^4 + 8 y^2 - 6 x^2 y^2 + y^4, {-2, 0}, x, y]

In[71]:= fm2 = Factor[PointMinForm[f, {2, 0}, x, y]]

Out[71]= PointMinForm[16 - 8 x^2 + x^4 + 8 y^2 - 6 x^2 y^2 + y^4, {2, 0}, x, y]
```

Since some critical points are points of the curve closest to the origin, we can use our simple procedure to find the point p of a curve closest to a given point q simply by translating the plane so that q is moved to the origin and then looking for the closest points to the origin.

An example using the previous curve:

```
In[71]:= closestPoint[f_, q_, x_, y_] := Module[{g, p, cp},
           g = Expand[f /. Thread[{x, y} → {x + q[[1]], y + q[[2]]}]];
           cp = criticalPoints[g, x, y];
           p = MinimalBy[cp, Norm[#] &][[1]];
           p + q];

In[72]:= p = closestPoint[f, {3, 3}, x, y]

Out[72]= {1.79714, 3.3654}

In[73]:= ContourPlot[f == 0, {x, -4, 4}, {y, -4, 4},
          Epilog → {{Red, PointSize[Large], Point[p]}, {Black, PointSize[Large], Point[{3, 3}]}}]
```

Out[73]=

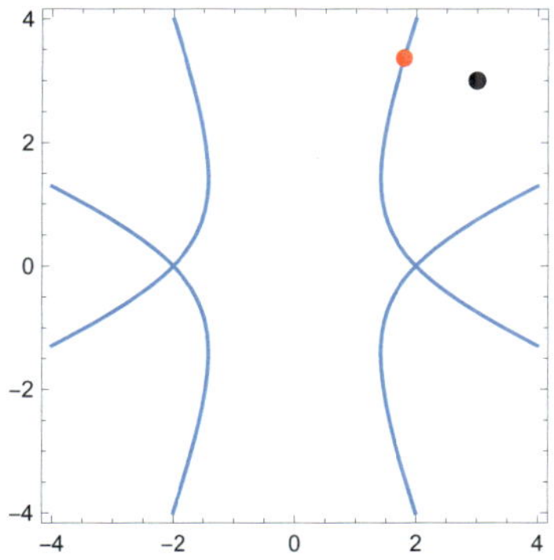

It should be noted that in a given case, there may be several points the minimal distance from q. In fact, if q were {0, 0} in the example above, then there are four points closest to q. For later use, this procedure only returns one arbitrarily chosen closest point.

More to Explore

For more information on Chebyshev polynomials, check out Wolfram MathWorld's entry (wolfr.am/t40IRnDb)

Check out this fun Wolfram Demonstration using Chebyshev polynomials (wolfr.am/t40S78Xh)

4 | Gauss Decomposition

In this section, we discuss our method for decomposing curves inside a circular region. The idea behind this comes from C. F. Gauss's fourth proof of the fundamental theorem of algebra (FTA) written in 1849. Let's start with an exploration of some of the topographical terminology.

4.1 Plane Topology

To avoid too much abstraction, we will work in the real Euclidean plane $\mathbb{R}^2$ viewed as a metric space. The *norm* of a point is $\| p \| = \| \{x_0, y_0\} \| = \sqrt{x_0^2 + y_0^2}$, and the distance between two points p, q will be denoted by $\| p - q \|$.

In the Wolfram Language, the distance from {3.2, −5.6} to {−1.3, 3.8} is one of the following:

```
In[1]:= Norm[{3.2, -5.6}-{-1.3, 3.8}]

Out[1]= 10.4216
```

```
In[2]:= EuclideanDistance[{3.2, -5.6}, {-1.3, 3.8}]

Out[2]= 10.4216
```

In this subsection, S will denote a subset of $\mathbb{R}^2$, often called a curve or a semialgebraic set. A semialgebraic set is a subset of the plane that is obtained by finite unions, complements and/or intersections of sets of the form $\{\{x, y\} \mid f(x, y) > 0\}$ or $\{\{x, y\} \mid f(x, y) \geq 0\}$ where $f[x, y]$ is a polynomial function in variables x, y—that is $f(x, y)$—is a curve. Mostly, we will just use the *atomic* semialgebraic set $\{\{x, y\} \mid f(x, y) > 0\}$. *Real Algebraic Geometry* (Bochnak, 1998), one of the most comprehensive texts in the eponymous field, points out that semialgebraic sets are very useful in studying algebraic curves, a fact that Gauss noted in his proof written in 1849.

A subset $\mathcal{U}$ of $\mathcal{S}$ is called *open* if, for every point p in $\mathcal{U}$, there is a number $\epsilon > 0$ so that every point q in $\mathcal{S}$ with $\|p - q\| < \epsilon$ is also in $\mathcal{U}$.

A point p of $\mathbb{R}^2$ is called a *boundary point* of $\mathcal{V} \subset \mathcal{S}$ if, given $\epsilon > 0$, there are points $u \notin \mathcal{V}$, $v \in \mathcal{V}$ with $\|u - p\| < \epsilon$ and $\|v - p\| < \epsilon$. Here, p itself may or may not be in $\mathcal{V}$. The set $\mathcal{V}$ is called *closed* in $\mathcal{S}$ if it contains all its boundary points in $\mathcal{S}$. A standard elementary result in topology is that $\mathcal{V}$ is closed in $\mathcal{S}$ if and only if the complement of $\mathcal{V}$ in $\mathcal{S}$ is open in $\mathcal{S}$. The set consisting of $\mathcal{V}$ and all its boundary points in $\mathcal{S}$ is called its *closure* in $\mathcal{S}$. A point in $\mathcal{S}$ that is not a boundary point is called an *interior point*.

1. The set $\mathcal{U} = \{\{x, y\} \mid x^2 + y^2 < 1\}$ is open in $\mathbb{R}^2$ while $\mathcal{V} = \{\{x, y\} \mid x^2 + y^2 \leqslant 1\}$ is closed in $\mathbb{R}^2$ and is the closure of $\mathcal{U}$.
2. The set $C = \{\{x, y\} \mid x^2 + y^2 = 1\}$ is the boundary of both $\mathcal{U}$ and $\mathcal{V}$ and is closed in $\mathbb{R}^2$.
3. While $\mathcal{W} = \{\{x, y\} \mid x^2 + y^2 = 1,\ y > 0\}$ is not open in $\mathbb{R}^2$, it is open in C with closure $\{\{x, y\} \mid x^2 + y^2 = 1,\ y \geqslant 0\}$.

Additional easy exercises in point set topology are to show an arbitrary union of open sets in $\mathcal{S}$ is open in $\mathcal{S}$ while finite unions of closed sets are closed. Using complements, finite intersections of open sets are open but infinite intersections of closed sets are closed. Both the empty set and $\mathcal{S}$ itself are open and closed in $\mathcal{S}$.

The key definition of this subsection is:

The set $\mathcal{S}$ is connected if and only if $\mathcal{S}$ is not the union of two disjoint open subsets of $\mathcal{S}$.

Alternatively, $\mathcal{S}$ is *connected* if and only if the only subsets of $\mathcal{S}$, which are both open and closed in $\mathcal{S}$, are the empty set and $\mathcal{S}$ itself. If $p \in \mathcal{S}$, then the union of all connected subsets of $\mathcal{S}$ containing p are called the *connected component* of p in $\mathcal{S}$.

Examples:

1. Let $C = \{\{x, y\} \mid x^2 + y^2 = 1\}$ be the circle curve. Then the complement of C in $\mathbb{R}^2$ is the union of two open sets $\{\{x, y\} \mid x^2 + y^2 < 1\} \cup \{\{x, y\} \mid x^2 + y^2 > 1\}$. Each of these is a connected component of the complement of C.
2. Another famous example of a disconnected open set of the plane comes from an analysis of the complex Newton method *basin of attraction*. Consider the complex polynomial $p[z] = z^3 - 1$, and let $\mathcal{U}$ be the set of points p in $\mathbb{R}^2$ such that Newton's method, starting from initial point p, converges. It can be shown that $\mathcal{U}$ is open in $\mathbb{R}^2$. Let $\mathcal{V}$ be the complement of $\mathcal{U}$, which is then closed.

Mathematica can draw this set:

In[3]:= `JuliaSetPlot[z - (z^3 - 3)/(3 z^2), z, ColorFunction → "RedBlueTones"]`

Out[3]=
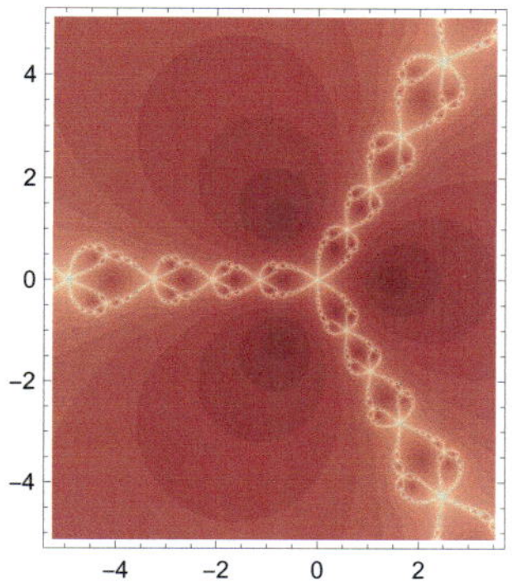

$\mathcal{U}$ consists of the points in the "RedBlueTones" colors, and $\mathcal{V}$ consists of the white points where Newton's method does not converge. If you could blow this picture up sufficiently, you would see that the complement $\mathcal{U}$ of $\mathcal{V}$ is the disjoint union of infinitely many connected components! It is also interesting to note that the white boundary of $\mathcal{V}$ is neither one- nor two-dimensional; it is what is known as a fractal.

Now let $\mathcal{S}, \mathcal{T}$ be plane subsets, perhaps one or both is the real line $\mathbb{R}$. A function $\phi : \mathcal{S} \longrightarrow \mathcal{T}$ is called *continuous* if for every open set $\mathcal{V} \subset \mathcal{T}$, the inverse image set $\phi^{-1}(\mathcal{V}) = \{p \in \mathcal{S} \mid \phi(p) \in \mathcal{V}\}$ is open in $\mathcal{S}$. In this case, both $\mathcal{S}, \mathcal{T}$ are real intervals, which is equivalent to the Calculus I definition of continuity. By simple set considerations, the definition is equivalent to the statement that the inverse image of every closed set is closed.

This last statement shows that if $f(x, y)$ is viewed as the polynomial function $f : \mathbb{R}^2 \longrightarrow \mathbb{R}$, then the point set $V(f) = \{p \mid f[p] = 0\}$ is closed in $\mathbb{R}^2$ since $\{0\}$ is closed in $\mathbb{R}$. Further, its complement $\{p \mid f(p) > 0 \text{ or } f(p) < 0\}$ is the union of two open sets in $\mathbb{R}^2$. Thus a main application of the topological considerations above can be summarized as follows:

A square-free curve, $f = 0$, as a point set $V(f)$ disconnects its complement into two (not necessarily connected) open sets $\{p \mid f(p) > 0\} \cup \{p \mid f(p) < 0\}$.

The reason f needs to be square-free is that otherwise, one of the two sets above could be empty. It is somewhat harder to prove that in the square-free case, the curve is the complete boundary of each of these semialgebraic sets in $\mathbb{R}^2$. We leave it to Gauss to make the pronouncement that this is true in the next subsection. It actually follows easily from the result in Chapter 5 that a square-free curve has a finite number of singular points.

We have already seen examples where the semialgebraic set $\{p \mid f(p) > 0\}$ is not connected. Nor is it true that the set $V(f)$ is connected; however, since $V(f)$ is a closed set of $\mathbb{R}^2$, each component of $V(f)$ is too.

For later use, the following theorem of point set topology is easy to prove in our context of open sets:

The continuous image of a connected set is connected.

In particular, the image of a parametrically defined curve $\phi(t)$ from a connected interval of $\mathbb{R}$ to $\mathbb{R}^2$ is connected. It is the main goal of this section to show that not only is each connected component of the boundary of each component of $\{p \mid f(p) > 0\}$ the image of such a continuous parametric curve, but to actually find a numerically accurate approximation of these boundary curves. Thus curves and semialgebraic sets are much better behaved than arbitrary sets (such as the Newton basin of attraction).

4.2 The Gauss Principle

D'Alembert gave what the French considered to be the first acceptable proof of the fundamental theorem of algebra in 1748, but Gauss received credit for said proof (in the general world) in 1799. At the time of his fourth proof, Gauss was in his 70s, one of many examples that creativity in math does not end at 40. Overall, this last proof was similar to the first but with one new idea clarifying his earlier argument. I will call this the Gauss Principle. I state this using modern mathematical terminology:

Let C be the point set given by the solution set of the square-free equation $f(x, y) = 0$, and let $\mathcal{D} = \{\{x, y\} \mid x^2 + y^2 \leq r^2\}$ be the disk of radius r. Then the complete boundary of the semialgebraic set $\mathcal{S} = \{p \in \mathcal{D} \mid f(p) > 0\}$ consists of $C \bigcap \mathcal{D}$ together with portions of the boundary $x^2 + y^2 = r^2$ of $\mathcal{D}$. $\mathcal{S}$ has finitely many connected components. Further, if $\mathcal{K}$ is any connected component of $\mathcal{S}$, then the connected components of the boundary of $\mathcal{K}$ are images of continuously parameterized curves, which can be parameterized in the direction so that as the parameter increases, the interior of $\mathcal{K}$ lies to the right, that is the regions are parameterized clockwise.

For low-degree or well-behaved nonsingular curves, it is easy to view this using RegionPlot. Note that $\{p \in \mathcal{D} \mid f(p) > 0\}$ is the darker orange-blue region. For convenience, in the rest of this chapter, we define the utility function gaussPlot.

For this function, *f* will be a polynomial in *x*; *y*, *r* will be a positive real number and *pts* will be a possibly empty set of points of norm less than or equal to *r*:

```
In[4]:= gaussPlot[f_, r_, pts_] := Module[{plta, pltb, pltc},
          plta := RegionPlot[x^2+y^2 < r^2, {x, -1.1 r, 1.1 r},
             {y, -1.1 r, 1.1 r}, PlotStyle → LightOrange, BoundaryStyle → Orange];
          pltb := RegionPlot[{f > 0&& x^2+y^2 < r^2}, {x, -r, r},
             {y, -r, r}, BoundaryStyle → Blue];
          pltc := ListPlot[Table[Labeled[pts[[n]], n], {n, Length[pts]}],
             PlotStyle → Red, PlotMarkers → Automatic];
          Show[plta, pltb, pltc, ImageSize → Small]];
```

We now give several examples of the Gauss Principle:

In[5]:=
```
{gaussPlot[y^2 - x^3 + x, 3, {}], gaussPlot[-1 - 3 x^2 + x^3 + 3 y^2 - 3 x y^2, 6, {}],
  gaussPlot[-1 + x - 0.2*x^2 + 0.008*x^3 + y + x*y -
    0.2*x^2*y - 0.2*y^2 - 0.2*x*y^2 + 0.008*y^3, 15, {}]}
```

Out[5]=

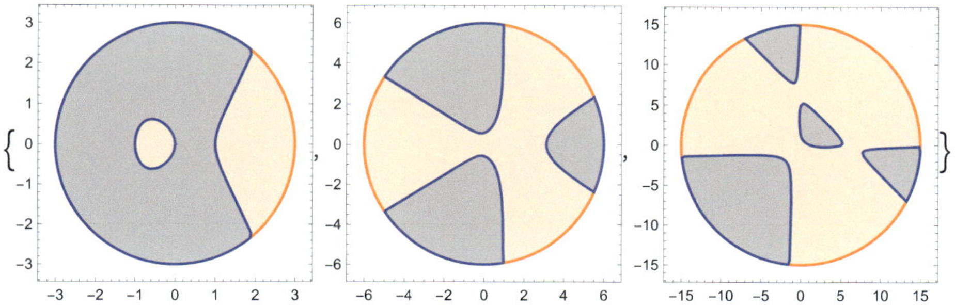

Note that there are one, three and four components of $\{p \in \mathcal{D} \mid f[p] > 0\}$, respectively.

Here is another simple example using RegionPlot, a union of four lines that clearly shows Gauss's decomposition:

In[6]:=
```
f = (y - .75 x - 4) (y + .75 x + 4) (y - 2 x) (y + 2 x);
P0 := RegionPlot[x^2 + y^2 < 100, {x, -10, 10},
    {y, -10, 10}, PlotStyle → LightOrange, BoundaryStyle → None];
P1 := RegionPlot[y > .75 x + 4 && y > 2 x && y > -2 x && x^2 + y^2 < 100,
    {x, -10, 10}, {y, -10, 10}, PlotStyle → LightGreen, BoundaryStyle → Green];
P2 := RegionPlot[y < .75 x + 4 && y < 2 x && y > -2 x && y > -.75 x - 4 && x^2 + y^2 < 100,
    {x, -10, 10}, {y, -10, 10}, PlotStyle → LightBlue, BoundaryStyle → Blue];
P3 := RegionPlot[y < .75 x + 4 && y < -2 x && y > -.75 x - 4 && y > 2 x && x^2 + y^2 < 100,
    {x, -10, 10}, {y, -10, 10}, PlotStyle → LightGray, BoundaryStyle → Gray];
P4 := RegionPlot[y < -.75 x - 4 && y > .75 x + 4 && x^2 + y^2 < 100, {x, -10, 10},
    {y, -10, 10}, PlotStyle → LightPurple, BoundaryStyle → Purple];
P5 := RegionPlot[y < 2 x && y < -.75 x - 4 && y < -2 x && x^2 + y^2 < 100, {x, -10, 10},
    {y, -10, 10}, PlotStyle → LightMagenta, BoundaryStyle → Magenta];
P6 = ContourPlot[x^2 + y^2 == 100, {x, -10, 10}, {y, -10, 10}, ContourStyle → Orange];
{ContourPlot[f == 0, {x, -10, 10}, {y, -10, 10}], Show[P0, P1, P2, P3, P4, P5, P6]}
```

Out[14]=

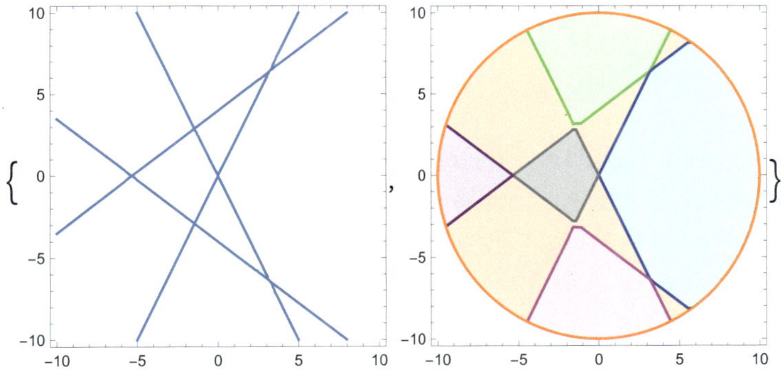

Let's take a closer look:

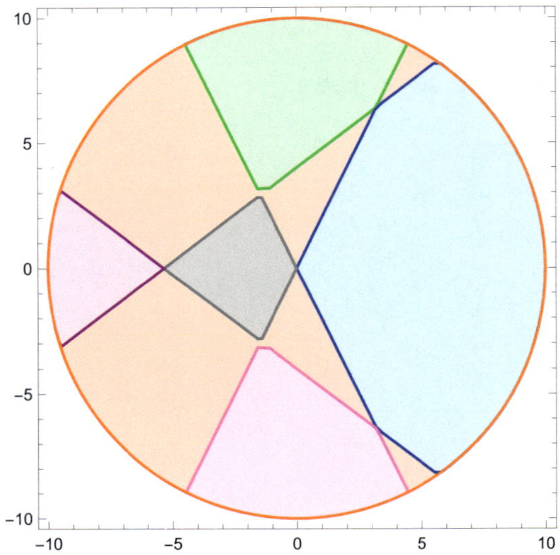

Note that the light orange region is $f < 0$ while the other colored regions are the five connected components of the region $f > 0$.

An important consequence of Gauss's Principle is that the complement of a square-free curve f can be two-colored as it is the union of the two disjoint nonempty sets $\{f > 0\}$ and $\{f < 0\}$.

Note that Abhyankar's illustration (also in Chapter 1) violates this in several ways. More generally, there cannot be an odd number of curve segments meeting at a point.

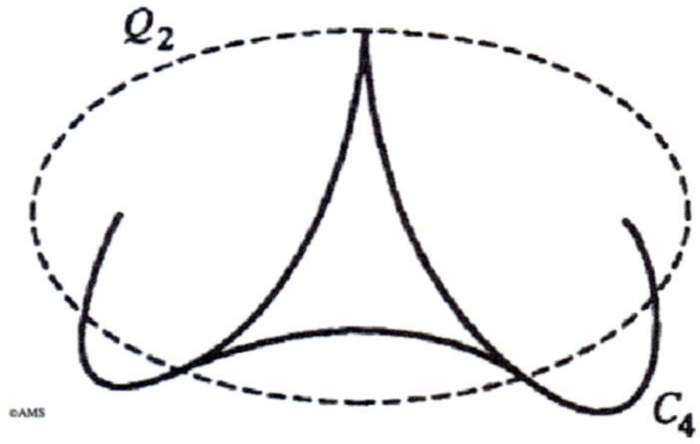

Following the practice from Euclid until the end of the nineteenth century, Gauss most likely did not attempt to prove this topological statement as it was meant to be self evident. I am unaware of the original source for this statement; all modern accounts go back to the discussion in Appendix 1 of Uspensky (1948). I will not attempt a proof in this book but will give many examples and several algorithms illustrating this principle.

Here are some examples using Gauss curves, which Gauss had already introduced in his first proof. We use them to illustrate Gauss's fourth proof.

We recall the definition:

```
In[15]:= gaussCurve[p_, x_, y_, z_] := ComplexExpand[Re[p /. {z → x+I*y}]];
```

```
In[16]:= f = gaussCurve[z^3-3 z +1, x, y, z]
         g = gaussCurve[z^3-1, x, y, z]
         h = gaussCurve[z^3, x, y, z]
         {gaussPlot[f, 4, {}], gaussPlot[g, 4, {}], gaussPlot[h, 4, {}]}
```

Out[16]= $1-3x+x^3-3xy^2$

Out[17]= $-1+x^3-3xy^2$

Out[18]= x^3-3xy^2

Out[19]=
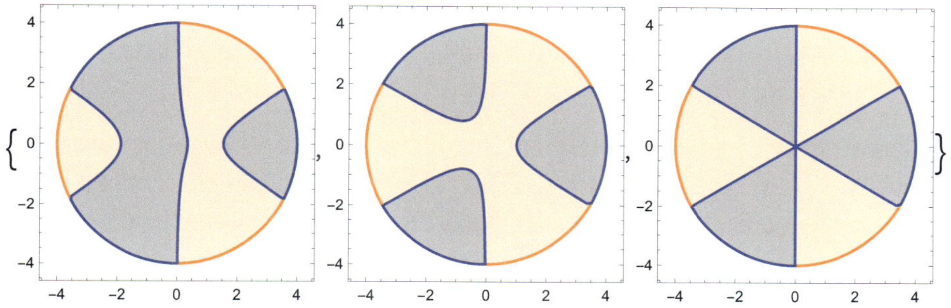

Gauss noticed that for complex polynomial $p(z)$ with leading term z^n, the intersections of different Gauss curves with the circle of large radius r were nearly independent of the lower-degree terms and the plots near the circle were essentially the same.

Note that if we white out the center of each of the previous plots and indicate points on the boundary in the direction of Gauss's infinite points, our plots are now all the same:

```
In[20]:= WP := RegionPlot[x^2+y^2 < 9, {x, -4, 4},
              {y, -4, 4}, PlotStyle → White, BoundaryStyle → LightGray];
         sol = {x, y} /. NSolve[{x^2+y^2-15, h}];
         bp = SortBy[sol, ArcTan@@# &];
         {Show[gaussPlot[f, 4, bp], WP],
           Show[gaussPlot[g, 4, bp], WP], Show[gaussPlot[h, 4, bp], WP]}
```

Out[23]=
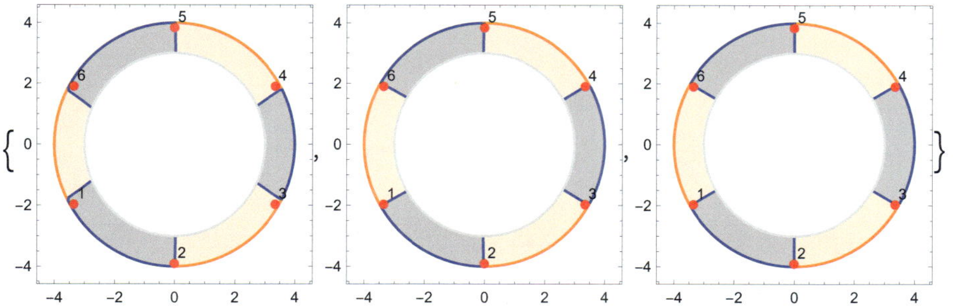

Focusing on the picture for $p(z) = z^n$ (right plot), we see $p(x_j + y_j \boldsymbol{i})$ for $j = 1, 3, 5$ has real part 0 and negative imaginary part while $p(x_j + y_j \boldsymbol{i})$ for $j = 2, 4, 6$ also has real part 0 (the definition of a Gauss curve) but imaginary part positive. If you extrapolate from the full pictures above to the numbered pictures in the previous output, we see that following the curve inwards from 1, 3 or 5 exits at one of the points 2, 4 or 6 (although which one depends on the lower-degree curve we see. In any case, Gauss observes that the imaginary part of $p(x + yi)$ goes continuously from a negative value to a positive value along the trip. However, the image of the connected path to $\mathbb{R}$ is connected, which means that it must pass through 0. At that point, $p(x + yi) = 0$, so Gauss has found a zero of $p(z)$ on the Gauss curves. In fact, since there are n-curves he has found n-zeros. This is his fourth proof of the FTA. I remind the reader that the argument in this paragraph only holds for Gauss curves of monic complex polynomials, but the Gauss Principle holds for all square-free curves.

There is a slight problem with the $p(z) = z^n$ case because of the singularity at the origin, but the curve is removed, so there are still three components of $\{p \mid f(p) > 0\}$. Therefore, this is essentially like the middle picture. More generally, using the Gauss Principle for singular curves, we can use *perturbation* to approximate the curve for RegionPlot.

In classical theory, a plane curve singularity must satisfy the three equations in two unknowns: $f[x, y] = 0$, $D[f, x] = 0$, $D[f, y] = 0$, but an over-determined system tends to be very unstable, so any small change (perturbation) in the system causes a singular point to no longer be a point and/or singular. In our current semialgebraic point of view, adding a small positive constant to f expands the region $f(p) > 0$ while subtracting a small positive constant contracts it. In either case, a pre-existing singular point is no longer on the perturbed curve.

We illustrate this with Gauss's original example, which, among other things, indicates that Gauss was certainly considering the case where the Gauss curve might be singular:

```
In[24]:= p = z^5 - 5 z^4 + 9 z^3 - 5 z^2;
f = -gaussCurve[p, x, y, z]
f1 = f - .1;
f2 = f + .1;
sol = {x, y} /. NSolve[{x^2 + y^2 - 25, f}];
bp = Reverse[SortBy[sol, ArcTan @@ # &]];
{gaussPlot[f, 5, bp], gaussPlot[f1, 5, bp], gaussPlot[f2, 5, bp]}
```

Out[25]= $5x^2 - 9x^3 + 5x^4 - x^5 - 5y^2 + 27xy^2 - 30x^2y^2 + 10x^3y^2 + 5y^4 - 5xy^4$

Out[30]=

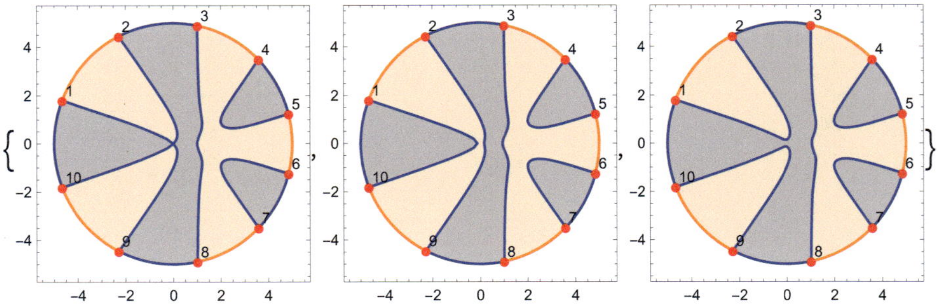

Gauss's example is on the left. The paths that Gauss is taking (that is, his decomposition of the left-hand singular system is approximated by the middle picture), so starting from 1, Gauss ends up at 10; from 3, he ends at 8; from 5, he ends at 4; from 7, he ends at 6; and from 9, he ends at 2. The picture on the right would give an alternate way to construct the paths; for example, from 1, he would end at 2. In fact, this is the way Gauss would decompose the curve $-f = 0$. In either case, the original curve on the left would be the union of these paths.

The example shows that for singular curves, the Gauss Principle gives different decompositions for the curve f and $-f$. While this may cause a bit of confusion, the important thing is that the Gauss decomposition does allow for the decomposition of singular curves.

4.3 Strategy for Gauss Decomposition of Curves in a Disk

Based on the ideas of the last section, what we need to do to decompose an algebraic curve into continuously parameterized arcs (parameterized curve, that is, $1 - 1$) is the following:

1. Find critical points on the curve that include the points where the curve intersects the boundary of the disk, singular points and at least two points on every topological connected component.
2. Connect these critical points with continuously parameterized arcs parameterized in the correct direction.

The critical points from Section 3.5 take care of most of the requirements of Step 1, but there may be critical points that are not in our disk because they are too far from the origin. Ideally, the radius r of the disk will be large enough to include all critical points, but in later sections, we will see examples of curves with critical points that have very large norms.

We claim the following procedure will get us at least a preliminary set of critical points, and we will call these points Gauss gPoints (For a complete list of global functions, see Appendix 2):

```
In[31]:= gPoints[f_, r_, x_, y_] := Module[{bp, cp, sol},
           sol = NSolve[{f, x^2+y^2-r^2}, {x, y}, Reals];
           If[Length[sol] > 0, bp = {x, y} /. sol, bp = {}];
           bp = Reverse[SortBy[bp, ArcTan@@# &]];
           cp = DeleteDuplicates[ Select[{x, y} /. NSolve[{f, y D[f, x] - x D[f, y]}, {x, y}, Reals],
                 Norm[#] < r &], Norm[#1 - #2] < .1 &];
           Join[
             bp,
             cp]];
```

Note that for later convenience, we sort the boundary points clockwise around the circle and put them first on our list.

Our first example is the previously discussed Gauss curve. Later, we may need to add more points by the singularity (left) or use perturbation (right):

```
In[32]:= p = z^5 - 5 z^4 + 9 z^3 - 5 z^2;
         f = -gaussCurve[p, x, y, z]
         cpf = gPoints[f, 5, x, y];
         f1 = f - .2;
         cpf1 = gPoints[f1, 5, x, y];
         {gaussPlot[f, 5, cpf], gaussPlot[f1, 5, cpf1]}
```

Out[33]= $5x^2 - 9x^3 + 5x^4 - x^5 - 5y^2 + 27xy^2 - 30x^2y^2 + 10x^3y^2 + 5y^4 - 5xy^4$

Out[37]=

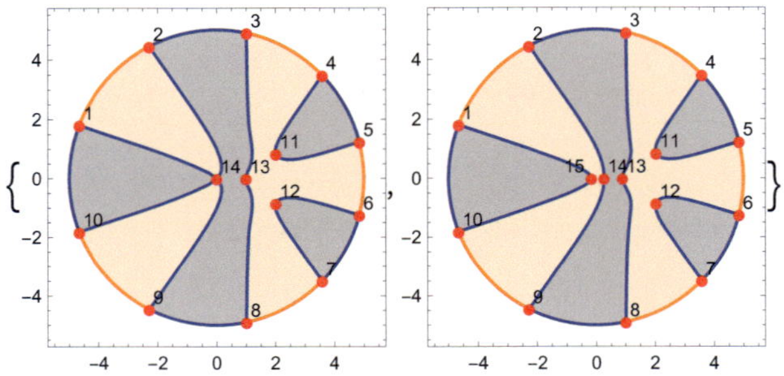

Here are some bounded examples. Note that we change signs so the exterior of the curve is negative. We get plenty of points:

```
In[38]:= g1 = -(x^4 + 4 x y + y^4 - 1/4); (* peanut curve *)
         cpg1 = gPoints[g1, 3, x, y];
         g2 = -(1 - 6 x^2 + x^4 - 5 y^2 + 5 x^2 y^2 + 4 y^4);
         cpg2 = gPoints[g2, 3, x, y];
         {gaussPlot[g1, 3, cpg1], gaussPlot[g2, 3, cpg2]}
```

Out[42]=

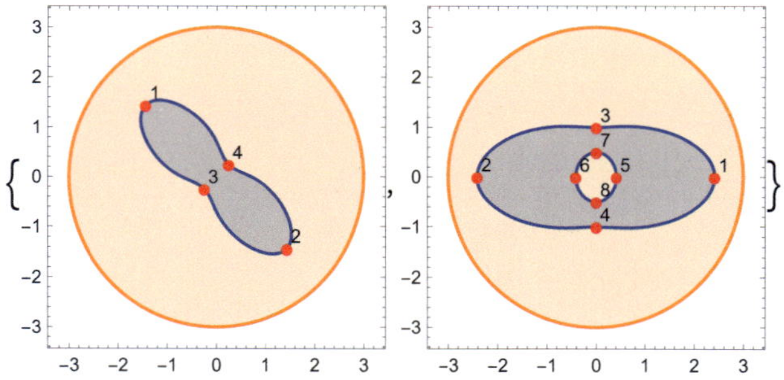

Here are some harder examples:

```
In[43]:= h1 = -(1 - 8 x^2 + 8 x^4 + 3 y - 4 y^3); (* Chebyshev quartic *)
         cph1 = gPoints[h1, 2, x, y];
         h2 = -(x^4 - x^2 y + y^3); (*bow curve*)
         cph2 = gPoints[h2, .5, x, y];
         {gaussPlot[h1, 2, cph1], gaussPlot[h2, .5, cph2]}
```

Out[47]=

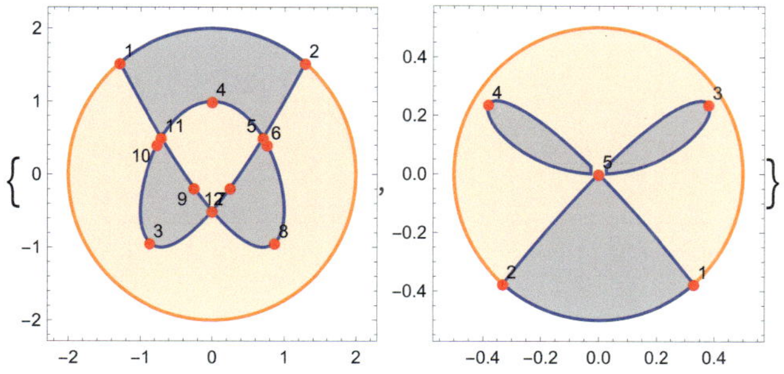

We see on the Chebyshev quartic that points 5 and 6 and points 10 and 12 are very close together.

We may want to delete 6 and 10 but leave the singular points, but on the bow curve, we may wish to add a few points:

```
In[48]:= cph1a = Delete[cph1, {{6}, {10}}];
         cph2a = Join[cph2, {x, y} /. NSolve[{h2, x^2 + y^2 - .02}, {x, y}, Reals]];
         {gaussPlot[h1, 2, cph1a], gaussPlot[h2, .5, cph2a]}
```

Out[50]=

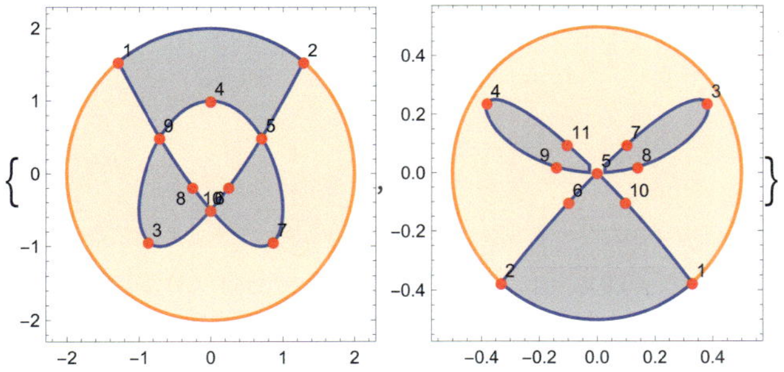

Finally, we recall that choosing the positive semialgebraic set determines a direction for the curve—we go clockwise around each component of the positive region inside the disk. An important observation is that of the two unit tangent vectors at any regular point on a plane algebraic curve, we can select the one going in the correct clockwise direction with the following simple Mathematica procedure.

Use Cross to get the vector cross and Thread to put *f* over the objects or arguments:

```
In[51]:= gTvec[f_, p_, x_, y_] :=
           Cross[Normalize[Chop[{D[f, x], D[f, y]} /. Thread[{x, y} → p], 1.*^-6]]];
```

Note that this procedure returns {0, 0} for a near-singular point; otherwise, it returns a vector of norm 1.

Here we use Table to Append the vector arrows to each curve:

```
In[52]:= f = gaussCurve[z^4-2z^3-z+1, x, y, z]
         cp = gPoints[f, 5, x, y];
         pltc = Graphics[Table[{Black, Arrow[{p, p+gTvec[f, p, x, y]}]}, {p, cp}]];
         Show[gaussPlot[f, 5, cp], pltc, ImageSize → Medium]
```

Out[52]= $1-x-2x^3+x^4+6xy^2-6x^2y^2+y^4$

Out[55]=

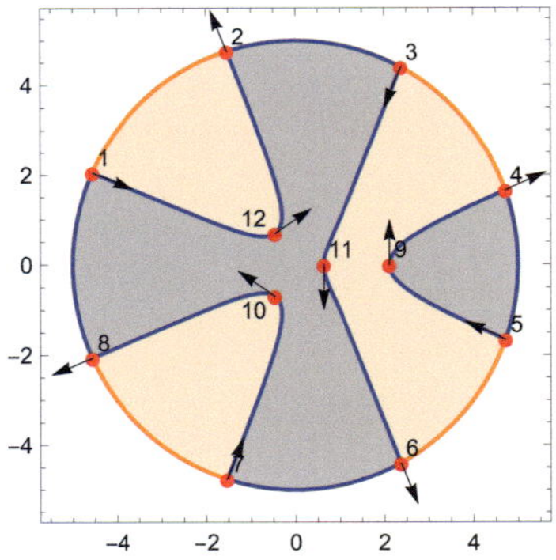

Now that we have critical points and a direction of travel to complete Step 2, we need to be able to travel from one regular critical point to the next. We can treat singular points by removing them with a small perturbation or by finding nearby nonsingular points in the direction of travel at start (such as we did with the bow curve) and we can get these by using a straight line from the singular point to these nearby points.

4.4 Naive Curve Tracing

Curve (or path) tracking is a basic tool in *numerical algebraic geometry* (see Sommes and Wampler [2005] for examples). The simple idea, borrowed from numerical differential equation solving, is to use the predictor/corrector method we discuss here. Mathematica has good numerical differential equation solvers and we could use them for curve tracking (we will discuss this in a later section).

However, for the predictor/corrector method, given a point on the curve, we can make use of our known direction of travel. So at point p of the curve, we choose a *step size* s (which, for now, will be fixed) and predict that the next point will be near q = p + s gTvec[f, p, x, y]. We correct this by finding a nearby point to q on the curve. The most common method is to find the point of intersection of the line through q orthogonal to the tangent vector with the curve, typically using the two-variable Newton's method. In practice, one or two iterations are a sufficient correction. In the Wolfram Language, we can use FindRoot.

Consider the curve $f = x^3 - y^2 + 3 = 0$ and point $p_0 = \{1, 2\}$.

In reference to the next plot, let the step size s = 1. A step size of .2, .1 or .05 will be more common in practice:

```
In[56]:= f = x^3+3-y^2;
         p0 = {1., 2.};
         tv = gTvec[f, p0, x, y]
Out[58]= {0.8, 0.6}

In[59]:= q = p0+1.*tv
Out[59]= {1.8, 2.6}
```

Find the line through *q* orthogonal to *tv* as follows using the **line** procedure from Chapter 1:

```
In[60]:= nv = Cross[tv]
Out[60]= {-0.6, 0.8}

In[61]:= l = line[q, Append[nv, 0], x, y]
Out[61]= -1.58201+0.421871 x+0.316403 y

In[62]:= lx = {x, y} /. NSolve[{l, x == 2}][[1]]
         ly = {x, y} /. NSolve[{l,  y == 2.9}][[1]]
Out[62]= {2., 2.33333}
Out[63]= {1.575, 2.9}

In[64]:= p1 = {x, y} /. FindRoot[{f, l}, {x, q[[1]]}, {y, q[[2]]}]
Out[64]= {1.67215, 2.77047}

In[65]:= p2 = closestPoint[f, q, x, y]
Out[65]= {1.63357, 2.71279}

In[66]:= P1 = RegionPlot[f > 0, {x, .8, 2}, {y, 1.7, 2.9}];
         P2 = Graphics[{{Black, PointSize[Medium], Point[{p0, p1, p2}]},
                {Black, Text[Style["p0", 16], p0+{.02, .08}]},
                {Black, Text[Style["p2", 16], p2+{-.07, .02}]}, {Black,
                  Text[Style["p1", 16], p1+{.04, .08}]}, {Black, Thick, Arrow[{p0, p0+{.8, .6}}]},
                {Black, Thick, Line[{lx, ly}]}, {Black, Text[Style["tv", 16], {1.5, 2.3}]}}];
         Show[P1, P2]
```

Out[68]=

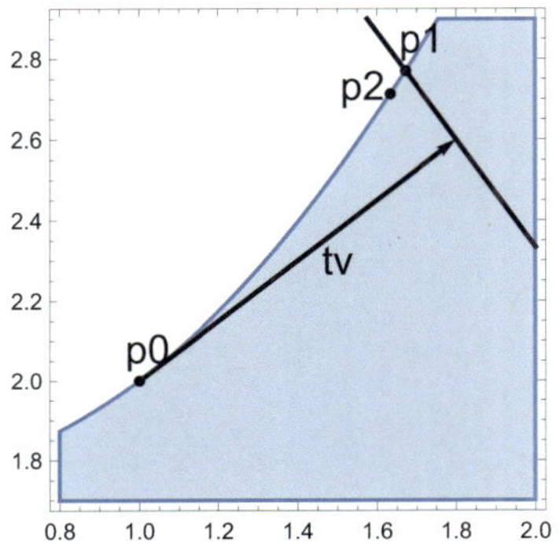

This method is implemented as pathFinderT in Appendix 2.

Although a bit slower, we will use our closestPoint procedure from Section 3.5. This works a little better and is easier to program:

```
In[69]:= p2 = closestPoint[f, q, x, y]

Out[69]= {1.63357, 2.71279}
```

We can now write a procedure that, given a point p on the curve, will attempt to find a list of points on the segment to the next critical point q in the direction of the curve, assuming it is known by the user. The step size s is one of the arguments, and it is assumed that all the points are more than $2\,s$ from each other.

This procedure returns a list of points on the curve starting from the given initial point p and, hopefully, ending with q. Since we may miss it, we limit the number of iterations (this limit maxit can be adjusted in the code).

By default, this does 30 iterations and quits if it has not gotten to within distance $2\,s$ of the endpoint. An option maxit → m will change the maximum number of iterations. Note that even if this function does not achieve it's objective, it will return a list of the the points it has found putting the endpoint q a the end of the list.

A warning is given to indicate the failed objective (For a complete list of global functions, see Appendix 2):

```
In[70]:= Options[pathFinder] = {maxit → 30};
pathFinder[f_, p_, q_, s_, x_, y_, OptionsPattern[]] :=
  Module[{L, p0, p1, u, mxt, k},
   mxt = OptionValue[maxit];
   p0 = p;
   L = Reap[Sow[p];
      k = 0;
      While[Norm[q - p0] > 2 s && k < mxt,
       u = p0 + s * gTvec[f, p0, x, y];
       p1 = closestPoint[f, u, x, y];
       Sow[p1];
       p0 = p1;
       k++];
      Sow[q]][[2, 1]];
   If[k ≥ mxt, Print[Style["Warning: iteration limit reached at", Orange], p1]];
   L];
```

We test this using the points in our previous example:

```
In[72]:= f = x^3+3-y^2;
         L = pathFinder[f, p0, p2, .2, x, y]
         Show[gaussPlot[f, 4, {p0, p2}], ListLinePlot[L, PlotStyle → Black]]
```

Out[73]= {{1., 2.}, {1.15222, 2.12831}, {1.2919, 2.27073}, {1.4212, 2.42292}, {1.63357, 2.71279}}

Out[74]=

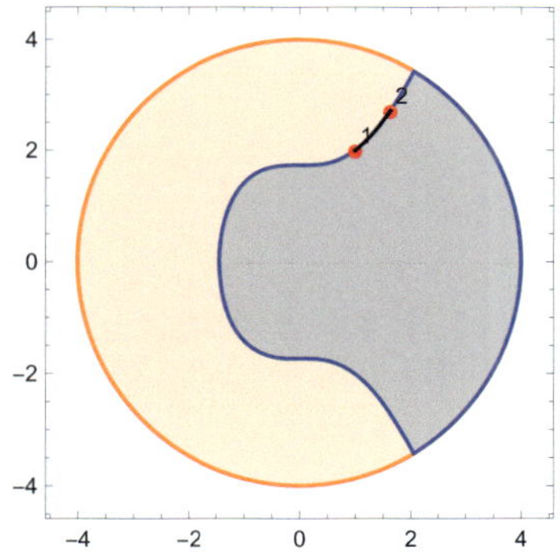

In general, path finding is difficult, especially in real cases. Constructing a black box program, which will decompose any real plane curve into paths, is very difficult. Receiving guidance from a person who knows which points lie on a path can be helpful. If we want the computer to reliably and quickly figure out the curve by using path tracing to find the next point, the naive approach used previously will not work well. We will attempt an improvement in the next section.

For example, to find Gaussian decomposition, set up the following Gauss curve using perturbation first:

```
In[75]:= f = gaussCurve[z^5-5z^4+9z^3-5z^2, x, y, z];
```

```
In[76]:= f1 = f-.1
```

Out[76]= $-0.1-5x^2+9x^3-5x^4+x^5+5y^2-27xy^2+30x^2y^2-10x^3y^2-5y^4+5xy^4$

```
In[77]:= gp = gPoints[f1, 5, x, y];
         gaussPlot[f1, 5, gp]
```

Out[78]=

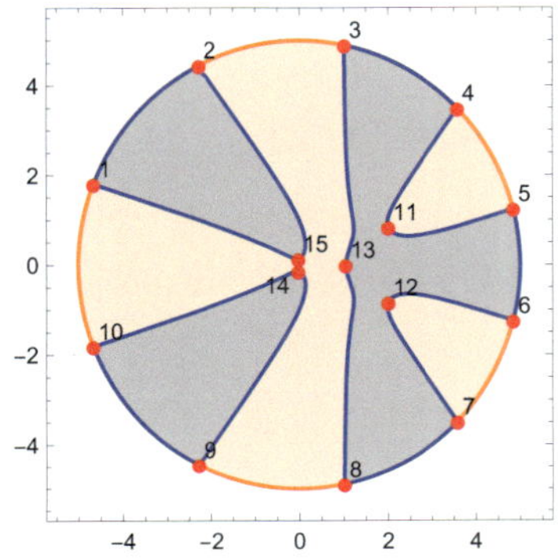

In[79]:=
```
Timing[
  L1 = pathFinder[f1, gp[[2]], gp[[15]], .2, x, y];
L2 = pathFinder[f1, gp[[15]], gp[[1]], .2, x, y];
L3 = pathFinder[f1, gp[[4]], gp[[11]], .2, x, y];
L4 = pathFinder[f1, gp[[11]], gp[[5]], .2, x, y];
L5 = pathFinder[f1, gp[[6]], gp[[12]], .2, x, y];
L6 = pathFinder[f1, gp[[12]], gp[[7]], .2, x, y];
L7 = pathFinder[f1, gp[[8]], gp[[13]], .2, x, y];
L8 = pathFinder[f1, gp[[13]], gp[[3]], .2, x, y];
L9 = pathFinder[f1, gp[[10]], gp[[14]], .2, x, y];
L10 = pathFinder[f1, gp[[14]], gp[[9]], .2, x, y];]
```

Out[79]= {19.7022, Null}

In[80]:=
```
Show[RegionPlot[f > 0 && x^2+y^2 < 25, {x, -5, 5}, {y, -5, 5}, BoundaryStyle → None],
  ListLinePlot[{Join[L1, L2], Join[L3, L4], Join[L5, L6], Join[L7, L8], Join[L9, L10]},
   PlotStyle → {Black, Red, Brown, Blue, Purple}]]
```

Out[80]=

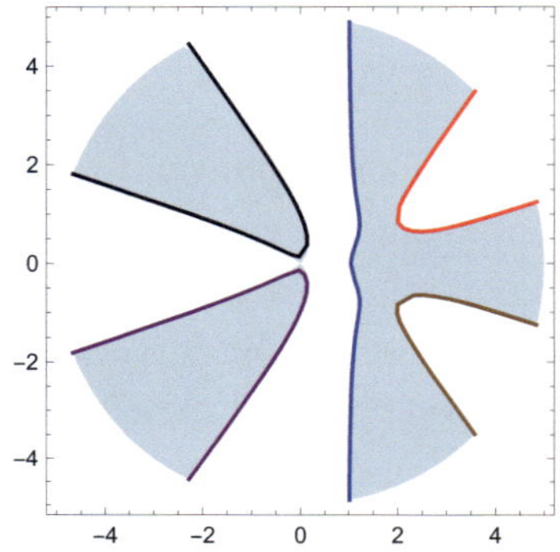

We got a good plot and decomposition, but it took a long time.

For singular curves, it is important to note that pathFinder can trace into a singular point but not out of one. Thus we should augment our critical points with points a little upstream from the singular points and trace into the critical point by replacing the curve equation with its negative.

As before, our code for the bow curve:

In[81]:=
```
h2 = -(x^4-x^2 y+y^3);
cph2 = Chop[gPoints[h2, .5, x, y]];
cph2a = Join[cph2, {x, y} /. NSolve[{h2, x^2+y^2-.02}, {x, y}, Reals]]
plta := gaussPlot[h2, .5, cph2a];
pltc =
   Graphics[Table[{Black, Arrow[{p, p+.1 gTvec[h2, p, x, y]}]}, {p, Delete[cph2a, 5]}]];
Show[
  plta,
  pltc]
```

Out[83]= {{0.330354, −0.375322}, {−0.330354, −0.375322}, {0.380892, 0.237985}, {−0.380892, 0.237985}, {0, 0}, {−0.0977421, −0.102208}, {0.10284, 0.0970767}, {0.139999, 0.0200083}, {−0.139999, 0.0200083}, {0.0977421, −0.102208}, {−0.10284, 0.0970767}}

Out[86]=

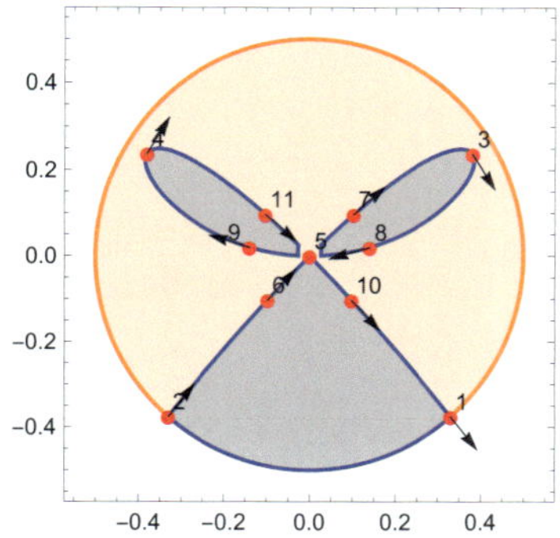

Thus we proceed as follows using a smaller step size going into the singular point:

In[87]:=
```
H1 = pathFinder[h2, cph2a[[2]], cph2a[[6]], .05, x, y];
H2 = pathFinder[h2, cph2a[[6]], cph2a[[5]], .025, x, y];
H3 = pathFinder[−h2, cph2a[[7]], cph2a[[5]], .025, x, y];
H4 = pathFinder[h2, cph2a[[7]], cph2a[[3]], .025, x, y];
H5 = pathFinder[h2, cph2a[[3]], cph2a[[8]], .025, x, y];
H6 = pathFinder[h2, cph2a[[8]], cph2a[[5]], .025, x, y];
H7 = pathFinder[−h2, cph2a[[9]], cph2a[[5]], .025, x, y];
H8 = pathFinder[h2, cph2a[[9]], cph2a[[4]], .02, x, y];
H9 = pathFinder[h2, cph2a[[4]], cph2a[[11]], .025, x, y];
H10 = pathFinder[h2, cph2a[[11]], cph2a[[5]], .025, x, y];
H11 = pathFinder[−h2, cph2a[[10]], cph2a[[5]], .025, x, y];
H12 = pathFinder[h2, cph2a[[10]], cph2a[[1]], .05, x, y];
```

In[98]:=
```
{Show[gaussPlot[h2, .5, cph2a],
   ListLinePlot[{H1, H2, H3, H4, H5, H6, H7, H8, H9, H10, H11, H12}]],
 ListLinePlot[{H1, H2, H3, H4, H5, H6, H7, H8, H9, H10, H11, H12},
   PlotStyle → Blue, Axes → False, Frame → True]}
```

Out[98]= {

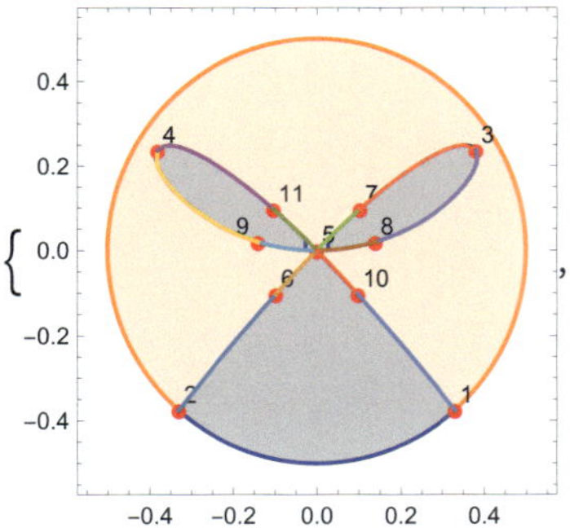

,

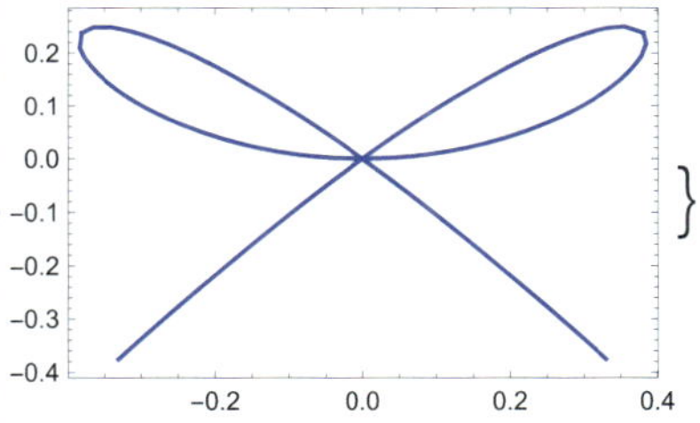

}

In[99]:=
```
Length[Join[H1, H2, H3, H4, H5, H6, H7, H8, H9, H10, H11, H12]]
```

Out[99]= 113

This takes some trial and error, including using different step sizes and some experimentation to generate these 113 points, but this gives the best plot of the bow curve.

4.5 The Affine Graph of a Curve

I have tried to use the word *plot* for a picture of a curve in this book. The word *graph* will be used in the same sense as it is used in the Graphs & Networks section of the Wolfram Language & System Documentation Center. Graphs provide a combinatorial way to view the topology of a curve in a simple way. We will build curve graphs using *critical points* as vertices and edges where there is a segment directly connecting one critical point to another. We will see that curve graphs have a special property that allows us to decompose curve graphs analogously to how we decompose curves.

Consider our plot from earlier in this section of the Chebyshev quartic but with direction vectors added:

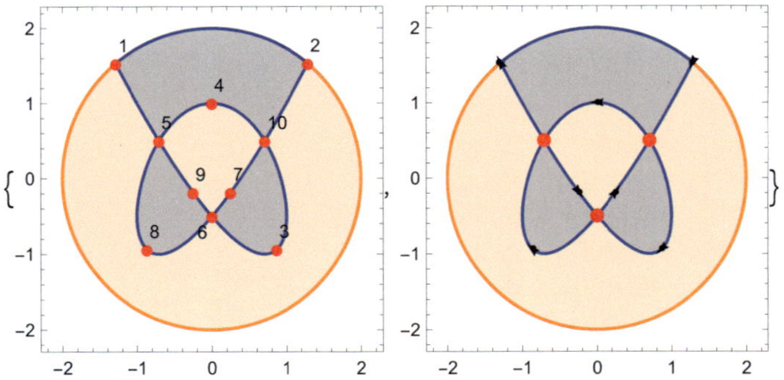

We can construct the *directed graph* as follows:

```
In[100]:= CGdata = {2 → 10, 10 → 3, 3 → 6, 6 → 7,
              7 → 10, 10 → 4, 4 → 5, 5 → 9, 9 → 6, 6 → 8, 8 → 5, 5 → 1};
          CG = Graph[CGdata, VertexLabels → "Name"]
```

Out[101]=

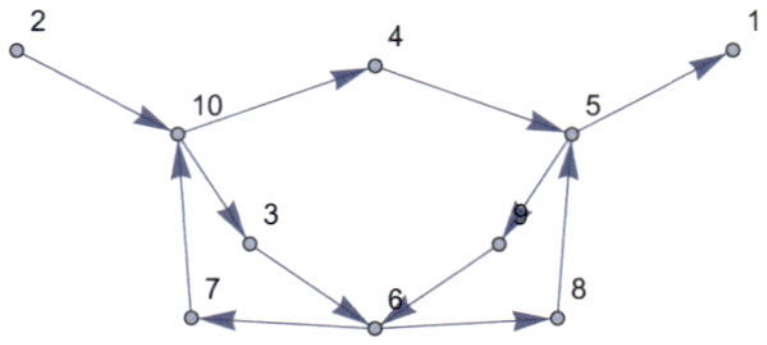

Or we can specify the graph more compactly by:

```
In[102]:= CGlist = EdgeList[CG]
```

```
Out[102]= {2 ⟶ 10, 10 ⟶ 3, 3 ⟶ 6, 6 ⟶ 7, 7 ⟶ 10, 10 ⟶ 4, 4 ⟶ 5, 5 ⟶ 9, 9 ⟶ 6, 6 ⟶ 8, 8 ⟶ 5, 5 ⟶ 1}
```

This last format is preferred because it distinguishes between directed edges and undirected edges.

Here are some useful graph tools:

```
In[103]:= e1 = DirectedEdge[1, 2]
          u1 = UndirectedEdge[1, 2]
          UG = UndirectedGraph[CG]
          EdgeList[UG]
```

Out[103]= 1 ⟶ 2

Out[104]= 1 ⟷ 2

Out[105]=

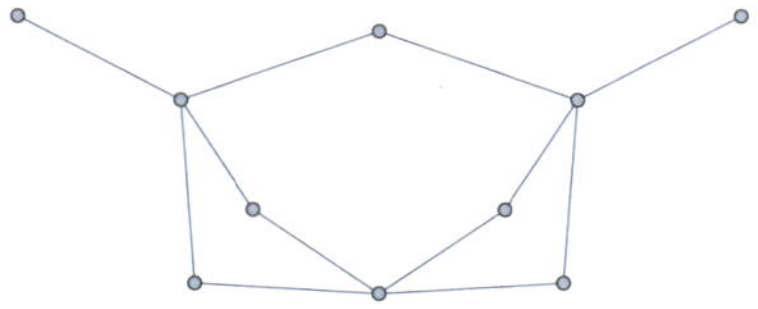

Out[106]= {2 ⟷ 10, 10 ⟷ 7, 10 ⟷ 3, 10 ⟷ 4, 3 ⟷ 6, 6 ⟷ 9, 6 ⟷ 7, 6 ⟷ 8, 4 ⟷ 5, 5 ⟷ 8, 5 ⟷ 9, 5 ⟷ 1}

An important idea in graph theory is vertex degree, which is the number of edges containing a given vertex:

```
In[107]:= {VertexDegree[CG, 2], VertexDegree[CG, 4], VertexDegree[CG, 10]}
```

Out[107]= {1, 2, 4}

Gaussian decomposition shows a very important fact about curve graphs. We call vertices not on the boundary of our disk *interior vertices*.

Every interior vertex has an even degree.

Like curves, graphs have connected components.

We are most interested in the built-in function WeaklyConnectedComponents:

```
In[108]:= WeaklyConnectedComponents[CG]
```

Out[108]= {{2, 10, 3, 6, 7, 4, 5, 9, 8, 1}}

This graph is connected, but as you see, it has two components:

```
In[109]:= c2 = y^2 - x^3 + x;
          gpc2 = gPoints[c2, 3, x, y];
          {gaussPlot[c2, 3, gpc2],
            GC2 = Graph[{4 → 5, 5 → 4, 1 → 3, 3 → 2}, VertexLabels → "Name"]}
```

Out[111]= {

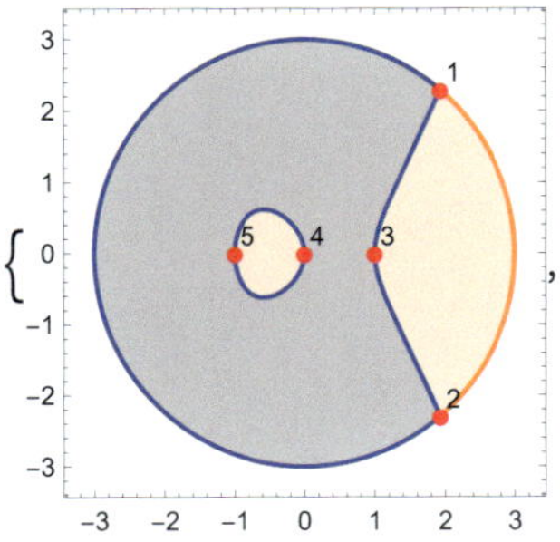

,

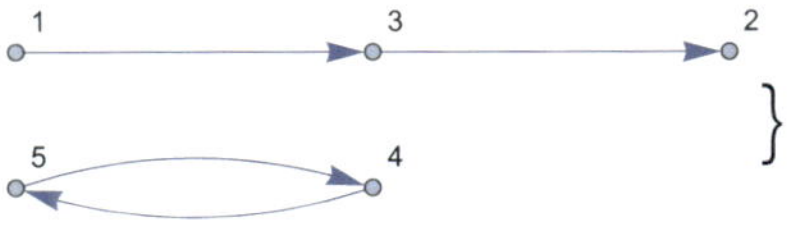

}

```
In[112]:= WeaklyConnectedComponents[GC2]

Out[112]= {{2, 3, 1}, {5, 4}}
```

We will see the theoretical importance of a curve graph in Chapter 5. One immediate use is to feed information on a graph to our pathFinder algorithm to automatically find a piecewise linear approximation to a curve component. For the next procedure defined by plotPath, we will take a curve $f = 0$ and a set of points (that is, vertices V), including enough boundary and critical points to identify segments for pathFinder.

The parameter *s* is the step size for pathFinder (For a complete list of global functions, see Appendix 2):

```
In[113]:= plotPath[f_, G_, V_, x_, y_, s_] := Module[{n, L},
              n = Length[G];
              L = Flatten[Table[pathFinder[f, V[[G[[i, 1]]]], V[[G[[i, 2]]]], s, x, y], {i, n}], 1];
              Append[L, V[[G[[n, 2]]]]]]
```

First, consider the example $c2 = y^2 - x^3 + x$ above with:

```
In[114]:= V = gpc2

Out[114]= {{1.93114, 2.2958}, {1.93114, −2.2958}, {1., 0.}, {0., 0.}, {−1., 0.}}
```

For component path graphs, the graph rules are sufficient:

```
In[115]:= G1 = {1 → 3, 3 → 2};
          G2 = {4 → 5, 5 → 4};

In[117]:= L1 = plotPath[c2, G1, V, x, y, .1];
          L2 = plotPath[c2, G2, V, x, y, .1];

In[119]:= Show[RegionPlot[c2 > 0 && x^2+y^2 < 9, {x, −3, 3}, {y, −3, 3},
              BoundaryStyle → {None, Pink}], ListLinePlot[{L1, L2}, PlotStyle → {Black, Orange}],
            ListPlot[Table[Labeled[V[[n]], n], {n, 5}], PlotStyle → Red, PlotMarkers → Automatic]]
```

Out[119]=

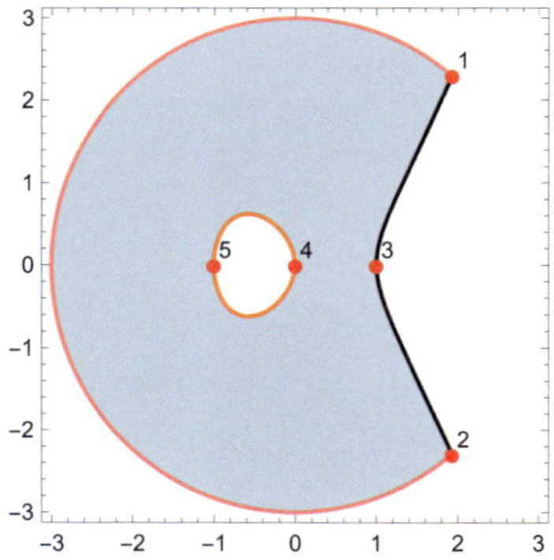

A second example is the difficult Newton Hyperbola 376, which has a component with several twists and turns:

In[120]:=
```
f = newtonHyperbola[376, x, y]
cp376 = Join[criticalPoints[f, x, y], {x, y} /. NSolve[{f, -x + y - 20}, {x, y}, Reals]];
```

Out[120]= $1 + x - 0.2\,x^2 - 0.008\,x^3 + y - x\,y + 0.2\,x^2\,y - 0.2\,y^2 - 0.2\,x\,y^2 + 0.008\,y^3$

In[122]:=
```
cp376 = {{-5.28009203049686`, -17.206102954932472`},
    {11.659821180891406`, 3.1684250826109506`},
    {5.167252947836531`, 0.5278467102679588`},
    {1.4273504023589225`, 2.0419148559610174`},
    {-0.39812614887988745`, -0.3862418222579013`},
    {-19.645226575920613`, 0.3547734240793867`},
    {0.05173064096128144`, 20.05173064096128`}}
```

Out[122]= {{-5.28009, -17.2061}, {11.6598, 3.16843}, {5.16725, 0.527847}, {1.42735, 2.04191}, {-0.398126, -0.386242}, {-19.6452, 0.354773}, {0.0517306, 20.0517}}

In[123]:=
```
gaussPlot[f, 20, cp376]
```

Out[123]=

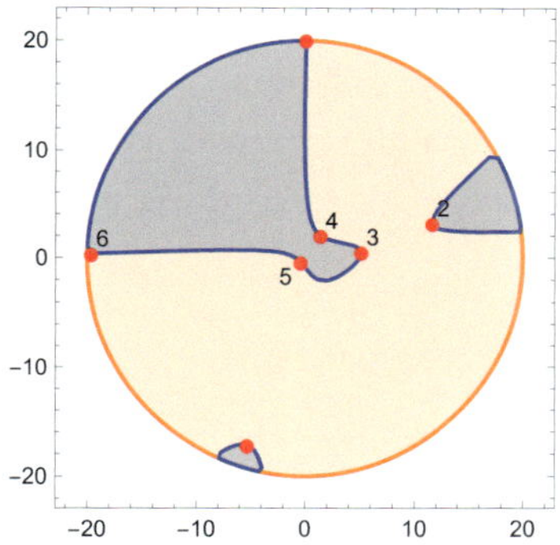

In[124]:=
```
G = {7 → 4, 4 → 3, 3 → 5, 5 → 6};
L = plotPath[f, G, cp376, x, y, .8];
ListLinePlot[L, Axes → False, Frame → True]
```

Out[126]=

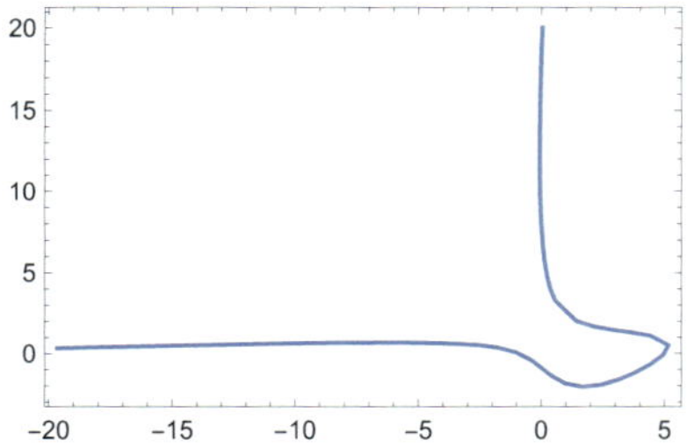

Note that this example will reappear in the next section.

4.6 Curve Graphs and Plots with a Differential Equation Solver

Ideally, we could develop a black box algorithm to find a graph of an affine algebraic curve and then plot it like we did in Section 4.4, but as was mentioned in Chapter 2, this is difficult when working numerically. Moreover, in Chapters 8 and 9, we will be more interested in plotting specific components rather than the whole curve, so we will go partway, letting several procedures do much of the work, but they will still require some direction from the user. Our main goal in this section is to give a randomized black box procedure, which will suggest a graph. The user can accept the result or try again. (The reader who is not interested in the details should jump to the summary at the end of this section.)

Our first step is to introduce a faster plotter than our naive plotter, which will allow smaller step sizes and give more chance of our software finding the next critical point. We have already seen gTvec and can now give our basic routine deTrace using the Wolfram Language's NDSolveValue.

Here, f is the curve equation, p is the initial point, s is the maximum Euclidean distance from the initial point that can be traced and r is the diameter of a circle around the origin containing the bounded part of the curve to be traced. We also note that there is a randomization, so running this procedure several times with the same input will give different outputs.

This function will return a pair of the Wolfram Language's built-in functions, InterpolatingFunction and ParametricPlot:

```
deTrace[f_, p_, x_, y_, s_, r_] := Module[{X, Y, F, u, t, tv, ttv, XS, YS},
    F = f /. Thread[{x, y} → {X[t], Y[t]}];
    tv = gTvec[f, p, x, y];
    u = Normalize[RandomReal[{.8, 1.2}, 2]];
    ttv = u.tv;
    NDSolveValue[{u.{X'[t], Y'[t]} == ttv, D[F, t] == 0, {X[0], Y[0]} == p,
        WhenEvent[X[t]^2 + Y[t]^2 > r^2 ||
            u[[1]] * X'[t] + u[[2]] * Y'[t] == 0 || (X[t] - p[[1]])^2 + (Y[t] - p[[2]])^2 > s^2,
          "StopIntegration"]}, {X, Y}, {t, 0, 20}]
  ]
```

The disadvantage to this, compared to pathFinder, is that it is hard to tell NDSolveValue exactly where to stop. As we will see, this method does not like to go around sharp curves and may stop on its own. Also, it will stop when its machine-defined parameter hits 20. On the other hand, deTrace is much faster than pathFinder and, therefore, can use a machine-defined smaller step size and a higher order of interpolation.

We illustrate with Newton Hyperbola 376. We will start at cp376[[7]] and hope to get to cp376[[6]] as in the previous example:

```
In[130]:= f = newtonHyperbola[376, x, y];
          p = cp376[[7]]
          {X, Y} = deTrace[f, p, x, y, 20, 20]

Out[131]= {0.0517306, 20.0517}

Out[132]= {InterpolatingFunction[ Domain: {{0., 16.4}} Output: scalar ],
           InterpolatingFunction[ Domain: {{0., 16.4}} Output: scalar ]}
```

This ended at:

```
In[133]:= dom = deEnd[{X, Y}][[2]]

Out[133]= 16.3794
```

```
In[134]:= ParametricPlot[{X[t], Y[t]}, {t, 0, dom},
            PlotRange → {{-20, 6}, {-5, 20}}, Axes → False, Frame → True]
```

Out[134]=

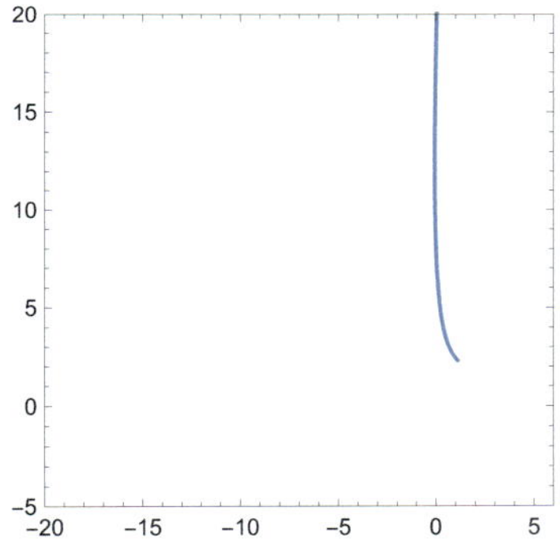

Here we barely got started around the bend when it stopped, so it is useful to have a procedure to tell us where and when (with t as parameter) these charts end.

deEnd uses the endpoint and maximum value for the domain of the interpolating function pair from deTrace:

```
In[128]:= deEnd[{X_, Y_}] := Module[{t1},
              t1 = X[[1, 1, 2]];
              {{X[t1], Y[t1]}, t1}]
```

From above:

```
In[128]:= deEnd[{X, Y}]

Out[128]= {{1.19091, 2.23588}, 16.3794}
```

We can run this several more times. Note we should probably allow *s* to be larger since:

```
In[135]:= Norm[cp376[[6]] - cp376[[7]]]

Out[135]= 27.8557
```

```
In[136]:= deEnd[deTrace[f, p, x, y, 30, 20]]
          deEnd[deTrace[f, p, x, y, 30, 20]]
          deEnd[deTrace[f, p, x, y, 30, 20]]

Out[136]= {{1.05022, 2.38671}, 16.0585}

Out[137]= {{1.16434, 2.26195}, 16.3236}

Out[138]= {{1.02992, 2.41134}, 16.0066}
```

We don't get anywhere near our desired endpoint in any of these examples. In fact, we don't get to negative *x* at all.

We can solve this problem by simply running deTrace again from the previous endpoint:

```
In[139]:= p = cp376[[7]]
          {X1, Y1} = deTrace[f, p, x, y, 30, 20];
          e1 = deEnd[{X1, Y1}]
          {X2, Y2} = deTrace[f, e1[[1]], x, y, 30, 20];
          e2 = deEnd[{X2, Y2}]
          {X3, Y3} = deTrace[f, e2[[1]], x, y, 30, 20];
          e3 = deEnd[{X3, Y3}]
          {X4, Y4} = deTrace[f, e3[[1]], x, y, 30, 20];
          e4 = deEnd[{X4, Y4}]
          {X5, Y5} = deTrace[f, e4[[1]], x, y, 30, 20];
          e5 = deEnd[{X5, Y5}]
          {X6, Y6} = deTrace[f, e5[[1]], x, y, 30, 20];
          e6 = deEnd[{X6, Y6}]
          cp376[[6]]

Out[139]= {0.0517306, 20.0517}

Out[141]= {{1.10545, 2.32358}, 16.1923}

Out[143]= {{1.30315, 2.13606}, 20.}

Out[145]= {{5.06612, 0.768387}, 10.1998}

Out[147]= {{5.06724, 0.766999}, 0.000890418}

Out[149]= {{1.3906, -2.00675}, 20.}

Out[151]= {{0.528923, -1.47735}, 0.45974}

Out[152]= {-19.6452, 0.354773}
```

My experiment took six applications to get to the desired endpoint. The reader with Mathematica can rerun the experiment, but often it will take even more applications. However, the user will see that this experiment is still much faster than running plotPath once!

Unfortunately, InterpolatingFunction does not combine nicely. Since the domains are different, they cannot be used in the same ParametricPlot statement. You can construct Piecewise functions out of them, but this is also messy.

Therefore, we will stick to the endpoints when plotting:

```
In[153]:= ListLinePlot[{p, e1[[1]], e2[[1]], e3[[1]], e4[[1]], e5[[1]], e6[[1]]},
            Axes → False, Frame → True, PlotRange → All]
```

Out[153]=

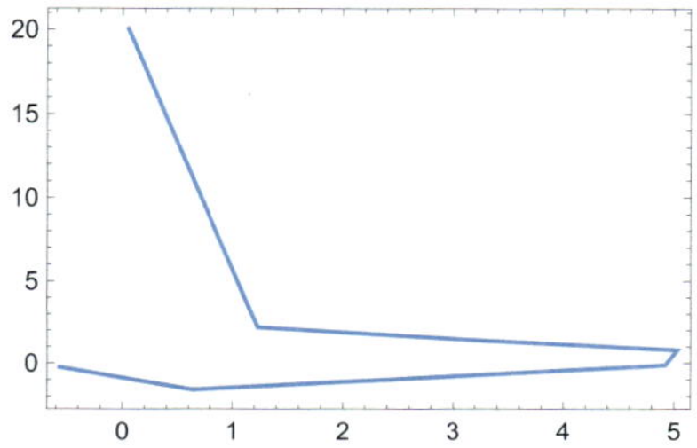

Code the last example as follows (note that we are turning off the common error message):

```
In[154]:= pathFinderDE[f_, p_, q_, r_, x_, y_, eps_] := Module[{i, e, X, Y, s},
              Off[NDSolveValue::ndsz];
              e = p;
              s = Norm[p - q];
              i = 1;
              Reap[Sow[p];
                  While[s > eps && Norm[e] < 2*r && i < 60,
                    {X, Y} = deTrace[f, e, x, y, s*eps, r];
                    e = deEnd[{X, Y}][[1]];
                    Sow[e];
                    s = Norm[e - q];
                    i++];
                  Sow[q]][[2, 1]]];
```

This example can plot the entire component of our NH376 example while in the last section we needed more critical points and plotPath. As an experiment, we compare performance:

```
In[155]:= f = newtonHyperbola[376, x, y];
          p = cp376[[7]]
          q = cp376[[6]]
          G = {7 → 4, 4 → 3, 3 → 5, 5 → 6}
          {td, L1} = Timing[pathFinderDE[f, p, q, 20, x, y, .05]];
          {tp, L2} = Timing[plotPath[f, G, cp376, x, y, .5]];
          {td, tp}
          {Length[L1], Length[L2]}
          ListLinePlot[{L1, L2}, PlotStyle → {Red, Blue},
            Frame → True, Axes → False, PlotRange → All]
```

Out[156]= {0.0517306, 20.0517}

Out[157]= {-19.6452, 0.354773}

Out[158]= {7 → 4, 4 → 3, 3 → 5, 5 → 6}

Warning: iteration limit reached at{0.219283, 5.06141}

Warning: iteration limit reached at{-15.1303, 0.508344}

Out[161]= {0.204629, 1.86518}

Out[162]= {61, 89}

Out[163]=

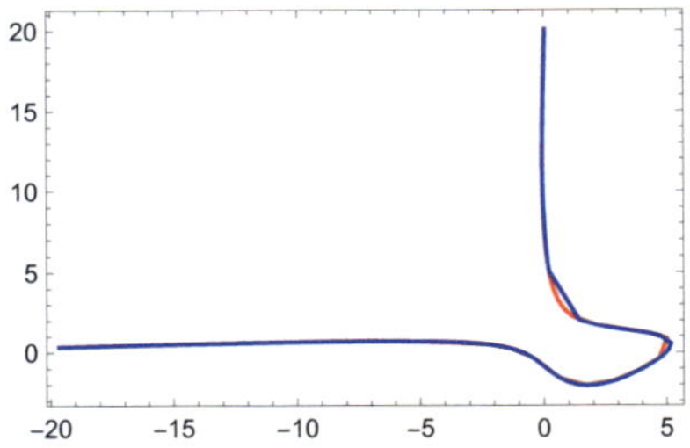

Note that plotPath took about 10 times longer with about 1.5 times the number of points, but the resulting plot is almost the same (see the difference marked in red).

Of course, if we used pathFinderDE instead of pathFinder, we could speed up plotPath:

```
In[164]:= plotPathDE[f_, G_, V_, x_, y_, s_] := Module[{r, n, L, LL, PF},
    n = Length[G];
    r = Max[Norm/@V];
    L = Flatten[Table[pathFinderDE[f, V[[G[[i, 1]]]], V[[G[[i, 2]]]], r, x, y, s], {i, n}], 1];
    Append[L, V[[G[[n, 2]]]]]]
```

```
In[165]:= {tdp, L3} = Timing[plotPathDE[NH376, G, cp376, x, y, .25]];
tdp
Length[L3]
ListLinePlot[{L1, L3}, PlotStyle → {Red, Blue},
  Frame → True, Axes → False, PlotRange → All]
```

Out[166]= 0.032386

Out[167]= 13

Out[168]=

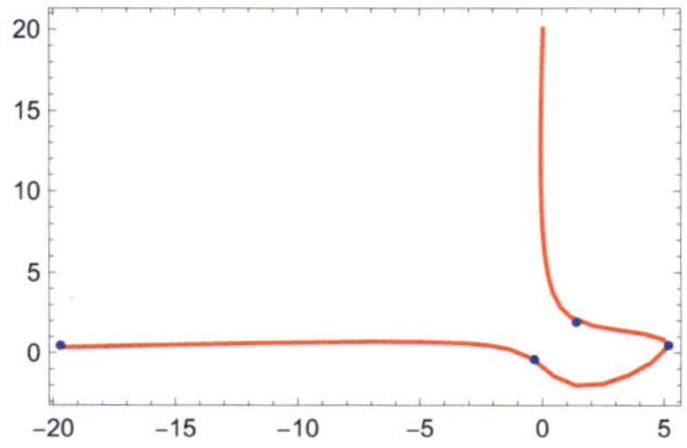

The difference here, marked in red, is much smaller. The next step is to wean pathFinderDE from needing the endpoint. The preferred state would be pathFinderDE having the choice of endpoints—from a critical point set CP to being able to find the next endpoint in the Gauss direction of the curve, but for nonsingular curves, this is just a bookkeeping problem. We replace q by the critical point set CP and s by the minimum distance to a point in CP. We do have one technical problem: we first need to delete the starting point from the critical point set, but in Mathematica, using Drop and Delete on lists requires a position, so we will use a position instead. However, the Wolfram Language Position function does not work well with machine numbers, so the following procedure finds the number of the point in CP following the nth point in CP.

If CP[[*n*]] is a point with no point of larger magnitude as the ending point, then the procedure produces 0 (For a complete list of global functions, see Appendix 2):

```
In[169]:= endFinder[f_, n_, CP_, x_, y_, eps_] := Module[{r, i, k, e, X, Y, s, CPD},
    Off[NDSolveValue::ndsz];
    r = Max[Norm /@ CP];
    e = CP[[n]];
    CPD = CP;
    CPD[[n]] = Nothing;
    s = Norm[e - First[Nearest[CPD, e]]];
    i = 1;
    e = Catch[
        While[s > eps && Norm[e] < 1.01 * r && i < 60,
          {X, Y} = deTrace[f, e, x, y, s, r];
          e = deEnd[{X, Y}][[1]];
          If[Norm[e] > r, Throw[e]];
          s = Norm[e - First[Nearest[CPD, e]]];
          i++];
        Throw[e]];
    If[Norm[e] > r, Return[0]];
    k = 1;
    While[k ≤ Length[CP] && Norm[e - CP[[k]]] > s, k++];
    If[k <= Length[CP], k, Fail]]
```

As before, our example is NH376:

```
In[170]:= NH376 = newtonHyperbola[376, x, y];
    gaussPlot[NH376, 21, cp376]
```

Out[171]=

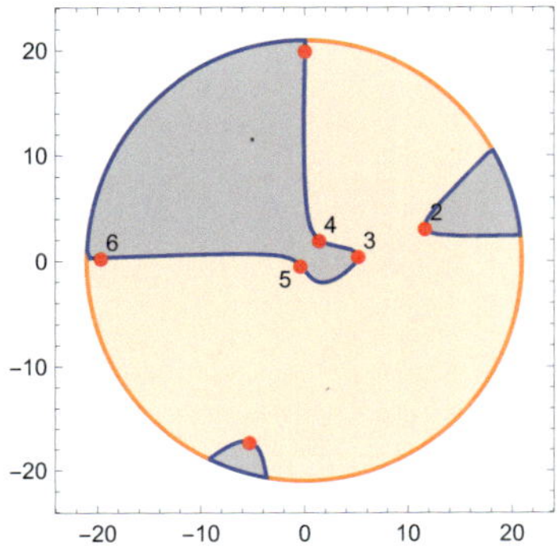

```
In[172]:= {endFinder[NH376, 7, cp376, x, y, .2],
            endFinder[NH376, 4, cp376, x, y, .2],
            endFinder[NH376, 3, cp376, x, y, .2],
            endFinder[NH376, 5, cp376, x, y, .2]}

Out[172]= {4, 3, 5, 6}
```

We can now use this to find graphs for nonsingular curves:

```
In[173]:= gaussGraphNS[f_, CP_, x_, y_, eps_] := Module[{n, G},
              n = Length[CP];
              G = Select[Table[k -> EndFinder[f, k, CP, x, y, eps],
                    {k, n}], #[[2]] > 0 &];
              graph[G, VertexLabels → "Name", ImageSize → Small]];
```

Here are some examples using Newton hyperbolas:

```
In[174]:= gaussGraphNS[NH376, cp376, x, y, .2]

Out[174]= graph[{}, VertexLabels → Name, ImageSize → Small]
```

```
In[175]:= g = newtonHyperbola[425, x, y];
          ggp = gPoints[g, 10, x, y];
          {gaussGraphNS[g, ggp, x, y, .2],
            Show[ContourPlot[g == 0, {x, -10, 12}, {y, -10, 12}],
              ListPlot[Table[Labeled[ggp[[n]], n], {n, Length[ggp]}]]]}
```

Out[177]= {graph[{}, VertexLabels → Name, ImageSize → Small],

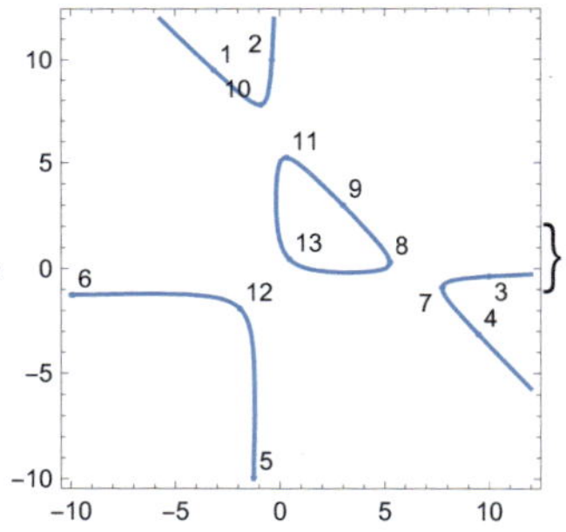

}

For examples of degree 5 or less, use random numbers under 2 000 000:

```
In[178]:= n = 26 536;
          f = newtonHyperbola[n, x, y]
          CP1 = gPoints[f, 15, x, y]
          gaussGraphNS[f, CP1, x, y, .4]
          r = 1.1 * Max[Norm[#] &/@CP1];
```

Out[179]= $1 + x - 0.2\,x^2 + 0.008\,x^3 - 0.000064\,x^4 + y + x\,y - 0.2\,x^2\,y + 0.008\,x^3\,y - 0.2\,y^2 - 0.2\,x\,y^2 + 0.2\,x^2\,y^2 - 0.008\,y^3 - 0.008\,x\,y^3 - 0.000064\,y^4$

```
Out[180]= {{-14.7873, 2.51732}, {-0.696342, 14.9838}, {2.01645, 14.8638},
            {14.9952, 0.37911}, {14.9964, -0.330793}, {1.26206, -14.9468},
            {-0.586659, -14.9885}, {-14.9819, -0.736523}, {8.04047, -0.0120441},
            {-0.00501908, 4.9753}, {-0.850408, -0.0000183323}, {-0.0000194407, -0.857838}}

Out[181]= graph[{}, VertexLabels → Name, ImageSize → Small]
```

In[183]:=
```
P1 = plotPathDE[f, {2 → 10, 10 → 3}, CP1, x, y, .1];
P2 = plotPathDE[f, {4 → 9, 9 → 5}, CP1, x, y, .1];
P3 = plotPathDE[f, {6 → 12, 12 → 7}, CP1, x, y, .1];
P4 = plotPathDE[f, {8 → 11, 11 → 1}, CP1, x, y, .1];
{Show[RegionPlot[f > 0, {x, −r, r}, {y, −r, r}],
    ListPlot[Table[Labeled[CP1[[n]], n], {n, Length[CP1]}]]],
  Show[RegionPlot[f > 0, {x, −r, r}, {y, −r, r}, BoundaryStyle → None],
    ListLinePlot[{P1, P2, P3, P4}, Axes → False, Frame → True, PlotRange → All],
    ListPlot[Table[Labeled[CP1[[n]], n], {n, Length[CP1]}]]]}
```

Out[187]=

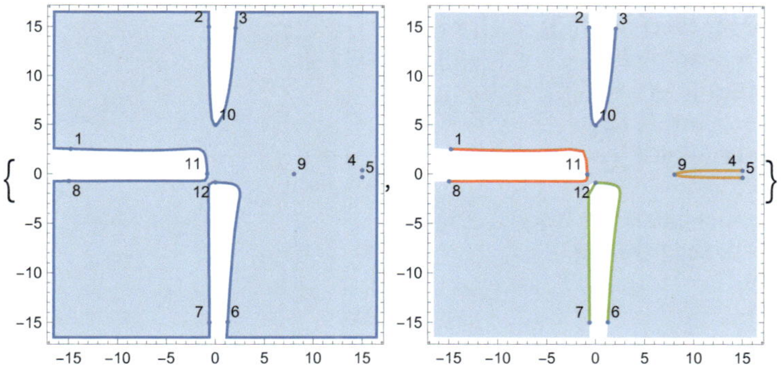

In this example, the graph out-performs the plot!

If there is a singular (or nearly singular) point, then gaussGraphNS does a decent job of tracing into the point.

Note that it cannot leave that point, often creating a loop:

In[188]:=
```
CP2 = {{0, 0}, {1, 0}, {0, 1}, {−1, 0}, {0, −1}};
gaussGraphNS[x y, CP2, x, y, .4]
```

Out[189]=
```
graph[{}, VertexLabels → Name, ImageSize → Small]
```

To rectify this situation, we use a weak criterion for singularity, Norm[gTvec[*f*, *p*, *x*, *y*]] < eps/2 and do not allow this point as a starting point.

We add to the critical point set nearPoints, which are points near the singular point with the Gauss direction pointing away from the singularity (For a complete list of global functions, see Appendix 2):

In[190]:=
```
nearPoints[f_, r_, x_, y_, pt_, ep_] := Module[{np, stnpts},
    np = {x, y} /. NSolve[{f, (x − pt[[1]]) ^ 2 + (y − pt[[2]]) ^ 2 − ep ^ 2}, {x, y}, Reals];
    np = Select[np, Norm[# − pt + ep / 2 * Normalize[gTvec[f, #, x, y]]] > ep &];
    Select[np, Norm[#] < r &]];
```

For each near point, we add an edge to the graph from the singular point to the near point. This leads to our main (almost black box) routine to find the graph of a curve on a bounded disc. It assumes that a set of critical points (including boundary and singular points) has been calculated.

gaussGraph calculates the radius from the singular points and returns a new critical point set, including the near points and the graph (For a complete list of global functions, see Appendix 2):

```
In[191]:= gaussGraph[f_, gp_, x_, y_, eps_] := Module[{n, k, e, G, np, CP, r, S},
        CP = gp;
        r = Max[Norm[#] &/@gp];
        n = Length[CP];
        G = {};
        k = 1;
        S = Reap[While[k ≤ n,
              If[Norm[gTvec[f, CP[[k]], x, y]] < eps/2,
                np = nearPoints[f, r, x, y, CP[[k]], 1.25 * eps];
                Do[Sow[k → n + j], {j, Length[np]}];
                CP = Join[CP, np];
                n = Length[CP];
                e = 0,
                e = endFinder[f, k, CP, x, y, eps]];
              If[e > 0, Sow[k → e]];
              k++]];
        G = S[[2, 1]];
        Echo[{Show[ContourPlot[f == 0, {x, -1.2 r, 1.2 r}, {y, -1.2 r, 1.2 r}],
              ListPlot[Table[Labeled[CP[[j]], j], {j, Length[CP]}]]],
            Graph[G, VertexLabels → "Name"]}];
        {CP, G}];
```

For a first example, consider the standard cubic node:

```
In[192]:= f = y^2 - x^3 - x^2;
          gp = gPoints[f, 3, x, y]
          gaussGraph[f, gp, x, y, .1]
```

Out[193]= {{1.58454, 2.54739}, {1.58454, −2.54739}, {−1., 0.}, {0., 0.}}

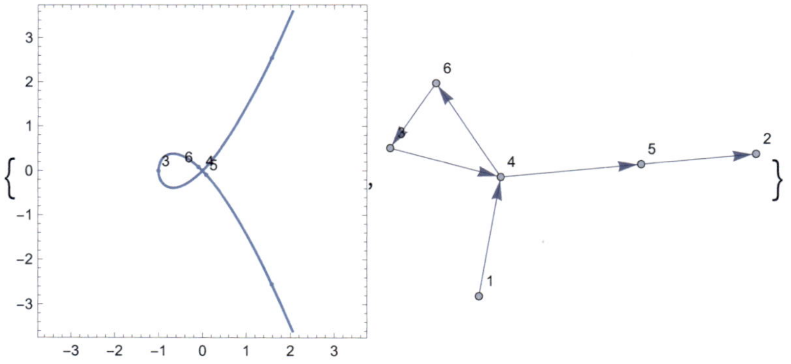

Out[194]= {{{1.58454, 2.54739}, {1.58454, −2.54739}, {−1., 0.}, {0., 0.}, {0.086536, −0.0902026}, {−0.0904577, 0.0862694}}, {1 → 4, 3 → 4, 4 → 5, 4 → 6, 5 → 2, 6 → 3}}

As a second example, consider the degree 4 Chebyshev curve:

In[195]:=
```
ch4 = 1 - 8 x^2 + 8 x^4 + 3 y - 4 y^3;
gp = gPoints[ch4, 2, x, y];
gaussGraph[ch4, gp, x, y, .1]
```

{ 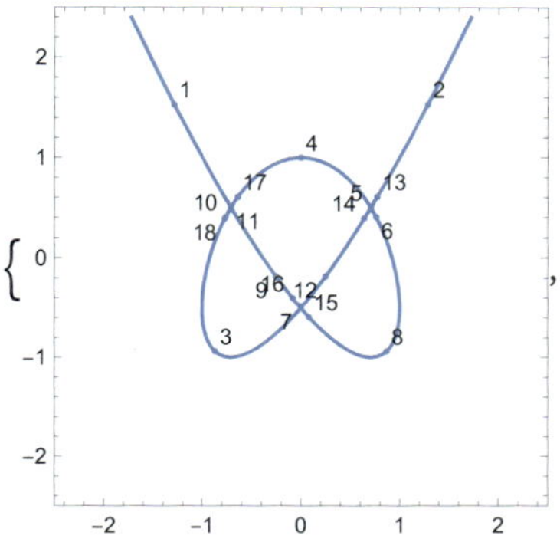

, 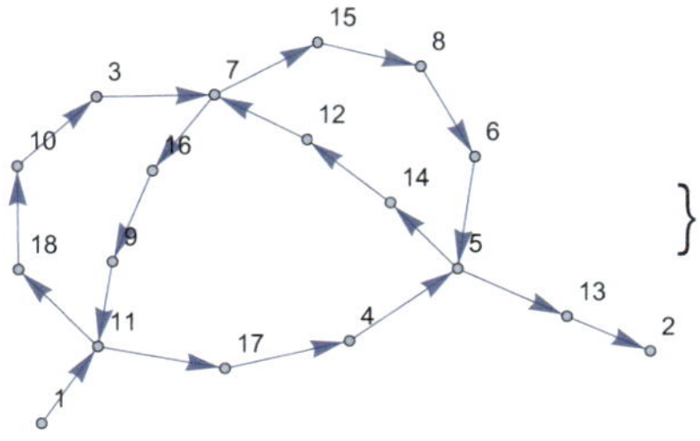 }

Out[197]= {{{−1.28789, 1.53014}, {1.28789, 1.53014}, {−0.867229, −0.938588}, {0., 1.}, {0.707107, 0.5}, {0.761401, 0.404868}, {0., −0.5}, {0.867229, −0.938588}, {−0.250427, −0.185031}, {−0.761401, 0.404868}, {−0.707107, 0.5}, {0.250427, −0.185031}, {0.771894, 0.6069}, {0.64131, 0.393719}, {0.083395, −0.593114}, {−0.0804368, −0.404319}, {−0.637826, 0.604044}, {−0.768609, 0.391177}},
{1 → 11, 3 → 7, 4 → 5, 5 → 13, 5 → 14, 6 → 5, 7 → 15, 7 → 16, 8 → 6, 9 → 11, 10 → 3, 11 → 17, 11 → 18, 12 → 7, 13 → 2, 14 → 12, 15 → 8, 16 → 9, 17 → 4, 18 → 10}}

Let's take a closer look at point 6. Looking at the Gauss points here:

In[198]:=
```
Show[ContourPlot[ch4 == 0, {x, -2, 2}, {y, -2, 2}],
  ListPlot[Table[Labeled[gp[[n]], n], {n, Length[gp]}]]]
```

Out[198]=

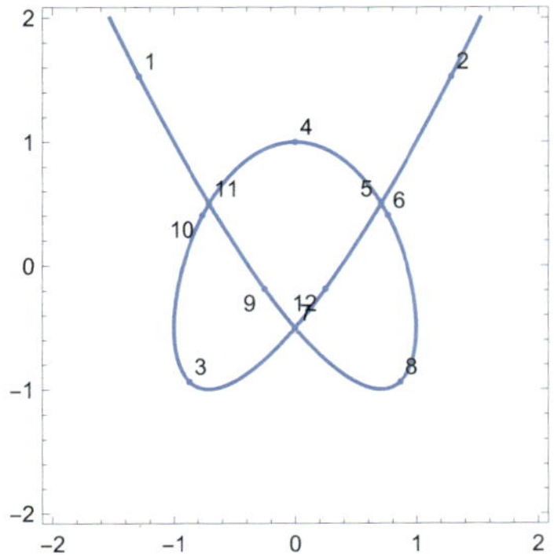

We see that points 6 and 10 are very close to 5 and 12, respectively, so we delete these points, removing point 10 first so that point 6 is still in the sixth position:

In[199]:=
```
gp1 = Delete[Delete[gp, 10], 6]
{ch4pts, ch4G} = gaussGraph[ch4, gp1, x, y, .1]
```

Out[199]=
```
{{-1.28789, 1.53014}, {1.28789, 1.53014}, {-0.867229, -0.938588},
  {0., 1.}, {0.707107, 0.5}, {0., -0.5}, {0.867229, -0.938588},
  {-0.250427, -0.185031}, {-0.707107, 0.5}, {0.250427, -0.185031}}
```

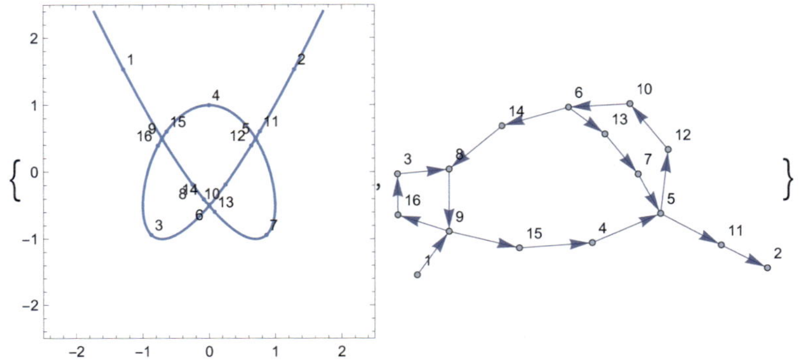

Out[200]=
```
{{{-1.28789, 1.53014}, {1.28789, 1.53014}, {-0.867229, -0.938588}, {0., 1.}, {0.707107, 0.5},
   {0., -0.5}, {0.867229, -0.938588}, {-0.250427, -0.185031}, {-0.707107, 0.5},
   {0.250427, -0.185031}, {0.771894, 0.6069}, {0.64131, 0.393719}, {0.083395, -0.593114},
   {-0.0804368, -0.404319}, {-0.637826, 0.604044}, {-0.768609, 0.391177}},
  {1 → 9, 3 → 8, 4 → 5, 5 → 11, 5 → 12, 6 → 13, 6 → 14, 7 → 5, 8 → 9, 9 → 15,
   9 → 16, 10 → 6, 11 → 2, 12 → 10, 13 → 7, 14 → 8, 15 → 4, 16 → 3}}
```

Due to the complexity of this software, it is possible that the labels may change when this is run on different sessions, versions of Mathematica and machines.

For ease of use and definiteness, we give a permanent version of the previous output as an Association:

In[201]:=
```
Ch4A = <|1 → {-1.2878893413505916`, 1.5301441253802013`},
     2 → {1.2878893413505876`, 1.5301441253802006`},
     3 → {-0.8672290665235644`, -0.938587657300242`},
     4 → {0.`, 1.0000000000004285`}, 5 → {0.7071067811865476`,
        0.5000000000000014`}, 6 → {0.`, -0.49999999999999856`},
     7 → {0.8672290665235052`, -0.9385876573001991`},
     8 → {-0.2504274405673725`, -0.18503072151060296`},
     9 → {-0.707106781186551`, 0.500000000000037`},
     10 → {0.2504274405674338`, -0.18503072151063518`},
     11 → {0.7718939492398669`, 0.6069000601292057`},
     12 → {0.6413097819106776`, 0.3937185110835575`},
     13 → {0.08339503632735218`, -0.5931142734276572`},
     14 → {-0.08043678659376786`, -0.4043186362844401`},
     15 → {-0.6378263144263652`, 0.604044302705585`},
     16 → {-0.7686094142232938`, 0.391177088214161`}|>;
```

We see that our directed graph breaks down to the union of the two directed subgraphs:

```
In[202]:= G1 = {1 → 9, 9 → 16, 16 → 3, 3 → 6, 6 → 13, 13 → 7, 7 → 5, 5 → 11, 11 → 2};
G2 = {9 → 15, 15 → 4, 4 → 5, 5 → 12, 12 → 10, 10 → 6, 6 → 14, 14 → 8, 8 → 9};
{Labeled[Graph[G1, VertexLabels → "Name"], "G1"],
  Labeled[Graph[G2, VertexLabels → "Name"], "G2"]}
```

Out[204]=

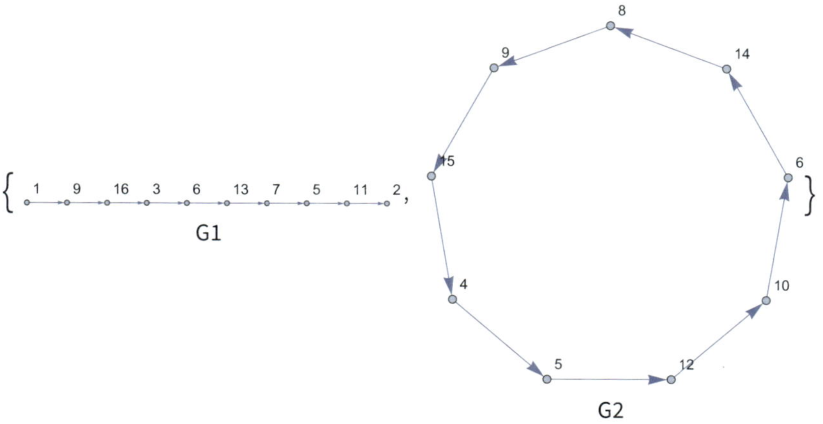

These subgraphs traverse the boundaries of our two connected components of the complement of the curve ch4 = 0. This is the decomposition of the curve proposed by Gauss.

In Chapters 5 and 8, we will learn how to extend these bounded graphs to global graphs. For now, we finish by extending our plotPath algorithms to this singular case.

We plot the edge as a line segment from the singular point to the near point:

```
In[205]:= plotPathS[f_, G_, V_, x_, y_, s_] := Module[{n, L, r},
    n = Length[G];
    r = Max[Norm /@ V];
    L = Flatten[
        Table[If[Norm[gTvec[f, V[[G[[i, 1]]]], x, y]] < s/2, {V[[G[[i, 1]]]], V[[G[[i, 2]]]]},
            pathFinder[f, V[[G[[i, 1]]]], V[[G[[i, 2]]]], s, x, y]],
          {i, n}], 1];
    AppendTo[L, V[[G[[n, 2]]]]];
    L]
```

```
In[206]:= plotPathDES[f_, G_, V_, x_, y_, eps_] := Module[{r, n, L},
    n = Length[G];
    r = Max[Norm[#] & /@ V];
    L = Flatten[
        Table[If[Norm[gTvec[f, V[[G[[i, 1]]]], x, y]] < eps/2, {V[[G[[i, 1]]]], V[[G[[i, 2]]]]},
            pathFinderDE[f, V[[G[[i, 1]]]], V[[G[[i, 2]]]], r, x, y, eps]],
          {i, n}], 1];
    AppendTo[L, V[[G[[n, 2]]]]];
    L]
```

As a different example, consider the lemniscate:

```
In[207]:= f = x^4 - 4 x y + y^4;
          gp = Chop[gPoints[f, 2.5, x, y]]

Out[208]= {{-1.41421, -1.41421}, {1.41421, 1.41421}, {0, 0}}

In[209]:= {CP, GG} = gaussGraph[f, gp, x, y, .05]
```

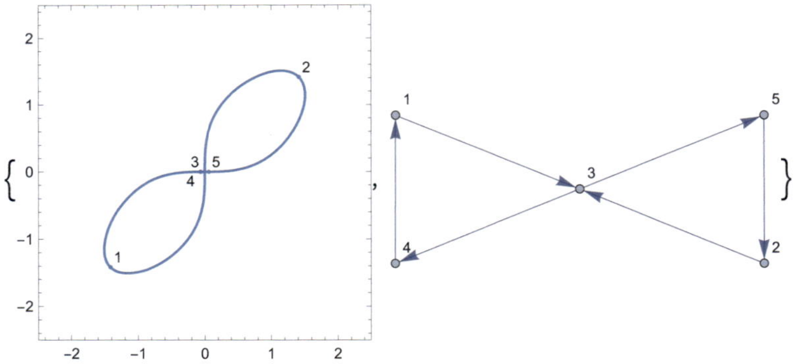

```
Out[209]= {{{-1.41421, -1.41421}, {1.41421, 1.41421}, {0, 0}, {-0.0625, -0.0000610351},
           {0.0625, 0.0000610351}}, {1 → 3, 2 → 3, 3 → 4, 3 → 5, 4 → 1, 5 → 2}}
```

Note the output of edge rules that GG is not a path. Since this is a closed graph (it begins and ends at the same point), we can use the following:

```
In[210]:= P = FindEulerianCycle[GG][[1]]

Out[210]= {1 ⟶ 3, 3 ⟶ 5, 5 ⟶ 2, 2 ⟶ 3, 3 ⟶ 4, 4 ⟶ 1}

In[211]:= {ListLinePlot[plotPathS[f, P, CP, x, y, .1], Axes → False, Frame → True],
           ListLinePlot[plotPathDES[f, P, CP, x, y, .2], Axes → False, Frame → True]}
```

Out[211]=

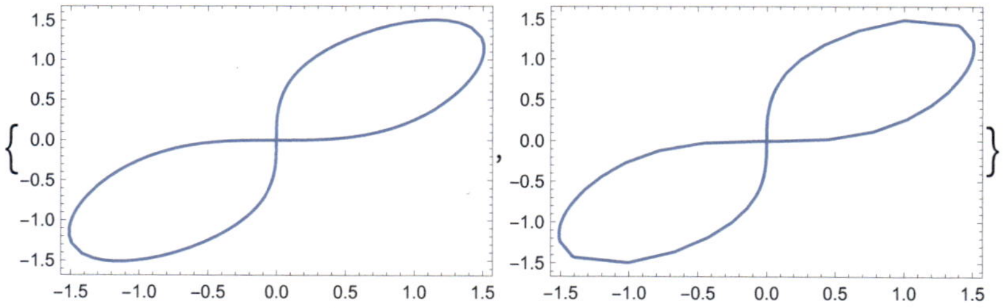

plotPathS produces a better curve, but **plotPathDES** runs more quickly. Note that the linear approximation near the singular point is not the problem in the right image.

Using the data above, we can now give a nice picture of the Gauss decomposition of the Chebyshev curve:

In[213]:=
```
G1 = {1 → 9, 9 → 16, 16 → 3, 3 → 6, 6 → 13, 13 → 7, 7 → 5, 5 → 11, 11 → 2};
G2 = {9 → 15, 15 → 4, 4 → 5, 5 → 12, 12 → 10, 10 → 6, 6 → 14, 14 → 8, 8 → 9};
Show[RegionPlot[ch4 > 0, {x, -2, 2}, {y, -2, 2},
    BoundaryStyle -> None, PlotStyle → LightOrange],
  ListLinePlot[plotPathS[ch4, G1, Values[Ch4A], x, y, .1], PlotStyle → Blue],
  ListLinePlot[plotPathS[ch4, G2, Values[Ch4A], x, y, .1], PlotStyle → Green],
  Axes → False, Frame → True]
```

Out[215]=

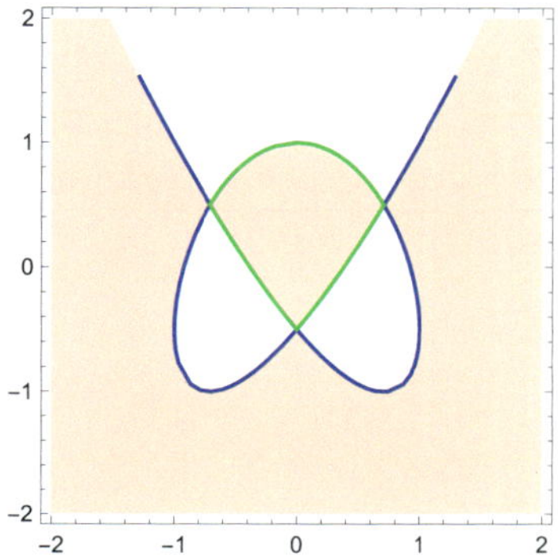

4.7 Paradigm for Analyzing Curves in a Bounded Circle about the Origin

In the social sciences, which are not as precise as mathematics, the analog of an algorithm is a paradigm. Since analyzing a curve numerically is not a precise science either, we will summarize our approach by a paradigm.

First, we choose a radius r and find boundary and critical points using, for example, gPoints.

For now, the region should be a finite disk centered at the origin. Later in Chapter 8, we show how to use a more general, possibly infinite, region bounded by a conic. It is important that our points contain all the points on the boundary of the region and that all singular points are present. Also, each connected component of the curve, as well as each algebraic component, should have at least two points represented. gPoints will accomplish this but may give too many or too few points.

Next, we find the curve graph, in this case using gaussGraph. *Separate into weak components.*

If it is a simple curve, gaussGraph should easily find the graph, but note that this algorithm is sensitive to the choice of r. If incorrect results are returned, try a larger or smaller r. Deleting unnecessary points or adding additional points to the critical point set may help. As a last resort, a RegionPlot with critical points labeled can help you give a graph manually. If the curve has more than one algebraic component, manual graph finding may be the best option—better even than algebraically decomposing the curve. The topology of the curve is easiest found from the picture given by Mathematica, but unfortunately, the decomposition tools do not work as well as preserving the paths and directions do, in the picture.

Optionally, plot using plotPathS *or* plotPathDES.

It is important to have actual paths—that is, the directed graph should be given by edges with the initial point of one edge being the terminal point for the preceding edge. In particular, EdgeList and EdgeRules are not always in path order. In theory, each path should begin and end at border points or should be a closed path. In the first case, use FindPath to convert EdgeList or EdgeRules to a path starting and ending in a boundary point. In the second case, you can use FindEulerianCycle to get a path. Paths should always be in the Gauss direction—clockwise about the positive set of the equation. Again (as indicated in the previous step), the picture, if correct, gives the decomposition into paths we want. For a relatively small critical point set, manually working from the picture is often the easiest strategy. However, if there are singular points, make sure that the paths go around connected components of the positive set $f > 0$.

As our last example, we find the components of the curve and positive set for Gauss's original example:

```
In[216]:= f = gaussCurve[z^5-5z^4+9z^3-5z^2, x, y, z]
```

Out[216]= $-5x^2+9x^3-5x^4+x^5+5y^2-27xy^2+30x^2y^2-10x^3y^2-5y^4+5xy^4$

After an unsuccessful attempt with $r = 5$, we try again with $r = 4$:

```
In[217]:= gp = Chop[gPoints[f, 4, x, y]]

Out[217]= {{-3.7105, 1.49404}, {-1.69429, 3.62345}, {1.02568, 3.86626},
  {2.99319, 2.65345}, {3.88591, 0.948515}, {3.88591, -0.948515},
  {2.99319, -2.65345}, {1.02568, -3.86626}, {-1.69429, -3.62345},
  {-3.7105, -1.49404}, {2.01524, 0.841511}, {2.01524, -0.841511}, {1., 0}, {0, 0}}
```

```
In[218]:= {cp, G} = gaussGraph[f, gp, x, y, .2]
```

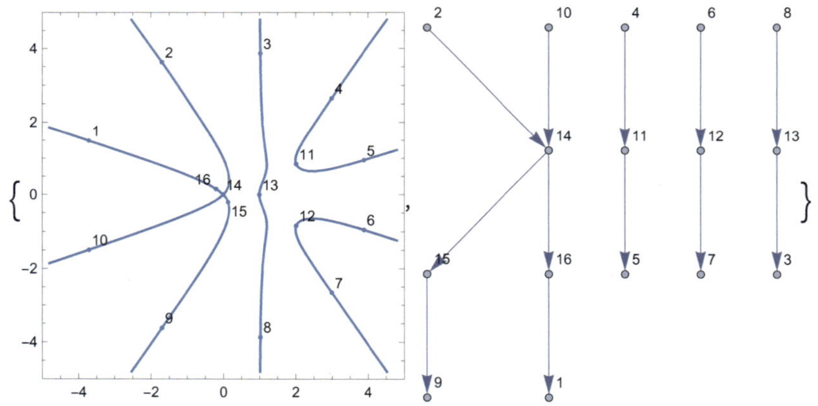

```
Out[218]= {{{-3.7105, 1.49404}, {-1.69429, 3.62345}, {1.02568, 3.86626}, {2.99319, 2.65345},
  {3.88591, 0.948515}, {3.88591, -0.948515}, {2.99319, -2.65345}, {1.02568, -3.86626},
  {-1.69429, -3.62345}, {-3.7105, -1.49404}, {2.01524, 0.841511}, {2.01524, -0.841511},
  {1., 0}, {0, 0}, {0.136872, -0.209203}, {-0.196973, 0.153953}}, {2 → 14, 4 → 11,
  6 → 12, 8 → 13, 10 → 14, 11 → 5, 12 → 7, 13 → 3, 14 → 15, 14 → 16, 15 → 9, 16 → 1}}
```

Again, before going on, we make a permanent version of this data and draw a RegionPlot:

```
In[219]:= gOrigA = <|1 → {-3.710503137713464`, 1.4940436623534898`},
        2 → {-1.694287437347172`, 3.6234500244446064`},
        3 → {1.0256836809227585`, 3.8662608534200804`},
        4 → {2.993193894204237`, 2.653448758069135`},
        5 → {3.885912999933733`, 0.9485147109801363`},
        6 → {3.885912999933767`, -0.9485147109801597`},
        7 → {2.993193894204127`, -2.6534487580692407`},
        8 → {1.0256836809227194`, -3.866260853420112`},
        9 → {-1.6942874373471617`, -3.623450024444607`},
        10 → {-3.7105031377133324`, -1.4940436623535953`},
        11 → {2.015235448916427`, 0.8415109240162972`},
        12 → {2.015235448916427`, -0.8415109240162972`}, 13 → {1.`, 0},
        14 → {0, 0}, 15 → {0.13687218865500828`, -0.209203259947802`},
        16 → {-0.19697325992704043`, 0.15395302814077477`}|>;

In[220]:= Show[RegionPlot[f > 0, {x, -4, 4}, {y, -4, 4}], ListPlot[Table[Labeled[gOrigA[i], i], {i, 16}]]]
```

Out[220]=

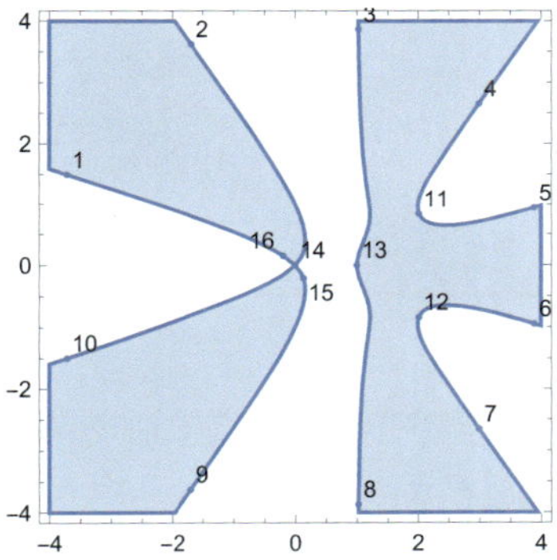

This shows we have four connected components of the curve and one singular path. We have paths:

```
In[221]:= P1 = {4 → 11, 11 → 5};
        P2 = {6 → 12, 12 → 7};
        P3 = {8 → 13, 13 → 3};
        P4 = {2 → 14, 14 → 16, 16 → 1};
        P5 = {10 → 14, 14 → 15, 15 → 9};
```

To plot our findings, we first find the paths. We use plotPathS, which takes a while but will give the better picture:

```
In[226]:= PP1 = plotPathS[f, P1, cp, x, y, .2];
        PP2 = plotPathS[f, P2, cp, x, y, .2];
        PP3 = plotPathS[f, P3, cp, x, y, .2];
        PP4 = plotPathS[f, P4, cp, x, y, .2];
        PP5 = plotPathS[f, P5, cp, x, y, .2];
```

Seeing no warnings, we can assume things went well. We can now plot our final picture:

In[231]:=
```
Labeled[Show[
    RegionPlot[{f > 0 && x^2+y^2 < 16}, {x, -4, 4}, {y, -4, 4}, BoundaryStyle → None],
    ContourPlot[x^2+y^2 == 16, {x, -4, 4}, {y, -4, 4}, ContourStyle → Pink],
    ListLinePlot[{PP1, PP2, PP3, PP4, PP5}], Axes → False, Frame → True],
  "Components of gaussCurve[z^5-4z^4+9z^3-5z^2]"]
```

Out[231]=

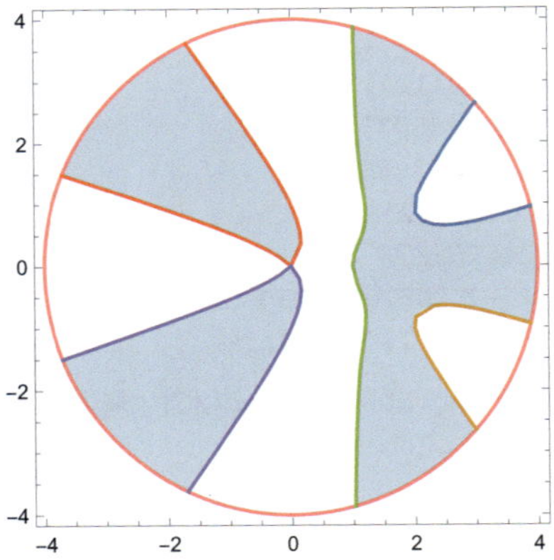

Components of gaussCurve[$z^5-4z^4+9z^3-5z^2$]

More to Explore

For more on the fundamental theorem of algebra, check out Wolfram MathWorld (wolfr.am/snvXr6BO)

To see the original German verison of Carl Friedrich Gauss's first proof, check it out online (wolfr.am/snw5oXcV)

For more on graphs and networks and their definitions, check out the Documentation Center (wolfr.am/vls3xhM3)

5 | A Classical Interlude

So far, we have been emphasizing how our approach differs from classical algebraic geometry. In this chapter, we will sketch some ideas from classical algebraic geometry that will be useful or interesting in the following chapters. In particular, we explore curve theory in the complex projective plane.

5.1 The Projective Plane in Homogeneous Coordinates

In Chapter 1, we introduced the idea of an infinite point on a line that could be a slope $\{1, m, 0\}$ or a direction vector $\{u_1, u_2, 0\}$. For the latter, we noted that direction vectors were homogeneous; that for any nonzero r, $\{r\,u_1, r\,u_2, 0\}$ is still a direction vector for the same line. For the purposes of the line algorithm, an affine point $\{x_0, y_0\}$ was considered to be a triple $\{x_0, y_0, 1\}$ so that it would satisfy the equation $a\,x_0 + b\,y_0 + c\,z_0 = a\,x_0 + b\,y_0 + c = 0$. Then multiplying the entire triple $\{x_0, y_0, 1\}$ by a constant r to get $z = r$, point $\{r\,x_0, r\,y_0, r\}$ gives another solution to the equation $a\,x_0 + b\,y_0 + c\,z_0 = 0$. Classical algebraic geometers prefer to stick to rational numbers anyway, and a point $\left\{\frac{3}{5}, \frac{5}{12}, 1\right\} = \{36, 25, 60\}$ seems nicer to them as an integer point. Thus we can consider the form $\{x_0, y_0, 1\}$ to be a homogeneous version of an affine point.

Now that we have consistency between affine and infinite points, we define the *real projective plane* to be the set of all homogeneous triples $\{x_0, y_0, z_0\}$ for which x_0, y_0, z_0 are not all zero. If $z_0 \neq 0$, we can, by the previous paragraph, consider this to be the (real) fractional point $\left\{\frac{x_0}{z_0}, \frac{y_0}{z_0}, 1\right\}$, so the infinite line consists of all the projective real fractions with denominator 0.

Unlike affine points, projective points cannot double as vectors; they cannot be added or scalar multiplied.

The real projective plane is typically denoted by $\mathbb{P}^2_{\mathbb{R}}$ or, if we allow our points to have complex coordinates, $\mathbb{P}^2_{\mathbb{C}}$.

We have set up our projective plane so that lines have the equation $a\,x_0 + b\,y_0 + c\,z_0 = 0$. Now it makes perfect sense to talk about the line $c\,z = 0$ when $c \neq 0$; it is the line at infinity.

Given two projective points p, q, we can find the line containing them with a simplification of our procedure from Chapter 1.

Assuming p, q have length 3 (For a complete list of global functions, see Appendix 2):

```
In[1]:= pLine[p_, q_, x_, y_, z_] := Module[{a, b, c, M},
          M = {p, q, RandomReal[{-1, 1}, 3]};
          ({a, b, c} /. (Solve[M.{a, b, c} == {0, 0, 1}][[1]])).{x, y, z}
        ];
```

Find the projective line through the affine point {3, 2} and the infinite point {2, −1, 0}:

```
In[2]:= l = pLine[{3, 2, 1}, {2, -1, 0}, x, y, z]

Out[2]= -2.81816 x - 5.63631 y + 19.7271 z
```

Use Thread to apply the variables:

```
In[3]:= l /. Thread[{x, y, z} → {9, 6, 3}]

Out[3]= 0.
```

The important idea here is that the equation of a projective line is *homogeneous*, that is, each monomial has the same total degree. Alternatively, if $\{x_0, y_0, z_0\}$ is a solution, so is $\{r\,x_0, r\,y_0, r\,z_0\}$ for any $r \neq 0$.

5.2 Homogenization and Specialization

An algebraic curve in the projective plane is now given by a *homogeneous polynomial* in three variables x, y, z. The total degree of a monomial $x^i\,y^j\,z^k$ is $i + j + k$, and a polynomial in three variables is homogeneous if each monomial has the same degree. No constant terms are allowed except 0. Each affine curve $f(x, y)$ of degree d gives rise to a projective curve $F(x, y, z)$, called the homogenization, in one of the following ways:

1. Each monomial $x^i\,y^j$ is replaced by $x^i\,y^j\,z^{d-i-j}$, or

2. $F(x, y, z) = z^d\,f\left(\frac{x}{z}, \frac{y}{z}\right)$

Homogenization gives a $1 - 1$ correspondence between affine polynomials of degree d in x, y and homogeneous polynomials of degree d in x, y, z. The projective curve is the set of all projective points such that $F[x, y, z] = 0$. Note that $f[x, y] = 0$ is equivalent to $F[x, y, 1] = 0$, so all affine points are, in particular, points of the homogenization. On the other hand, infinite solutions of F are when $F[x, y, 0] = 0$. By the first method of homogenization, all monomials of degree less than d have z raised to a positive power so become 0 in $F(x, y, 0)$, but monomials $x^i\,y^j$ for $i + j = d$ appear in the homogenization without a power of z. In other words, the maximal form $f_{\max} = F[x, y]$, so infinite points of $F(x, y, z)$ are just the infinite points of $f[x, y]$, which we calculated in Chapter 3. Therefore, the solution set of the homogenization of $f(x, y)$ are just the affine and infinite points of $f(x, y)$.

Find the homogenization with a function developed for this book (Evaluate the global functions notebook to ensure the code works properly):

```
In[4]:= homog[f_, x_, y_, z_] := Expand[z^tDeg[f, x, y] (f /. Thread[{x → x/z, y → y/z}])];

In[5]:= F = homog[x^2 - y^2 - 1, x, y, z]

Out[5]= x^2 - y^2 - z^2
```

Note that pLine is the same as homog[line[p, q, x, y], x, y, z].

The opposite of homogenization is *specialization*. Given a homogeneous polynomial, we can replace any one of the variables by 1 to obtain a bivariate polynomial. Points with that coordinate nonzero specialize by dividing the two remaining coordinates by the coordinate specialized.

For example, here are three common specializations:

Specialization by *z*:

```
In[6]:= F
        f = F /. {z → 1}

Out[6]= x^2 - y^2 - z^2

Out[7]= -1 + x^2 - y^2

In[8]:= F /. Thread[{x, y, z} → {a, b, c}]

Out[8]= a^2 - b^2 - c^2

In[9]:= f /. Thread[{x, y} → {a/c, b/c}]

Out[9]= -1 + a^2/c^2 - b^2/c^2
```

Specialization by *x*:

```
In[10]:= g = F /. {x → 1}

Out[10]= 1 - y^2 - z^2

In[11]:= F /. Thread[{x, y, z} → {a, b, c}]

Out[11]= a^2 - b^2 - c^2
```

In[12]:=
```
g /. Thread[{y, z} → {b / a, c / a}]
```

Out[12]= $1-\frac{b^2}{a^2}-\frac{c^2}{a^2}$

Specialization by *y*:

In[13]:=
```
h = F /. {y → 1}
```

Out[13]= $-1+x^2-z^2$

In[14]:=
```
F /. Thread[{x, y, z} → {a, b, c}]
```

Out[14]= $a^2-b^2-c^2$

In[15]:=
```
h /. Thread[{x, z} → {a / b, c / b}]
```

Out[15]= $-1+\frac{a^2}{b^2}-\frac{c^2}{b^2}$

We notice in all cases that although the value of the specialization function is different from the value of the homogeneous function at the specialized point, that it is a nonzero multiple, that is, $F(a, b, c) = 0$ if and only if $f(a, b) = 0$ when $c \neq 0$. Note also that $f = 0$ is the original curve, a hyperbola. $g = 0$ (in y-z plane) is different; it is a circle. Note also that $\{1, 1, 0\}$ is an infinite point of $f = 0$, $F(1, 1, 0) = 0$, so it is a point on the projective curve. Specializing at x sends this infinite point to the finite point $\{1, 0\}$ on the circle g.

In Chapter 6, we will see that we can specialize at any line. Moreover, we will be able to work entirely in the affine case, so we will not need homogeneous functions later in this book. In Chapter 7, we will discuss in more detail what happens to conics when homogenizing and specializing.

Generalizing our example above, homogenizing hyperbolas and specializing elsewhere show that, topologically, open sets around infinite points look just like open sets about affine points. This gives a topology for the projective plane that locally makes infinite points just like affine points. Globally, the topology of the projective plane is quite different from the affine plane. We will examine this in the second part of Chapter 8.

Note that homogenization has the property that homogenizing *f*, *g* to *F*, *G* gives the homogenization of *f*∗*g* as *F*∗*G*:

In[16]:=
```
f = 3 + x - x^2 y + y^4;
g = -1 - y + x y^2 - x^4;
```

In[18]:=
```
H1 = homog[f * g, x, y, z]
```

Out[18]= $-x^4 y^4+x^6 y z+x y^6 z-x^3 y^3 z^2-x^5 z^3-y^5 z^3-3 x^4 z^4+ 2 x^2 y^2 z^4-y^4 z^4+x^2 y z^5+3 x y^2 z^5-x y z^6-x z^7-3 y z^7-3 z^8$

```
In[19]:= H2 = Expand[homog[f, x, y, z]*homog[g, x, y, z]]
```

Out[19]= $-x^4 y^4 + x^6 y z + x y^6 z - x^3 y^3 z^2 - x^5 z^3 - y^5 z^3 - 3 x^4 z^4 + 2 x^2 y^2 z^4 - y^4 z^4 + x^2 y z^5 + 3 x y^2 z^5 - x y z^6 - x z^7 - 3 y z^7 - 3 z^8$

```
In[20]:= H1 - H2
```

Out[20]= 0

In particular, if h is irreducible, so is its homogenization and conversely.

5.3 Bézout's Theorem

Bézout's theorem is one of the most important and useful theorems for curves in $\mathbb{P}^2_{\mathbb{C}}$; it can be considered a generalization of the fundamental theorem of algebra. In particular, the existence part of this theorem is not true in the real affine plane. The theorem was first stated by Isaac Newton in his *Principia*. Later, Étienne Bézout gave what is now considered a heuristic proof in 1799. Since our emphasis is not on theory, we will not prove it here, but we do need to state this theorem and explain it in this long section. The first thing is to explain the concept of *intersection multiplicity*.

We now want to consider the *complex projective plane* $\mathbb{P}^2_{\mathbb{C}}$. As in the real projective plane, points are triples $\{x_0, y_0, z_0\}$, but the components x_0, y_0, z_0 are now complex numbers, at least one of which is not zero. Two such triples are equivalent, i.e. the same point, if they differ by multiplication by a complex number. Of course, real numbers are also complex numbers, and multiplying a real triple, not all zero, by a complex number c can only be real if c is real; hence, the equivalence relation inside $\mathbb{P}^2_{\mathbb{C}}$ for these points is compatible with that in $\mathbb{P}^2_{\mathbb{R}}$. A curve in $\mathbb{P}^2_{\mathbb{C}}$ is the zero set of a homogeneous complex polynomial in three variables that can arise as the homogenization of a two-variable complex polynomial. The main change is the topology, $\mathbb{P}^2_{\mathbb{C}}$ is, from a real point of view, a 4D surface, and a complex curve is a 2D surface inside $\mathbb{P}^2_{\mathbb{C}}$.

Our line-finding functions, `line` and `pLine`, work in $\mathbb{P}^2_{\mathbb{C}}$. We can also use `NSolve`. In fact, solvers tend to work better with complex numbers rather than real numbers. Our one problem in working numerically with complex numbers is deciding if numbers are real or not, but the Wolfram Language's solvers are generally reliable about picking out reals from a set of complex solutions.

Intersection Multiplicity

When two projective curves $F = 0$, $G = 0$ meet at a point p in $\mathbb{P}^2_{\mathbb{C}}$, that point is a singularity, see the next plot, of the union curve FG. However, these kinds of singularities work differently from a typical singularity in an irreducible curve. Intersection multiplicity $\text{mult}_p(F, G)$ measures how complicated this intersection point is, or if we give curves names like C_1, C_2, $\text{mult}_p(C_1, C_2)$. However, each point of $\mathbb{P}^2_{\mathbb{C}}$ lies in some affine complex specialization, and since intersection multiplicity is a local property, we can study the concept in the affine case $\mathrm{mult}_p(f, g)$ where f, g are bivariate polynomials that are specializations of F, G respectively.

Fortunately, for our purposes, I along with Professor Zhonggang Zeng gave an algorithm in 2005 that solves the general intersection multiplicity problem using numerical linear algebra. This algorithm uses an idea from F. S. Macaulay in 1916 that had been largely forgotten because it involves analyzing large matrices, which was impossible in practice back then and impractical even up through the year 2000 when modern PCs began working at gigahertz speeds with gigabytes of memory. This algorithm uses numerical linear algebra and will be discussed in Appendix 1. For those using the notebook version of this book, in the Wolfram Language the code for the plane curve case is hidden below. For more information about the algorithm I suggest the introduction to our expanded 2011 paper (Dayton, Li, and Zeng).

If the two affine curves $f = 0, g = 0$ do not meet at p in the xy plane, then $\mathrm{mult}_p(f, g) = 0$, otherwise the multiplicity is an integer greater than zero. When a numerical algorithm returns an integer, there will always be situations where the answer is ambiguous; the use of a tolerance helps resolve this problem, see Appendix 1. If f, g have small integer coefficients, then the default tolerance dTol should work. For simplicity, our software assumes the varibles are x, y.

Here are some examples of multiplicities:

```
In[21]:= {Labeled[ContourPlot[{y == x^2-1, y^2-x^2 == 5}, {x, 0, 4}, {y, 1, 5}], "Multiplicity 1"],
  Labeled[ContourPlot[{y^2-x^2 == 5, x-1.5 y == -2.5, y == -1/3 x^2+2 x+1/3},
    {x, 0, 4}, {y, 0, 4}, ContourStyle → {Blue, Dashed, Green}], "Multiplicity 2"],
  Labeled[ContourPlot[{y == x^3, y == 0}, {x, -2, 2}, {y, -8, 8}], "Multiplicity 3"]}
```

Out[21]=

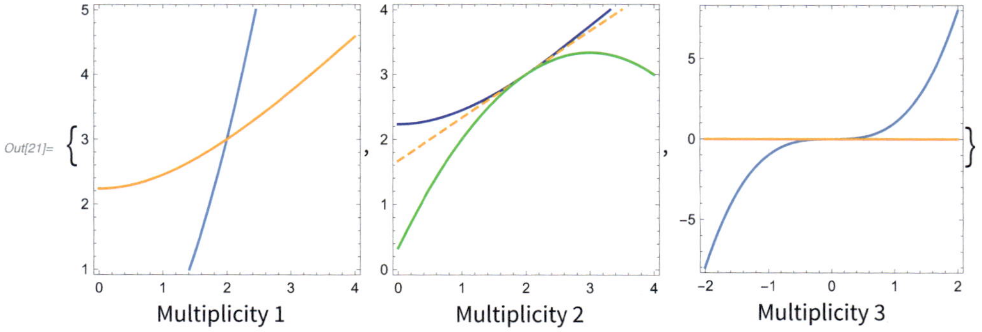


```
In[22]:= intersectionMultiplicity[y-x^2+1, y^2-x^2-5, {2, 3}, dTol]

Out[22]= 210

In[23]:= intersectionMultiplicity[y+1/3 x^2-2 x-1/3, y^2-x^2-5, {2, 3}, dTol]

Out[23]= 210
```

```
In[24]:= intersectionMultiplicity[y-x^3, y, {0, 0}, dTol]

Out[24]= 210
```

In the first case, we have a *normal crossing* with multiplicity 1, but the tangent directions are different. In the second case, both curves are tangent to the same line (dashed). In the third case, the curve has an inflection point at the point of tangency.

Some other examples with multiplicity greater than 1 at {0, 0}:

```
In[25]:= {Labeled[ContourPlot[{y-x == x^3, y-x == y^3}, {x, -1, 1}, {y, -1, 1}], "Multiplicity 5"],
   Labeled[ContourPlot[{y^2 == x^3, y-3 x}, {x, -1, 1}, {y, -1, 1}], "Multiplicity 2"],
   Labeled[ContourPlot[{x^2+y^2 == 0, y == 3 x}, {x, -1, 1}, {y, -1, 1}], "Multiplicity 2"]}
```

Out[25]=

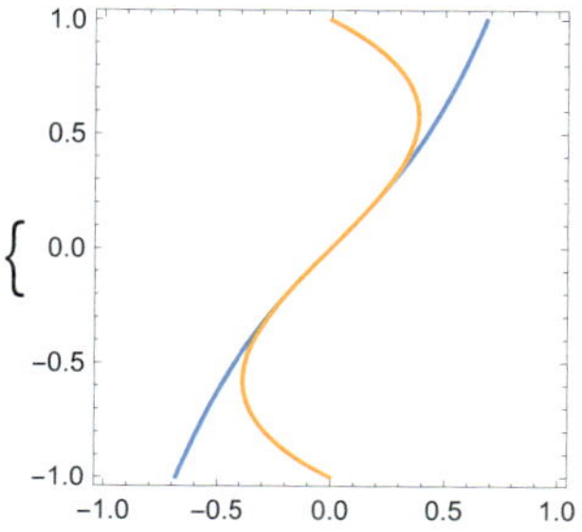

Multiplicity 5

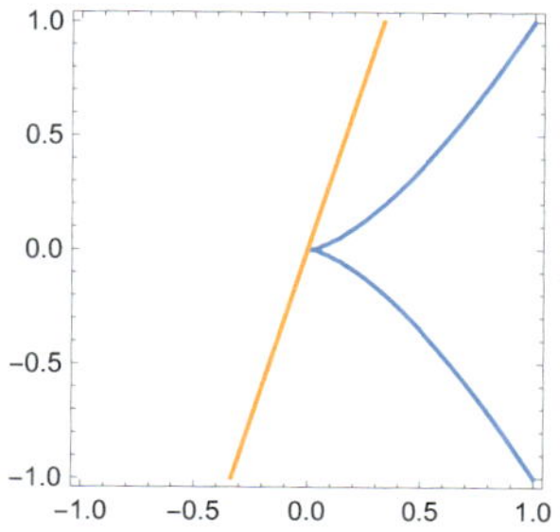

Multiplicity 2

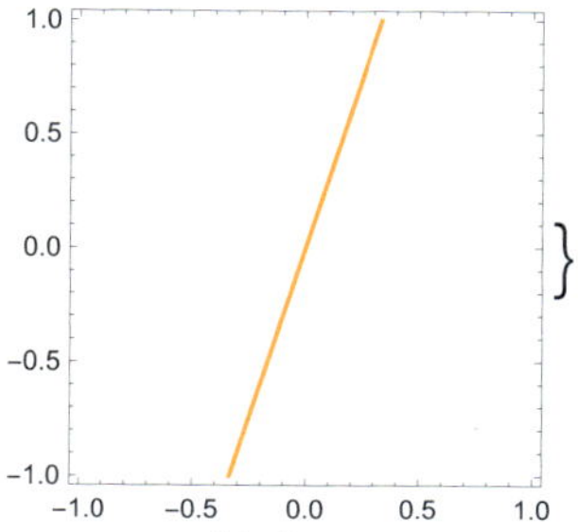

Multiplicity 2

```
In[26]:= intersectionMultiplicity[y-x-x^3, y-x-y^3, {0, 0}, dTol]

Out[26]= 210

In[27]:= intersectionMultiplicity[y^2-x^3, y-3 x, {0, 0}, dTol]

Out[27]= 210

In[28]:= intersectionMultiplicity[x^2+y^2, y-3 x, {0, 0}, dTol]

Out[28]= 210
```

In fact, determining when the intersection multiplicity is 1 is easy; $f = 0$ and $g = 0$ and the Jacobian determinant $\mathcal{J}(f, g) = \mathrm{Det}[D[\{f, g\}, \{\{x, y\}\}]] = \frac{\partial f}{\partial x}\frac{\partial g}{\partial y} - \frac{\partial f}{\partial y}\frac{\partial g}{\partial x}$ is not zero at the point. If this determinant is zero, then the intersection multiplicity is higher but more complicated to calculate. Note that in the right two examples in the previous graphs, we have a line intersecting a curve at a singularity, with the rightmost singularity being an isolated point, which is invisible to a contour plot. This illustrates an important general fact:

If p is a singular point of $f = 0$, then any line through p has intersection multiplicity ≥ 2 with f.

We will discuss complex projective singular points a bit later in this chapter.

Another important fact about intersection multiplicity is that if $f = 0, g = 0$ are reducible curves with a common component $h = 0$ and p is a point on h, then the intersection multiplicity of f, g at p is infinite. To discover common components, one takes greatest common divisors using `PolynomialGCD` if f, g are given by integer coefficients. Otherwise, use the `nGCD` algorithm in Appendix 2.

Bézout's Theorem

We now come to the most important theorem in complex plane algebraic geometry, Bézout's theorem:

Let $F(x, y, z) = 0$, $G(x, y, z) = 0$ be two projective curves, that is, two homogeneous three-variable polynomials of degrees m, n, respectively. Assume F, G have no common nonconstant component. Then

$$\sum_p mult_p(F, G) = m\,n,$$

where the sum ranges over all intersection points in the complex projective plane.

In simpler language, this says two curves without a common component intersect in exactly m, n points counted according to multiplicity. This is similar to the fundamental theorem of algebra in that to get all zeros, one must look at complex as well as real points. In addition, here we also have to consider infinite points. If one is only interested in the affine case, then this says there are at most m, n solutions. Newton was interested in finding more precise estimates in this latter case as he did not acknowledge complex or infinite points extending to higher dimensions, and finding algorithms for his results is a big area of modern computational algebra.

In the affine curve case, the Wolfram Language command would be:

```
In[29]:= NSolve[{f, g}, {x, y}, Reals]

Out[29]= {}
```

This will generally correctly answer this question although to get the correct multiplicity, one may need to count nearby points as a multiple intersection. Again, our algorithm in the appendix may be helpful. Remember that solving for multiple zeros accurately is difficult, so you might need to increase the tolerance in the code for that procedure.

Unfortunately, there are no easy proofs of Bézout's theorem. Abhyankar's work (1990) gives a heuristic argument; the proof in Fischer's work (2001) is as direct as any but requires methods from an appendix. Computationally, this can be checked by computing affine intersection points in each of the three specializations above and converting to projective notation remembering that triples that are scalar multiples are the same projective point.

Here is a simple example:

```
In[30]:= f = x ^2 y - 1;
         g = y - x ^2 - 2;
         ContourPlot[{f == 0, g == 0}, {x, -3, 3}, {y, -1, 5}]
```

Out[32]=

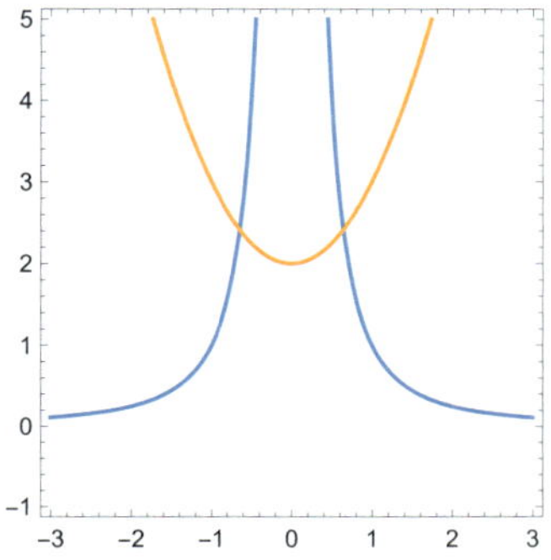

There appear to be two real affine intersections. Bézout claims six complex projective solutions:

```
In[33]:= sol0 = NSolve[{f, g}]

Out[33]= {{x → 0.643594, y → 2.41421}, {x → -0.643594, y → 2.41421},
          {x → 0. - 1.55377 ⅈ, y → -0.414214}, {x → 0. + 1.55377 ⅈ, y → -0.414214}}
```

We now see two more complex intersections. We write these in homogeneous coordinates:

```
In[34]:= {x, y, 1} /. sol0

Out[34]= {{0.643594, 2.41421, 1}, {-0.643594, 2.41421, 1},
          {0. - 1.55377 ⅈ, -0.414214, 1}, {0. + 1.55377 ⅈ, -0.414214, 1}}
```

Homogenizing and specializing at $x = 1$, we get the following:

```
In[35]:= fh = x^2 y - z^3;
         gh = y z - x^2 - 2 z^2;

In[37]:= fh1 = fh /. {x → 1}

Out[37]= y - z^3

In[38]:= gh1 = gh /. {x → 1}

Out[38]= -1 + y z - 2 z^2

In[39]:= sol1 = NSolve[{fh1, gh1}]
         ({1, y, z}/z) /. sol1

Out[39]= {{y → 3.75114, z → 1.55377}, {y → -3.75114, z → -1.55377},
          {y → 0. + 0.266585 ⅈ, z → 0. - 0.643594 ⅈ}, {y → 0. - 0.266585 ⅈ, z → 0. + 0.643594 ⅈ}}

Out[40]= {{0.643594, 2.41421, 1}, {-0.643594, 2.41421, 1},
          {0. + 1.55377 ⅈ, -0.414214 + 0. ⅈ, 1}, {0. - 1.55377 ⅈ, -0.414214 + 0. ⅈ, 1}}
```

We see we get the same four projective solutions. Specializing the homogenous equation at $y = 1$, we get:

In[41]:=
```
fh2 = fh /. {y → 1}
```

Out[41]= $x^2 - z^3$

In[42]:=
```
gh2 = gh /. {y → 1}
```

Out[42]= $-x^2 + z - 2 z^2$

In[43]:=
```
sol2 = NSolve[{fh2, gh2}]
```

Out[43]=
```
{{x → 0. - 3.75114 ⅈ, z → -2.41421},
  {x → 0. + 3.75114 ⅈ, z → -2.41421}, {x → 0.266585, z → 0.414214},
  {x → -0.266585, z → 0.414214}, {x → 0., z → 0.}, {x → 0., z → 0.}}
```

Now we have six solutions, counting the repeat. The last two homogenize as a double solution {0, 1, 0}, and the first four give:

In[44]:=
```
({x, 1, z}/z) /. Take[sol2, 4]
```

Out[44]=
```
{{0. + 1.55377 ⅈ, -0.414214, 1}, {0. - 1.55377 ⅈ, -0.414214, 1},
  {0.643594, 2.41421, 1}, {-0.643594, 2.41421, 1}}
```

These are the four previously known solutions, so counting complex and infinite solutions by multiplicity, we get six as required by Bézout's theorem.

In general, there are specializations that contain all intersection points, although not necessarily one of the three so far described. In Chapter 6, we will find such a specialization and find all intersection points with one calculation.

If f, g are real of degree m, n, we can say a bit more. Imaginary, nonreal intersections must come in pairs with the same multiplicity. Thus the number of real intersections, including multiplicity and infinite real intersection points, must differ from mn by an even number. In particular, if m, n are both odd, there must be at least one real intersection point.

If g is a line, then one does not need the full force of Bézout's theorem since counting intersections can be done by reduction to one variable by substitution. But Bézout makes clear that the one-variable multiplicity is the same as intersection multiplicity in this case. Hence, a line intersects a degree m curve in exactly m complex projective points counted by multiplicity. In particular, this applies to the infinite line, so a curve of degree m has m infinite points counting complex points and multiplicity. If m is odd, there must be at least one real infinite point that cannot be isolated since isolated points have even multiplicity. So, in fact, odd-degree real curves must contain infinitely many real points and, as we will see in Chapter 7, a least one asymptote.

5.4 Complex Singularities

In some situations, even if our interest is in real curves, we need information available by considering the curve in the complex affine or projective plane. Two cases are:

A plane algebraic curve is nonsingular or regular if it has no complex projective singular points.

A plane algebraic curve is irreducible if it cannot be factored as a product of nonconstant complex bivariate polynomials.

Not only does this make our definitions consistent with classical definitions, it allows us to separate out curves that may have unexpected properties.

Two simple examples are as follows:

1. The first example:

In[45]:=
```
f = 4 - 5 x^2 + x^4 - 5 y^2 + 2 x^2 y^2 + y^4;
ContourPlot[f == 0, {x, -3, 3}, {y, -3, 3}]
```

Out[46]=

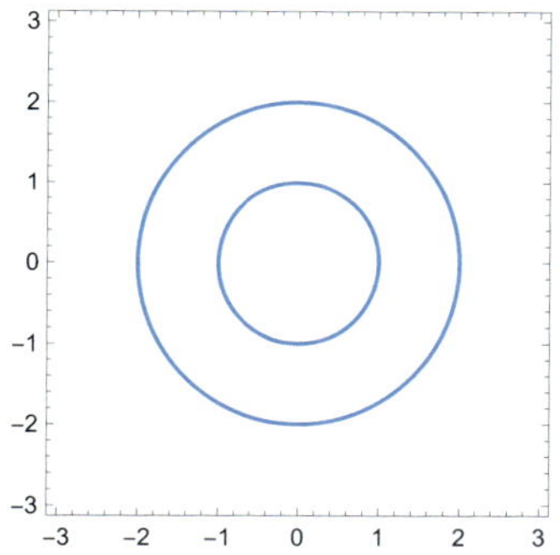

We will see in the second example that this has complex infinite singular points $\{1, \pm i, 0\}$, so although there are no real affine singularities, this curve is singular:

2. The second example:

In[47]:=
```
g = 18 - 24 x + 12 x^2 - 8 x^3 + 2 x^4 - 24 y + 32 x y - 8 x^2 y + 9 y^2 - 8 x y^2 + 3 x^2 y^2 - 4 y^3 + y^4;
ContourPlot[g == 0, {x, -2, 4}, {y, -2, 4}]
```

Out[48]=

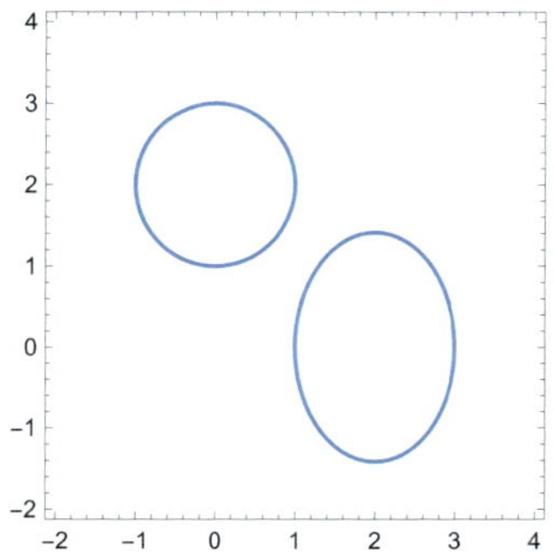

We will see the next section that this has four affine complex singular points, so again it is a singular curve.

Finding Complex Singular Points

From a statistics point of view, if we make the null hypothesis that a given curve is nonsingular, it is a Type II error that we want to avoid. Labeling a regular point as singular will not be as serious as labeling a singular curve as nonsingular, so in coding a probabilistic algorithm to find singular points, we will prefer to err on the side of calling a regular point singular.

Recall that in Chapter 3, we define an affine point p to be singular if point p is a solution to the system for all real a, b:

```
{ f = 0
{ a D[f, x] + b D[f, y] = 0
```

We now change this to requiring it to be true for all complex a, b.

We note that, again, such a p will be a multiple zero of this system. In practice, this could show up in NSolve by p appearing more than once in the list of solutions, which is the method of discovery in Chapter 3. The problem is that for severe singularities where the intersection multiplicity at p is large, NSolve will do a poor job and list the multiple solutions as different. The other alternative is we could use two random choices of a, b and look for common solutions. In order to make sure we don't miss singular points, we can easily combine both. We also use a very loose threshold for deciding two points are the same (For a complete list of global functions, see Appendix 2).

Our algorithm for finding complex affine singular points will be:

```
In[49]:= complexSingularPoints[f_, u_, v_] := Module[{a, b, sol1, sol2, solz},
           {a, b} = RandomComplex[{-1 - I, 1 + I}, 2];
           sol1 = {u, v} /. NSolve[{f, a D[f, u] + b D[f, v]}];
           {a, b} = RandomComplex[{-1 - I, 1 + I}, 2];
           sol2 = {u, v} /. NSolve[{f, a D[f, u] + b D[f, v]}];
           solz = Join[sol1, sol2];
           Flatten[Take[Select[Tally[solz, Norm[#1 - #2] < .001 &], #[[2]] > 1 &], All, 1], 1]]
```

Looking at Example 2 on the previous page:

```
In[50]:= g = 18 - 24 x + 12 x^2 - 8 x^3 + 2 x^4 - 24 y + 32 x y - 8 x^2 y + 9 y^2 - 8 x y^2 + 3 x^2 y^2 - 4 y^3 + y^4;
         complexSingularPoints[g, x, y]

Out[51]= {{6.88879 - 4.73523 i, 6.76932 + 6.83955 i}, {6.88879 + 4.73523 i, 6.76932 - 6.83955 i},
          {1.11121 - 0.517534 i, 1.23068 - 0.747525 i}, {1.11121 + 0.517534 i, 1.23068 + 0.747525 i}}
```

For Example 1 above, we must homogenize and take a specialization, find affine complex singular points there and homogenize the result to get the projective singular points:

```
In[52]:= f = 4 - 5 x^2 + x^4 - 5 y^2 + 2 x^2 y^2 + y^4;

In[53]:= fh = homog[f, x, y, z]

Out[53]= x^4 + 2 x^2 y^2 + y^4 - 5 x^2 z^2 - 5 y^2 z^2 + 4 z^4
```

In[54]:=
```
f1 = fh /. {x → 1}
```

Out[54]= $1+2y^2+y^4-5z^2-5y^2z^2+4z^4$

In[55]:=
```
c1 = complexSingularPoints[f1, y, z]
{1, #[[1]], #[[2]]} &/@c1
```

Out[55]= {{0. + 1. ⅈ, −9.00573 × 10^−6 + 0.0000410139 ⅈ}, {0. − 1. ⅈ, −0.0000165785 − 0.0000335048 ⅈ}}

Out[56]= {{1, 0. + 1. ⅈ, −9.00573 × 10^−6 + 0.0000410139 ⅈ}, {1, 0. − 1. ⅈ, −0.0000165785 − 0.0000335048 ⅈ}}

Since we are expecting infinite points, we can assume the third coordinates of the last answer are 0.

Here is a similar but even less obvious example:

In[57]:=
```
h = −1 + x^4 + 2 x^2 y^2 + y^4;
ContourPlot[h == 0, {x, −3, 3}, {y, −3, 3}]
```

Out[58]=

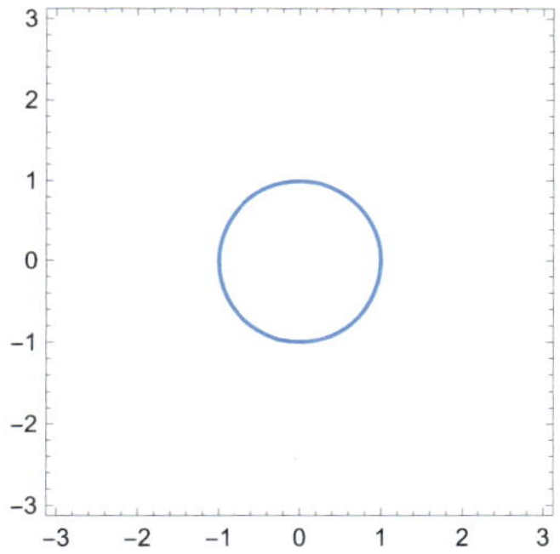

This does not look singular. In fact, it looks like a circle, but this is not expected with a 4th degree equation:

In[59]:=
```
complexSingularPoints[h, x, y]
```

Out[59]= {}

No affine singular points but:

In[60]:=
```
hh = x^4 + 2 x^2 y^2 + y^4 − z^4;
hh1 = hh /. {x → 1}
```

Out[61]= $1+2y^2+y^4-z^4$

In[62]:=
```
hh2 = hh /. {y → 1}
```

Out[62]= $1+2x^2+x^4-z^4$

In[63]:=
```
px = {1, #[[1]], #[[2]]} &/@complexSingularPoints[hh1, y, z]
```

Out[63]= {{1, 0. − 1. ⅈ, 0.000109249 − 0.000146589 ⅈ}, {1, 0. + 1. ⅈ, −0.000121456 + 0.000140766 ⅈ}}

In[64]:=
```
py = {#[[1]], 1, #[[2]]} &/@complexSingularPoints[hh2, x, z]
```

Out[64]= {{0. − 1. ⅈ, 1, −0.0000180499 − 0.000154631 ⅈ}, {0. + 1. ⅈ, 1, 0.0000382071 + 0.000150842 ⅈ}}

Again, we can check that $\{1, \pm i, 0\}$, $\{\pm i, 1, 0\}$ are the actual four complex infinite points of $h = 0$ that were discovered by our algorithm to be singular.

For example:

```
In[65]:= hh /. Thread[{x, y, z} → {1, I, 0}]
Out[65]= 0
```

It should be noted that instead of finding real singular points, we can make a tiny change in our algorithm to find nonsingular points. Compare with the above complexSingularPoints to see if you can find the difference other than we specify how many points to produce, if known.

Compare the previous code using complexSingularPoints with the following code using realRegularPoints (For a complete list of global functions, see Appendix 2):

```
In[66]:= realRegularPoints[f_, x_, y_, n_] := Module[{a, b, c, d, k},
           If[tDeg[f, x, y] < 2, Echo["Degree must be > 1"]; Return[{}]];
           k = 0;
          Reap[While[k < n,
                {a, b} = RandomReal[{-10, 10}, 2];
                {c, d} = closestPoint[f, {a, b}, x, y];
                If[
                  Abs[Det[D[{f, a x - b y}, {{x, y}}]] /. Thread[{x, y} → {c, d}]] > .01, Sow[{c, d}];
                  k++]]][[2, 1]]];

In[67]:= f = x^4 - 4 x y + y^4;
         realRegularPoints[f, x, y, 3]
Out[68]= {{1.44126, 1.38296}, {1.33897, 1.46855}, {-1.40886, -0.757537}}
```

The Singularity Theorem

A famous theorem in classical algebraic geometry is the following:

Let $f = 0$ be an irreducible curve of degree d. Then there are at most $\frac{(d-1)(d-2)}{2}$ complex projective singular points.

It is clear that this is true in degree 1 and 2 since we know that all irreducible curves are nonsingular.

We will mostly follow the proof in Fischer's book (2001), which is an indirect proof using Bézout's theorem. However, since in this book I emphasize algorithms rather than proofs, I will describe instead an algorithm to do the following:

Given an algebraic plane curve $f = 0$ of degree $d \geq 3$ with more than $\frac{(d-1)(d-2)}{2}$ singularities, produce a nonconstant factor.

We know from Chapter 2 that given $\frac{(n+2)(n+1)}{2} - 1$ sufficiently random points, there is a unique curve of degree n through these points, so $\frac{d(d-1)}{2} - 1$ points determine a curve of degree $d - 2$.

Now some simple algebra:

In[69]:=
```
Expand[((d) (d - 1) / 2 - 1) - ((d - 1) (d - 2) / 2 + 1)]
```

Out[69]= $-3 + d$

We see that if f has more than $\frac{(d-1)(d-2)}{2}$ singularities, then taking $\frac{(d-1)(d-2)}{2} + 1$ of them together with $d - 3$ nonsingular points of f, we get a curve g of degree $d - 2$ passing through these points. A technical issue is that if these points are not sufficiently random, or in algebraic geometry terms, in sufficiently general position, then the algorithm in Chapter 2 does not quite work since there could be several such curves, but we just choose one. In any case, we can get a curve of degree $d - 2$ passing through all the $\frac{(d-1)(d-2)}{2} + 1$ singular points we took as well as at least $d - 3$ additional points. However, we learned previously that the intersection multiplicity of two curves meeting at a singular point of one of them is at least 2.

The sum of the intersection multiplicities of f and g is at least as follows:

In[70]:=
```
Expand[2 ((d - 1) (d - 2) / 2 + 1) + (d - 3)]
```

Out[70]= $1 - 2 d + d^2$

Bézout says the sum of the multiplicities of all complex projective intersection points of f, g is only $d(d - 2) = d^2 - 2 d$, so Bézout fails because f, g have a nonconstant component since GCD[f, g] is not constant, so f, in particular, has a nonconstant component.

For degrees 3 and 4, this gives us a numerical factoring algorithm for any reducible bivariate real or complex polynomial. Unfortunately, for degree 5 and above, there are reducible curves with less than $\frac{(d-1)(d-2)}{2} + 1$ singular points. Generally, to get more singular points, a curve of degree 5 or larger needs to be the product of more than two irreducible curves. Careful analysis of this situation leads to the conclusion that a square-free reducible curve of degree d can have no more than $\frac{d(d-1)}{2}$ singularities. This is achieved by a union of d lines with no more than two meeting at any one point.

We should, however, note this important corollary of Bézout's theorem since the components of an irreducible curve meet in singular points:

A reducible curve must have at least one singular point, possibly complex and/or infinite. That means a reducible curve cannot be nonsingular even if there are no visible singularities on the plot.

Here is one worked-out example:

In[71]:=
```
f = Expand[(x^2 + 4 y^2 - 4) * (4 x^2 + y^2 - 4)]
```

Out[71]= $16 - 20 x^2 + 4 x^4 - 20 y^2 + 17 x^2 y^2 + 4 y^4$

In[72]:=
```
csg = complexSingularPoints[f, x, y]
```

Out[72]= {{0.894427, 0.894427}, {-0.894427, -0.894427}, {-0.894427, 0.894427}, {0.894427, -0.894427}}

This is obviously reducible with too many singular points to be nonsingular, so we take these four points and one random real regular point:

In[73]:= **rrp = realRegularPoints[f, x, y, 1][[1]]**

Out[73]= {−1.8517, −0.377888}

In[74]:= **allpts = Append[csg, rrp];**

In[75]:= **Labeled[Show[ContourPlot[f == 0, {x, −2, 2}, {y, −2, 2}], ListPlot[
Table[Labeled[allpts[[i]], i], {i, 5}], PlotMarkers → Automatic]], "singularities"]**

Out[75]=

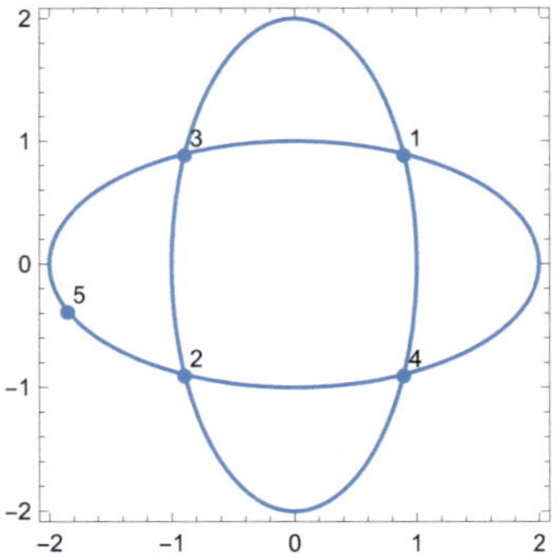

singularities

We now apply the function aCurve from Chapter 2 (or Appendix 2):

In[76]:= **h = aCurve[allpts, x, y]**

Out[76]= $1.10241 + 2.03993 \times 10^{-15}\, x - 1.10241\, x^2 + 4.85328 \times 10^{-15}\, y + 5.12933 \times 10^{-15}\, x\, y - 0.275601\, y^2$

In[77]:= **h = Chop[Expand[h / Coefficient[h, y ^ 2]]]**

Out[77]= $-4. + 4.\, x^2 + 1.\, y^2$

We then divide *f* by *h* to get the other factor *k*. Note the built-in function PolynomialReduce seems to work better in our context than PolynomialDivide:

In[78]:= **k = PolynomialReduce[f, h, {x, y}][[1, 1]]**

Out[78]= $1.06169 \times 10^{16} \left(-3.76757 \times 10^{-16} + 9.41893 \times 10^{-17}\, x^2 + 3.76757 \times 10^{-16}\, y^2\right)$

In[79]:= **Labeled[Show[ContourPlot[{h == 0, k == 0}, {x, −2, 2}, {y, −2, 2}],
ListPlot[Table[Labeled[allpts[[i]], i], {i, 5}], PlotMarkers → Automatic]],
"intersection points"]**

Out[79]=

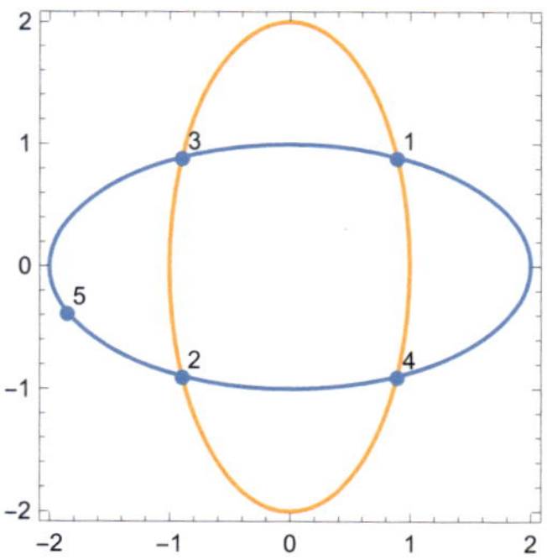

intersection points

The algorithm described above may be implemented in the Wolfram Language. The function singularFactor is given in Appendix 1 or can be accessed in Appendix 2:

In[80]:=
```
singularFactor[f, x, y, 1.*^-10]
```

Out[80]= $\{7.18023 - 7.18023\,x^2 - 1.79506\,y^2, \text{Fail}\}$

Before leaving this example, I need to make an important distinction about the two monst recent plots. In the first plot, the points labeled 1–4 are *singularities*. In the second, two irreducible components, points 1–4 are *intersection points*. Singularities and intersection points behave differently. If a curve with singularities is perturbed slightly, it is likely that the singularities will disappear. In other words, singularities are *unstable*. If the two component curves are each perturbed slightly, there will still be an intersection point nearby. More generally, the total intersection multiplicity in a small open set about the original intersection point will remain unchanged, so intersections are *stable*. This is why it is much easier to work with intersections numerically than to work with singularities. An important topic in classical algebraic geometry is assigning invariants to singularities. For an example, see Appendix 5 in Fischer's book (2001) because I will not cover it in this text.

In[81]:=
```
{Labeled[ContourPlot[f == -.09, {x, -2.5, 2.5}, {y, -2.5, 2.5}], "singularities perturbed"],
  Labeled[ContourPlot[{h == -.3, k == .3}, {x, -2.5, 2.5}, {y, -2.5, 2.5}],
   "intersections perturbed"]}
```

Out[81]= 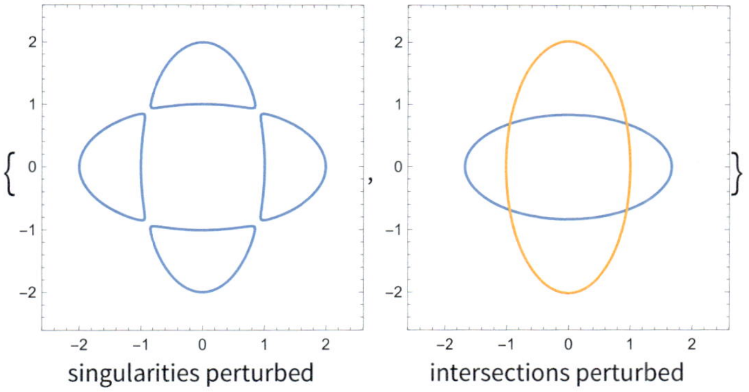

The curve $f = -.09$ is actually nonsingular, but the curve $(h + .3)\,(k - .3) = 0$ is still singular:

In[82]:=
```
complexSingularPoints[f+.09, x, y]
```

Out[82]= {}

5.5 Fundamental and Riemann–Roch Theorems

No book on algebraic curve theory is complete without at least mentioning the Riemann–Roch theorem, the most famous theorem in classical curve theory. The importance of this theorem is that it is basic to the notion of *genus*, a number that, among other things, describes the topology of a complex nonsingular projection curve. Since this book is about real curves, my main interest is to compare this theorem with my *fundamental theorem of real projective plane curves.*

Before we start, we need one definition. A set $\mathcal{X}$ will be called a *concrete topological space* if it is a subset of one of $\mathbb{R}^2$, $\mathbb{C}^2$, $\mathbb{P}^2_{\mathbb{R}}$, $\mathbb{P}^2_{\mathbb{C}}$. In particular, the point set $\mathcal{V}(f)$ of a projective curve qualifies. With such a set, we can define continuous functions to and from the set.

A continuous function Φ from a concrete topological space to a component of a projective curve C, real or complex, will be called a complete parameterization if for each nonsingular point q in C, there exists a unique p in $\mathcal{X}$ with $\Phi(p) = q$, and if q is a singular point in C, there is a positive number of points p_i with $\Phi(p_i) = q$.

Fundamental Theorem of Real Projective Curves

In the real projective plane $\mathbb{P}^2_{\mathbb{R}}$, what we have called infinite points are just like any other point, especially after specializing to the $x\,z$ or $y\,z$ affine plane. Thus the curve inside a large circle about the origin can, in theory, be described in terms of a graph, as in Chapter 4, where the infinite point is now an interior point. In particular, if we add that point to our critical points, it becomes an interior vertex of the graph and, hence, even. More generally, the boundary points of the circle are interior points of other circles, so they also give even vertices when viewed in the projective context. We will make this more precise in Chapter 8, where we explain how to get the projective graph with all even vertices in practice. Thus we obtain our:

Fundamental Theorem: Given a real projective plane curve C, there is a graph with vertices v_i representing points p_i on C satisfying the following:

1. *For each edge e from vertex v_i to vertex v_j, there is a closed continuous path contained in C connecting the points p_i to p_j so that each interior point of this path is nonsingular.*

2. *Every point of C is either a vertex p_k or an interior point of a path corresponding to a unique edge e, as in 1.*

3. *Every vertex is even.*

A graph with every vertex even is sometimes called an *Euler graph,* and we will use this nomenclature. It is well known that every Euler graph has the property that each connected component is an *Euler cycle*; that is, there is a closed path that includes each edge of the graph exactly once.

The Gauss curves give good examples. Gauss tells us that for a monic complex polynomial p of degree n in z, the Gauss curve of p intersects a large circle about the origin of radius r in approximately $2\,n$ points, $p_j = r\left(\text{Cos}\left[\frac{(2\,j+n-1)\,\pi}{2\,n}\right], \text{Sin}\left[\frac{(2\,j+n-1)\,\pi}{2\,n}\right]\right)$, $j = 1, \ldots, 2\,n$, (the indexes were changed to correspond to our indexing of the boundary points in gPoints). We can conclude that the infinite points are then $P_j = r\left\{\text{Cos}\left[\frac{(2\,j+n-1)\,\pi}{2\,n}\right], \text{Sin}\left[\frac{(2\,j+n-1)\,\pi}{2\,n}\right], 0\right\}$, $j = 1, \ldots, 2\,n$ with

$P_j = P_{j+n}$. This says that we can simply construct the projective graph of a Gauss curve by identifying the vertex v_{j+n} with v_j for $j = 1, \ldots, n$ in the affine graph using the Wolfram Language function VertexReplace.

In particular, consider Gauss's famous example (Initialize Appendix 2):

```
In[83]:= p = z^5 - 4 z^4 + 9 z^3 - 5 z^2;
```

```
In[84]:= g = gaussCurve[p, x, y, z]
```

Out[84]= $-5\,x^2 + 9\,x^3 - 4\,x^4 + x^5 + 5\,y^2 - 27\,x\,y^2 + 24\,x^2\,y^2 - 10\,x^3\,y^2 - 4\,y^4 + 5\,x\,y^4$

```
In[85]:= T5 = Table[N[20 {Cos[(2 j + 9) Pi / 10], Sin[(2 j + 9) Pi / 10]}], {j, 10}]
```

```
Out[85]= {{-19.0211, -6.18034}, {-11.7557, -16.1803}, {0., -20.},
  {11.7557, -16.1803}, {19.0211, -6.18034}, {19.0211, 6.18034},
  {11.7557, 16.1803}, {0., 20.}, {-11.7557, 16.1803}, {-19.0211, 6.18034}}
```

```
In[86]:= Show[ContourPlot[g == 0, {x, -22, 22}, {y, -22, 22}],
  ListPlot[Table[Labeled[T5[[j]], 11 - j], {j, 1, 10}]]]
```

Out[86]=

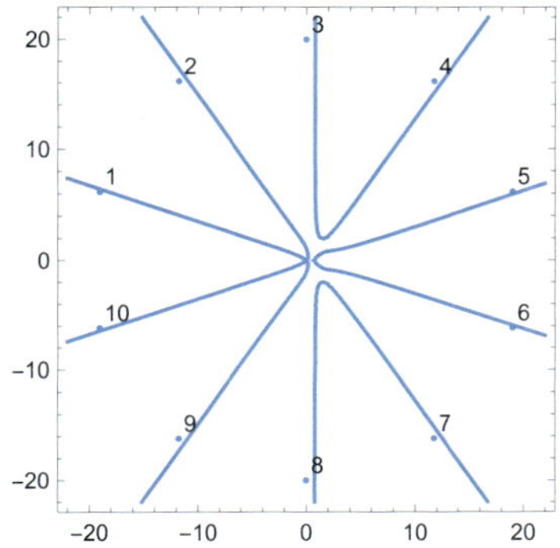


```
In[87]:= cp = Chop[gPoints[g, 5, x, y]]
```

```
Out[87]= {{-4.66433, 1.80113}, {-2.32319, 4.42751}, {0.855926, 4.92619},
  {3.34022, 3.72061}, {4.79137, 1.42927}, {4.79137, -1.42927}, {3.34022, -3.72061},
  {0.855926, -4.92619}, {-2.32319, -4.42751}, {-4.66433, -1.80113},
  {1.35091, 2.02496}, {0.766771, 0}, {1.35091, -2.02496}, {0, 0}}
```

The affine graph is given by the code below. If you are using the notebook version of this, initialize all global functions:

```
In[88]:= AG = gaussGraph[g, cp, x, y, .4];
```

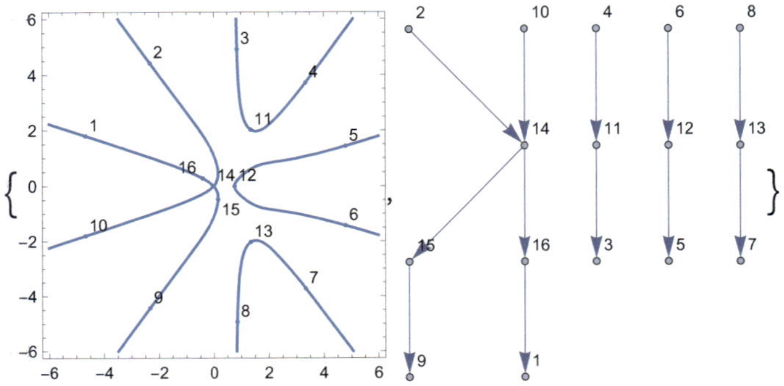

The projective graph is then:

```
In[89]:= PG = Graph[VertexReplace[AG[[2]], {6 → 1, 7 → 2, 8 → 3, 9 → 4, 10 → 5}],
           VertexLabels → "Name"]
```

Out[89]=

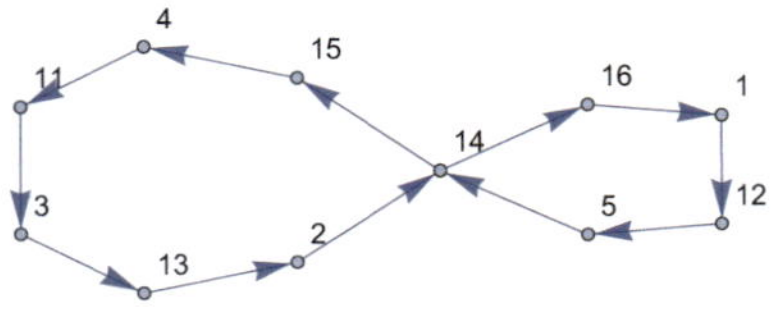

The commands in the above example, other than the check on Gauss's assertion, can be collected so it is only necessary to enter a new polynomial in z and possibly change some parameters.

As an example, we can do a generic polynomial in z:

```
In[90]:= p = -1.768628847250528` + 0.4539057530410755` z - 1.0579937564992576` z^2 -
           2.8339474110658296` z^3 + 0.35200766125787464` z^4 + z^5

Out[90]= -1.76863 + 0.453906 z - 1.05799 z^2 - 2.83395 z^3 + 0.352008 z^4 + z^5

In[91]:= g = Chop[gaussCurve[p, x, y, z]]

Out[91]= -1.76863 + 0.453906 x - 1.05799 x^2 - 2.83395 x^3 + 0.352008 x^4 + x^5 +
           1.05799 y^2 + 8.50184 x y^2 - 2.11205 x^2 y^2 - 10 x^3 y^2 + 0.352008 y^4 + 5 x y^4

In[92]:= cp = gPoints[g, 5, x, y]

Out[92]= {{-4.78227, 1.45942}, {-3.07112, 3.94566}, {-0.0732468, 4.99946},
           {2.9796, 4.01522}, {4.77103, 1.49575}, {4.77103, -1.49575}, {2.9796, -4.01522},
           {-0.0732468, -4.99946}, {-3.07112, -3.94566}, {-4.78227, -1.45942}, {1.73732, 0.},
           {-1.19202, -0.43087}, {-1.19202, 0.43087}, {0.279974, 0.67943}, {0.279974, -0.67943}}
```

```
In[93]:= AG = gaussGraph[g, cp, x, y, .5];
```

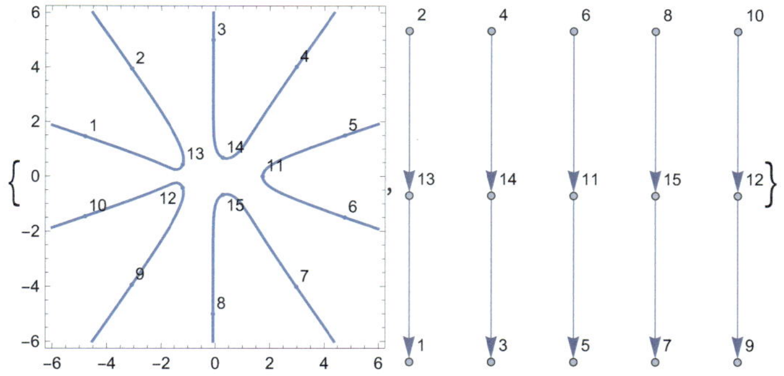

```
In[94]:= PG = Graph[VertexReplace[AG[[2]], {6 → 1, 7 → 2, 8 → 3, 9 → 4, 10 → 5}],
           VertexLabels → "Name"]
```

Out[94]=

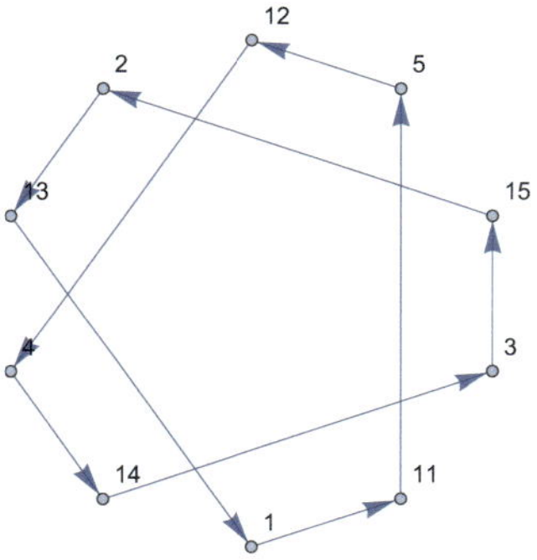

Note that this is just a cycle:

```
In[95]:= Graph[FindEulerianCycle[PG][[1]], VertexLabels → "Name"]
```

Out[95]=

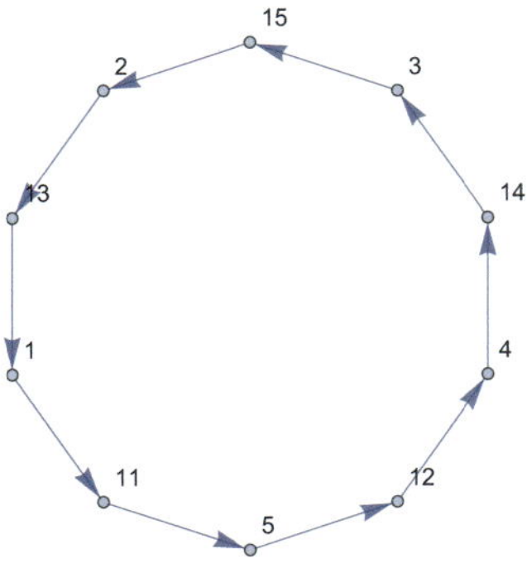

For those reading the Wolfram Notebook version of this text, try other examples of polynomials $p(z)$ in degree 5, possibly with complex coefficients. In general, one would obtain one or two connected components. In particular, you should try various placements of multiple zeros of $p(z)$ of various multiplicities. Each multiple zero at $a + bi$ will give a singularity at (a, b), but there may be additional singularities as well.

Two more examples of Gauss curves and projective graphs are:

In[96]:= **g1 = gaussCurve[z (z^2-4)^2, x, y, z]**

Out[96]= $16x-8x^3+x^5+24xy^2-10x^3y^2+5xy^4$

In[97]:=

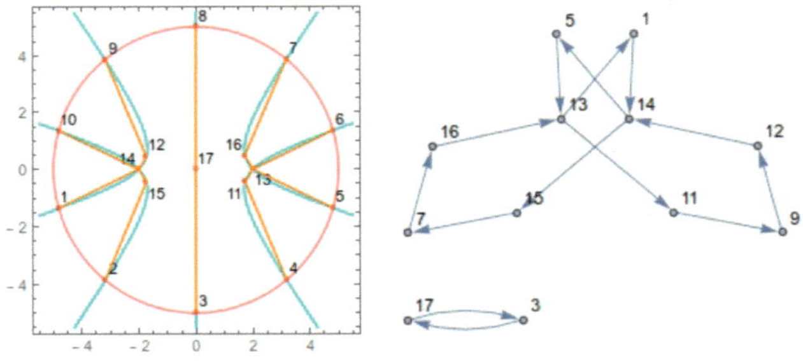

Out[97]=

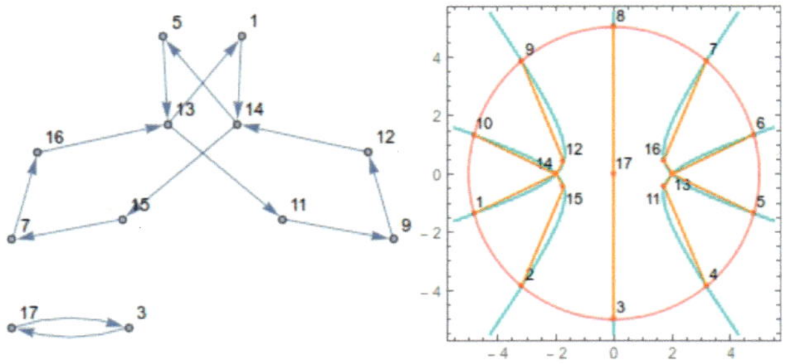

In[98]:= **g2 = gaussCurve[z^5, x, y, z]**

Out[98]= $x^5-10x^3y^2+5xy^4$

In[99]:=

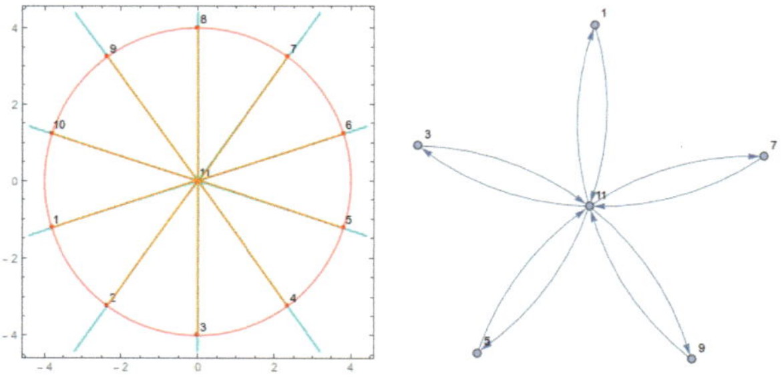

Out[99]=

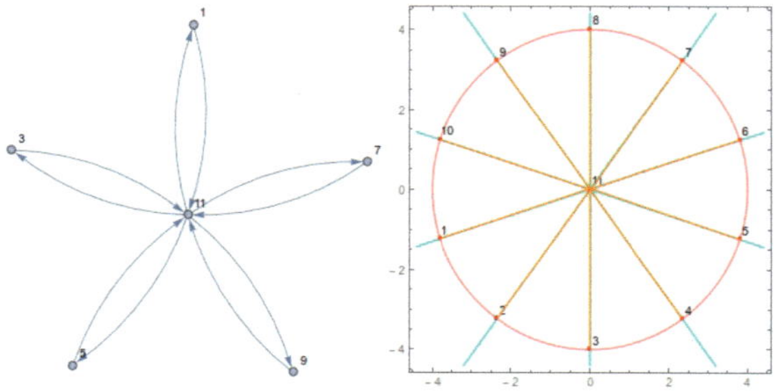

In the general case, we will show in Chapter 8 how to get the projective graph of any real curve.

Riemann–Roch

In the early 1800s, Abel and Jacobi worked on indefinite integration of elliptic integrals such as

$$\int \frac{dx}{\sqrt{x^3 + ax + b}}.$$

As a byproduct of their work, they surmised that the topology of the complex curve $y^2 = 4\,x^3 - ax - b$ where the right-hand side does not have repeated zeros, called *elliptic curve*, is equivalent to a torus. In the 1850s, Weierstrass confirmed this by explicitly parameterizing this curve by a function defined on a torus. (Look up WeierstrassP in the Wolfram Language & System Documentation Center or on Wolfram|Alpha or see Chapter 6 of my *Theory of Equations* book. For more information check out the More to Explore section at the end of this chapter.)

Parametrize the elliptic curve $y^2 = 4\,x^3 - 4\,x + 4$:

```
In[100]:= wp = WeierstrassP[x, {4, -4}];
          wpp = WeierstrassPPrime[x, {4, -4}];
          Plt1 = ContourPlot[y^2 == 4x^3-4x+4,
               {x, -2, 2}, {y, -5, 5}, Axes → True, ImageSize → Small];
          Plt2 = ParametricPlot[{wp, wpp}, {x, -4, 4}, AspectRatio → 1.25,
               PlotStyle → Red, ImageSize → Small];
          {Plt1, Plt2, Show[Plt1, Plt2, Axes → True, ImageSize → Small]}
```

Out[104]=

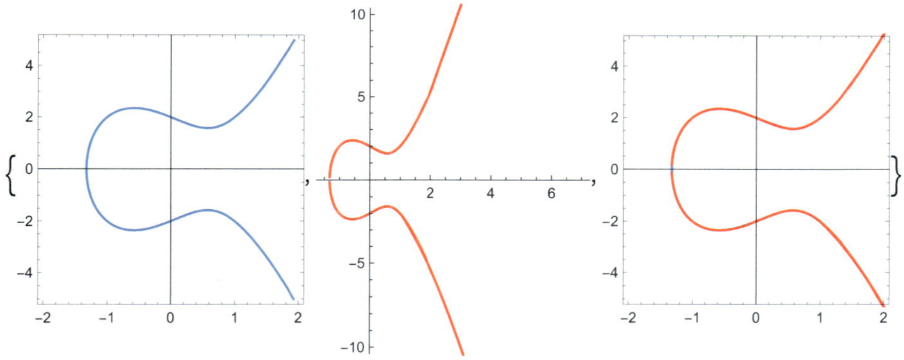

The results show a contour plot on the left, a parametric plot in the middle and a combined plot on the right showing that the plots coincide.

At the same time of Weierstrass's work, Riemann and his short-lived student Roch were working on understanding meromorphic (rational) functions on what are now called *Riemann surfaces*. Their work became known as the Riemann–Roch theorem. In the 1870s, geometers Felix Klein, Alfred Clebsch, and Max Noether noted that Riemann's work gave an abstract proof of the existence of not only Weierstrass's parameterization of elliptic curves but similar parameterization of all complex algebraic curves from Riemann surfaces. They called this result the Riemann–Roch theorem. These related versions of the Riemann–Roch theorem are now considered the fundamental theorems of complex curve theory:

An irreducible complex projective curve can be completely parameterized by a continuous function from a Riemann surface of genus g for some $g \geq 0$.

I will not discuss Riemann surfaces here, but note that they are variously described topologically as a *g-holed torus* or a *sphere with g handles*. The number g is called the *genus* of the Riemann surface or of a complex algebraic curve parameterized by such a surface. See Fischer's book (2001), or almost any book on algebraic curve theory, for more details. There is a well-known *Clebsch–Noether genus formula* for finding the genus from the degree of a curve together with certain *Clebsch numbers* describing each singularity. A nonsingular curve of degree d has genus $g = \frac{(d-1)(d-2)}{2}$, a number we are already familiar with, and Clebsch numbers are subtracted from this number. Each singularity has at least one Clebsch number leading to a quick proof of the singularity theorem.

A normal crossing has Clebsch number 1, but one must be careful about working from a plot as the following example shows:

In[105]:=
```
f = 3 x^4 - 10 x^3 y + 6 x^2 y^2 - 10 x y^3 + 3 y^4;
ContourPlot[f == 0, {x, -4, 4}, {y, -4, 4}]
```

Out[106]=

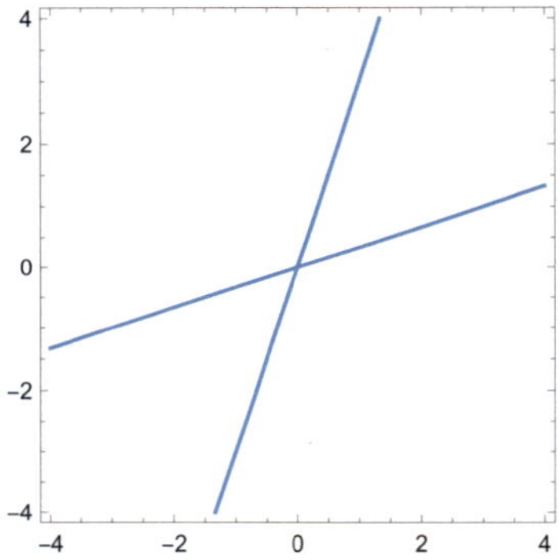

In[107]:=
```
g = Expand[{D[f, x], D[f, y]}.RandomReal[{-1, 1}, 2]]
```

Out[107]= $-4.12034\, x^3 + 11.3888\, x^2 y - 2.4466\, x y^2 + 3.12676\, y^3$

In[108]:=
```
NSolve[{f, g}]
```

Out[108]=
```
{{x → 0., y → 0.}, {x → 0., y → 0.}, {x → 0., y → 0.}, {x → 0., y → 0.},
 {x → 0., y → 0.}, {x → 0., y → 0.}, {x → 0., y → 0.}, {x → 0., y → 0.},
 {x → 0., y → 0.}, {x → 0., y → 0.}, {x → 0., y → 0.}, {x → 0., y → 0.}}
```

We expect multiplicity 4 for the zero, but we get multiplicity 12, so something is wrong. We actually have something unseen called an *embedded zero* at the origin as well as the obvious intersection point. Actually, this curve is reducible so Riemann–Roch does not apply.

As mentioned above, singularities are unstable numerically, so we will not attempt further discussion in this book.

Here are some pictures of some Riemann surfaces. These are called g-holed tori for $g = 1, 2, 3$. Other examples are *spheres with handles*:

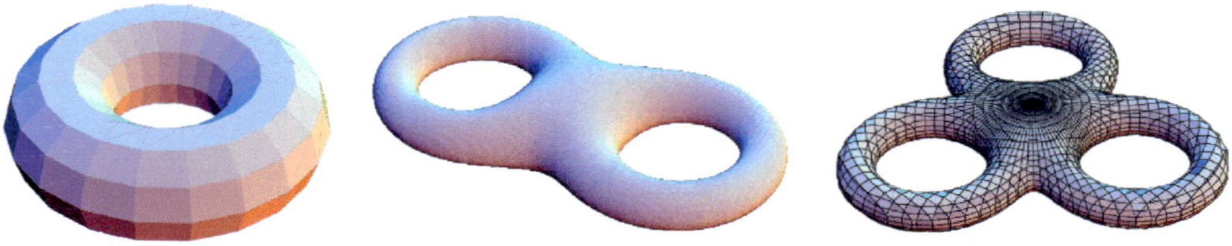

We can compare the fundamental theorem for real curves for the Riemann–Roch theorem for complex curves by looking at the following illustrations:

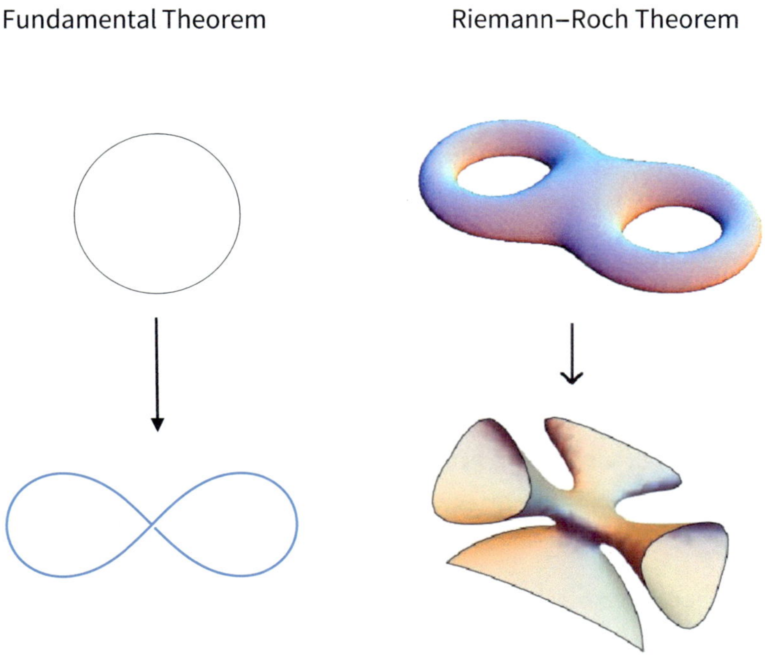

Tech Notes

- There is a form of classical geometry known as *projective geometry* that is different from *classical algebraic geometry* because it only considers points, lines and sometimes conics, but it does not, in general, consider algebraic curves. It generally starts with an axiomatic description and only later develops a coordinate version with points as above. In this case, lines are also denoted by a triple, say $[a, b, c]$, not all zeros. A main idea is *duality*, meaning two points determine a line but also two lines determine a point!

For example, consider lines:

```
In[109]:= l1 = 3 x - 2 y + 5 z;
          l2 = -x + 3 y - 2 z;
          l3 = pLine[{3, -2, 5}, {-1, 3, -2}, x, y, z]

Out[111]= pLine[{3, -2, 5}, {-1, 3, -2}, x, y, z]
```

```
In[112]:= p = {Coefficient[l3, x], Coefficient[l3, y], Coefficient[l3, z]}

Out[112]= {0, 0, 0}
```

```
In[113]:= l1 /. Thread[{x, y, z} → p]

Out[113]= 0
```

```
In[114]:= l2 /. Thread[{x, y, z} → p]

Out[114]= 0
```

So p is the point determined by lines [3, −2, 5] and [−1, 3, −2]. The same procedure to find a point through two lines also finds the projective point of intersection of two lines!

- A useful application of duality in projective geometry is that if one has a theorem about points and lines, then replacing the word *point* with the word *line* and vice versa results in one obtaining a new *dual* theorem. Two famous theorems of projective geometry are the *theorem of Desargues* and the *theorem of Pappus*.

More to Explore

For more on the Desargues theorem visit Wolfram|Alpha (wolfr.am/x1RlLDZz)

Also, check out several of the theorems that comprise Pappus's theorem on Wolfram MathWorld (wolfr.am/x1RpuJcD)

For a fun Riemann sum calculator, check out this Wolfram|Alpha widget submitted by Tiffanyj16 (wolfr.am/vT2UtWsc)

My book *Theory of Equations* (2002) is available as PDFs here (wolfr.am/vUzH0isW)

6 | Fractional Linear Transformations

Since the days of Felix Klein (around 1870), any discussion about any kind of geometry mentions the transformations that leave geometry invariant. (Abhyankar, 1990) has suggested that the proper group of transformations is the group of fractional linear transformations, or FLTs. While these generally work, we will keep the discussion simple by focusing on the plane invertible case.

This chapter includes half a dozen different functions for fractional linear transformations, all using the same basic function name, some variant of "FLT". For a full listing, see Appendix 2. I have also added a summary of these different functions and how they relate to the overall topic at the end of this chapter.

6.1 The Group of Fractional Linear Transformations

A fractional linear transformation is a transformation of the plane given by

$$F(x,\ y) = \left(\frac{a_{1,1}\ x + a_{1,2}\ y + \ a_{1,3}}{d_1\ x + d_2\ y + d_3},\ \frac{a_{2,1}\ x + a_{2,2}\ y + \ a_{2,3}}{d_1\ x + d_2\ y + d_3} \right),$$

where $a_{i,\,j}$, d_j are fixed real (or sometimes complex) numbers so that an inverse function of the same type exists.

Note that the equation $d_1\ x + d_2\ y + d_3\ = 0$ defines a line. Often $d_1,\ d_2\ = 0$, so this will be the infinite line. In either case, in the domain of the FLT, F, as shown above, will be the set of affine points with $d_1\ x + d_2\ y + d_3 \neq 0$.

The notation above suggests we describe the FLT compactly by the matrix

$$A = \begin{pmatrix} a_{1,1} & a_{1,2} & a_{1,3} \\ a_{2,1} & a_{2,2} & a_{2,3} \\ d_1 & d_2 & d_3 \end{pmatrix}.$$

This is more compact as well as useful because the FLT is actually given by this two-step procedure using matrix multiplication:

$$\{x,\ y\} \mapsto A\begin{pmatrix} x \\ y \\ 1 \end{pmatrix} = \{u,\ v,\ w\} \mapsto \left\{\frac{u}{w},\ \frac{v}{w}\right\}$$

In the Wolfram Language, this becomes:

```
In[1]:= flt[{x_, y_}, A_] := Module[{u, v, w},
          {u, v, w} = A.{x, y, 1};
          If[Abs[w] < 1.*^-6, {}, {u/w, v/w}]]
```

As an example:

```
In[2]:= A = {{1, 0, -2}, {0, -1, 0}, {0, 1, 1}};
        A // MatrixForm
```

Out[3]//MatrixForm=

$$\begin{pmatrix} 1 & 0 & -2 \\ 0 & -1 & 0 \\ 0 & 1 & 1 \end{pmatrix}$$

```
In[4]:= p = RandomReal[{-3, 3}, 2]

Out[4]= {1.81687, -1.33897}

In[5]:= q = flt[p, A]

Out[5]= {0.540235, -3.95007}

In[6]:= flt[q, Inverse[A]]

Out[6]= {1.81687, -1.33897}
```

This suggests that this FLT, as determined by matrix A, is invertible with inverse A^{-1}:

```
In[7]:= B = RandomInteger[{-5, 5}, {3, 3}]

Out[7]= {{-2, -4, -1}, {1, 0, 3}, {0, 2, 4}}

In[8]:= flt[p, B.A]
        flt[q, B]

Out[8]= {-3.51777, -0.907718}

Out[9]= {-3.51777, -0.907718}
```

We can generalize the following for invertible matrices A, B:

1. The composition of two FLTs is an FLT; in particular, the composition of the FLT given by matrix A with the FLT given by matrix B is the FLT given by the matrix product $B\,A$.
2. The FLT $\mathcal{I}$ defined by the identity matrix is the identity transformation; that is, $\mathcal{I}[x, y] = \{x, y\}$ for all $\{x, y\}$.
3. Each FLT has an inverse FLT, and the inverse of the FLT given by A is A^{-1}.

These facts can by defined as the set of FLTs in a group of transformations.

We have failed to mention a technicality: if we multiply a matrix by a nonzero constant c, that is

$$cA = \begin{pmatrix} c & 0 & 0 \\ 0 & c & 0 \\ 0 & 0 & c \end{pmatrix} A,$$

then we still get the same FLT due to the division at the end, so the group of FLTs is not the group of invertible 3×3 matrices but rather the group quotient by the nonzero constant diagonal matrices.

This group is generally known in mathematical literature as the projective general linear group PGL (3, ℝ):

```
In[10]:= A
         flt[{2, -1}, A]
Out[10]= {{1, 0, -2}, {0, -1, 0}, {0, 1, 1}}
Out[11]= {}
```

This means that the point $\{2, -1\}$ was not in the domain of the FLT defined by A. For this matrix, $d_1 = 0$, $d_2 = d_3 = 1$ is giving the line $d_1\, x + d_2\, y + d_3 = y + 1 = 0$ as the affine line not in the domain. The routine flt returns the empty list in this case.

In fact, these FLTs easily extend to the entire projective plane. The formula is actually simpler since the FLT given by A simply sends the projective point

$$\{x, y, z\} \mapsto A \begin{pmatrix} x \\ y \\ z \end{pmatrix}.$$

In this case, the fact that the point is only defined up to a nonzero constant multiple complements the fact that matrix A is also only defined up to a nonzero constant multiple.

Here is the code for projective FLT (For a complete list of global functions, see Appendix 2):

```
In[12]:= flti[p_, A_] := Module[{u, v, w},
           If[Length[p] == 2, {u, v, w} = A.Append[p, 1], {u, v, w} = A.p];
           If[Abs[w] < 1.*^-6, {u, v, 0}, {u/w, v/w}]]
```

Unlike the affine FLT, the projective FLTs are defined on the entire projective plane. The points of the affine plane, where the affine FLT was not defined, now go to the infinite line. In the previous example using flti, p could be either affine or projective. The output will be affine if the result is not infinite or projective otherwise.

Use the function flti, and not the other variants in this chapter, to get:

```
In[13]:= flti[{3, 2}, A]
```

Out[13]= $\left\{\frac{1}{3}, -\frac{2}{3}\right\}$

```
In[14]:= flti[{3, 2, 1}, A]
```

Out[14]= $\left\{\frac{1}{3}, -\frac{2}{3}\right\}$

```
In[15]:= flti[{2, -1}, A]
```

Out[15]= {0, 1, 0}

```
In[16]:= flti[{2, -1, 1}, A]
```

Out[16]= {0, 1, 0}

We will see that if, in the original formula, $d_1 = d_2 = 0$, then the FLT leaves the infinite line alone and simply transforms the affine plane. These transformations may be familiar to you from elementary transformational geometry.

Later in this chapter, we will also consider FLTs defined by matrices with complex entries. These work the same way, but now the group of transformations is known as PGL $(2, \mathbb{C})$. Unless noted otherwise, you can assume that we are talking about real FLTs.

6.2 The Action of FLT on Curves

The most important property of FLTs is that they act on curves as well as points. Again, we do the special plane invertible case only. One key is that the denominator in the formula (6.1) is the same for both coordinates, so passing to the projective case, we can use coordinate z for the denominator. Thus, projectively, the FLTs are polynomial transformations, not rational transformations.

Given an FLT $F(p)$ and a curve $f(x, y) = 0$, we want a new curve $g(x, y) = 0$ with the property that if each point p on $f = 0$, then $g(F(p)) = 0$; that is, $F(p)$ is on the curve $g = 0$, and conversely, if q is on the curve $g = 0$, then there is a unique point p on $f = 0$ with $F(p) = q$, but invertibility of F implies that $p = F^{-1}(q)$. Hence we conclude that our formula for g should be

$$g(x, y) = f\left(F^{-1}(x, y)\right).$$

Working affinely, we have the problem that since F^{-1} is a rational transformation, we will be getting denominators that are no longer the same, causing problems numerically since we cannot assume simplification, but we can avoid this by working projectively. Assume F is given by matrix A and the homogenization of the curve is the homogeneous polynomial $h(x, y, z)$. Then the new projective curve is given by

$$g(x, y, z) = h\left(A^{-1}(x, y, z)\right).$$

Or in the Wolfram Language, g (x, y, z} = FLTh[h, A, x, y, z] where (For a complete list of global functions, see Appendix 2):

```
In[17]:= FLTh[h_, A_, x_, y_, z_] := Expand h /. Thread[{x, y, z} -> Inverse[A].{x, y, z}];
```

Consider the projective circle of radius 2, $x^2 + y^2 - 4z^2 = 0$, and the FLT F, which translates the origin from {0, 0, 1} to {1, 2, 3}:

```
In[17]:= h = x^2+y^2-4z^2;
         A = {{1, 0, 2}, {0, 1, 3}, {0, 0, 1}};
         k = FLTh[h, A, x, y, z]
```

Out[19]= $x^2+y^2-4xz-6yz+9z^2$

Here, h is the homogenization of $x^2 + y^2 = 4$, but the specialization of k at $z = 1$ is just $(x-2)^2 + (y-3)^2 = 4$.

On the point level, we can use the following:

```
In[20]:= flth[p_, A_] := A.p;

In[21]:= flth[{2, 2, Sqrt[2]}, A]
```

Out[21]= $\left\{2+2\sqrt{2}, 2+3\sqrt{2}, \sqrt{2}\right\}$

As suggested by the example above, the affine case is now handled by homogenizing f, transforming by FLTh and finally specializing at $z = 1$.

We use homog from Chapter 5 or Appendix 2:

```
In[22]:= FLT[f_, A_, x_, y_] := Module[{fh, gh, AI, B, d, z},
         fh = homog[f, x, y, z];
         AI = Inverse[A];
         B = AI.{x, y, z};
         gh = Expand[fh /. Thread[{x, y, z} → B]];
         Chop[gh /. {z → 1}, 1*^-10]
         ];
```

Using A:

```
In[23]:= A = {{1, 0, 2}, {0, 1, 3}, {0, 0, 1}};
         g = FLT[x^2+y^2-4, A, x, y]
```

Out[24]= $9-4x+x^2-6y+y^2$

If an FLT takes curve $f = 0$ to curve $g = 0$, then we say f and g are *equivalent* projective curves. Although they may look different in the affine plane, the FLT is a polynomial map in the projective plane and hence continuous. Moreover, the fact that the projective FLT is invertible makes it 1-1 and onto. In particular, all topological properties are preserved. It is not difficult to show the projective graphs, as defined in Chapter 5, are also preserved. Thus, from a geometer's point of view, they are the same. We will see a number of examples in the next section.

6.3 Examples of FLTs

In this section, we will look at various subgroups of the FLT group; that is, sets of transformations that are closed under composition and taking inverses. Each subgroup defines a different kind of plane geometry. We will develop this material further in Chapter 7.

The Euclidean Group

An FLT defined by matrix A is Euclidean if the last row is $\{0, 0, 1\}$ and the determinant is ± 1. FLTs in the Euclidean group take the affine plane to the affine plane and the infinite line into the infinite line.

Translations: If we think of the affine plane as a vector space, a translation adds a constant vector to each point. A translation is defined by the image of a single point. The product of two translations is a translation, as is the inverse.

A nonidentity translation is not a linear transformation:

```
In[25]:= PQTranslation[p_, q_] := {{1, 0, q[[1]] - p[[1]]}, {0, 1, q[[2]] - p[[2]]}, {0, 0, 1}};
```

```
In[26]:= A = PQTranslation[{2, 3}, {5, 2}];
         A // MatrixForm
```

Out[27]//MatrixForm=
$$\begin{pmatrix} 1 & 0 & 3 \\ 0 & 1 & -1 \\ 0 & 0 & 1 \end{pmatrix}$$

```
In[28]:= Inverse[A] // MatrixForm
```

Out[28]//MatrixForm=
$$\begin{pmatrix} 1 & 0 & -3 \\ 0 & 1 & 1 \\ 0 & 0 & 1 \end{pmatrix}$$

Rotations: The standard clockwise rotation around the origin in θ radians is given by the matrix

$$\begin{pmatrix} \mathrm{Cos}[\theta] & -\mathrm{Sin}[\theta] & 0 \\ \mathrm{Sin}[\theta] & \mathrm{Cos}[\theta] & 0 \\ 0 & 0 & 1 \end{pmatrix}.$$

However, for our purposes, it is more useful to define a rotation by giving two intersecting affine lines, considering the rotation around the intersection point and taking the first line to the second line. The code makes use of the built-in Mathematica function RotationMatrix.

If we restricted ourselves to FLTs with rational matrices, we would not be able to construct these FLTs, even from lines defined by integer coefficients. This is perhaps the biggest single argument for doing curve theory numerically.

Note that this code returns a matrix with machine numbers (For a complete list of global functions, see Appendix 2):

```
In[29]:= klRotation[k_, l_, x_, y_] := Module[{p, a1, a2, b1, b2, T, R, A, B, m},
          a1 = Coefficient[k, x];
          b1 = Coefficient[k, y];
          a2 = Coefficient[l, x];
          b2 = Coefficient[l, y];
          If[Abs[a1*b2 - a2*b1] < 1.*^-9, Print["No affine center"];
            Return[Fail]];
          p = {x, y} /. NSolve[{k, l}][[1]];
          T = {{1, 0, -p[[1]]}, {0, 1, -p[[2]]}, {0, 0, 1}};
          R = N[RotationMatrix[{{a1, b1}, {a2, b2}}]];
          R = Join[Append[R, {0, 0}], {{0}, {0}, {1}}, 2];
          Inverse[T].R.T
        ]

In[30]:= R = klRotation[2 x + 3 y + 1, x - y + 2, x, y];

In[31]:= R // MatrixForm
```

Out[31]//MatrixForm=

$$\begin{pmatrix} -0.196116 & 0.980581 & -2.26291 \\ -0.980581 & -0.196116 & -0.655143 \\ 0. & 0. & 1. \end{pmatrix}$$

```
In[32]:= Inverse[R] // MatrixForm
```

Out[32]//MatrixForm=

$$\begin{pmatrix} -0.196116 & -0.980581 & -1.08621 \\ 0.980581 & -0.196116 & 2.09048 \\ 0. & 0. & 1. \end{pmatrix}$$

Again, since the intersection of the two lines is not the origin, this rotation matrix is also not a linear transformation of the plane. However, note that the upper left 2×2 submatrix in R^{-1} is the transpose of that in R.

Reflections: The reflection in the x-axis takes point $\{a, b\}$ to $\{a, -b\}$. This reflection is its own inverse. The best way to reflect about an arbitrary line is to use translations and rotations to move an arbitrary line to the x-axis, reflect and then move back. This combined operation is called conjugation in group theory.

Note that k must be an affine line, a x + b − c = 0, with either a or b being something other than zero (For a complete list of global functions, see Appendix 2):

```
In[33]:= kReflection[k_, x_, y_] := Module[{a, b, c, T, RX, RO},
    RX = {{1, 0, 0}, {0, −1, 0}, {0, 0, 1}};
    a = Coefficient[k, x];
    b = Coefficient[k, y];
    c = k /. Thread[{x, y} → {0, 0}];
    If[a == 0,
      T = PQTranslation[{0, 0}, {0, −c / b}];
      Return[T.RX.Inverse[T]],
      RO = klRotation[y, k, x, y];
      Return[RO.RX.Inverse[RO]];
    ]
  ]
```

```
In[34]:= A = kReflection[x − y, x, y]
```

```
Out[34]= {{0., 1., 0.}, {1., 0., 0.}, {0., 0., 1.}}
```

```
In[35]:= B = kReflection[3 x − 2 y − 1, x, y]
```

```
Out[35]= {{−0.384615, 0.923077, 0.461538}, {0.923077, 0.384615, −0.307692}, {0., 0., 1.}}
```

```
In[36]:= B // MatrixForm
         Inverse[B] // MatrixForm
```

Out[36]//MatrixForm=
$$\begin{pmatrix} -0.384615 & 0.923077 & 0.461538 \\ 0.923077 & 0.384615 & -0.307692 \\ 0. & 0. & 1. \end{pmatrix}$$

Out[37]//MatrixForm=
$$\begin{pmatrix} -0.384615 & 0.923077 & 0.461538 \\ 0.923077 & 0.384615 & -0.307692 \\ 0. & 0. & 1. \end{pmatrix}$$

Note that even though the line is not through the origin, the reflection FLT is its own inverse.

Generally speaking, any Euclidean transformation is a composition of the above three types. A Euclidean transformation is determined uniquely by taking the three noncolinear vertices of a triangle to corresponding vertices of a congruent triangle. In fact, it is enough to use two or three reflections. We will illustrate this in Chapter 7 by showing how to take any affine ellipse to an ellipse of the form $\frac{x^2}{a^2} + \frac{y^2}{b^2} - 1 = 0$.

If one includes similarity transformations with matrices of the form

$$\begin{pmatrix} r & 0 & 0 \\ 0 & r & 0 \\ 0 & 0 & 1 \end{pmatrix},$$

when $r > 0$, one gets the group of Euclidean similarities. With these, you can transform any triangle to a given similar triangle.

Affine Transformations

The word "affine" comes from "affine transformations," which are transformations that take Euclidean lines to Euclidean lines; specifically, an FLT defined by the matrix

$$A = \begin{pmatrix} a_{1,1} & a_{1,2} & a_{13} \\ a_{2,1} & a_{2,2} & a_{2,3} \\ 0 & 0 & 1 \end{pmatrix},$$

where $a_{1,1}\, a_{2,2} - a_{1,2}\, a_{2,1} \neq 0$. In particular, the group of affine transformations contains the Euclidean and similarity transformations. There are two new types that, with the Euclidean transformations, generate the entire group.

Shears: The standard shear leaves the x-coordinate fixed while shrinking or expanding the y-coordinate so that a given line l intersecting the x-axis is moved to the x-axis.

More general shears may be made by conjugating by a Euclidean transformation (For a complete list of global functions, see Appendix 2):

```
In[38]:= kShear[k_, x_, y_] := Module[{a, b, c, d, e},
           c = Coefficient[k, x];
           d = Coefficient[k, y];
           e = k /. {x → 0, y → 0};
           a = −e / d;
           b = −c / d;
           {{1, 0, 0}, {−b, 1, −a}, {0, 0, 1}}
         ]
```

Let's try generating a shear:

```
In[39]:= A = kShear[x + y, x, y];
         A // MatrixForm
```

Out[40]//MatrixForm=

$$\begin{pmatrix} 1 & 0 & 0 \\ 1 & 1 & 0 \\ 0 & 0 & 1 \end{pmatrix}$$

```
In[41]:= S = {{0, 0}, {1, 0}, {1, 1}, {0, 1}, {0, 0}};
         S1 = flt[#, A] & /@ S
         {Show[ListLinePlot[S, AspectRatio → 1,
               PlotRange → {{−.5, 1.5}, {−.5, 2.2}}, PlotStyle -> Thick],
             ListPlot[{Labeled[{0, 0}, "A", {Bottom, Left}], Labeled[{1, 0}, "B", {Bottom, Right}],
                Labeled[{1, 1}, "C", {Top, Right}], Labeled[{0, 1}, "D", {Top, Left}]}]],
           Show[ListLinePlot[flt[#, A] & /@ S, PlotRange → {{−.5, 1.5}, {−.5, 2.2}}, AspectRatio → 1,
               PlotStyle → Thick], ListPlot[{Labeled[flt[{0, 0}, A], "A'", {Bottom, Right}],
                Labeled[flt[{1, 0}, A], "B'", {Bottom, Left}], Labeled[flt[{1, 1}, A],
                 "C'", {Top, Left}], Labeled[flt[{0, 1}, A], "D'", {Top, Right}]}]]}

Out[42]= {{0, 0}, {1, 1}, {1, 2}, {0, 1}, {0, 0}}
```

Out[43]=

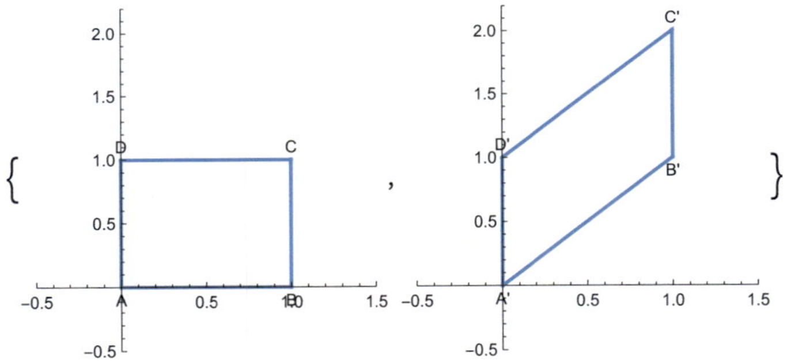

A shear takes the line $x = y$ to the y-axis, point A to point A', etc. In both examples, notice what happens to points B, C, and D; noting that point A stays fixed at the origin.

Homothety: Also in the affine group is the homothety given by the matrix

$$\begin{pmatrix} r & 0 & 0 \\ 0 & s & 0 \\ 0 & 0 & 1 \end{pmatrix},$$

where r and s are positive reals.

Here is my code for **homethety** (For a complete list of global functions, see Appendix 2):

```
In[44]:= homothety[r_, s_] := {{r, 0, 0}, {0, s, 0}, {0, 0, 1}};
```

Here is an example building on the one output above:

```
In[45]:= B = homothety[2, .5]
         S = {{0, 0}, {1, 0}, {1, 1}, {0, 1}, {0, 0}};

Out[45]= {{2, 0, 0}, {0, 0.5, 0}, {0, 0, 1}}

In[47]:= SS = flt[#, B] &/@S

Out[47]= {{0., 0.}, {2., 0.}, {2., 0.5}, {0., 0.5}, {0., 0.}}

In[48]:= ListLinePlot[{S, SS}, Axes → False, Frame → True, PlotStyle → {Blue, Orange}]
```

Out[48]=

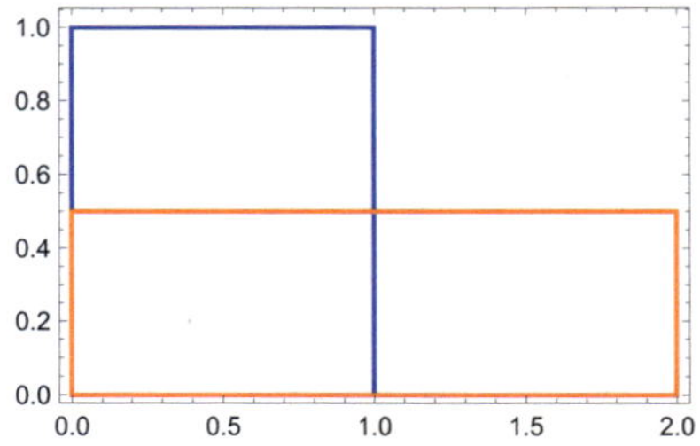

Now you can construct any affine transformation using the Euclidean, shear or homothety transformations.

Some Projective Transformations

As mentioned earlier, the full projective linear group consists of all FLTs. There are two that will generally be useful in the next two chapters that will be mentioned here. The first is the ip2z transform, which takes an infinite point and places it at the origin. The second, iTransform, places affine line ℓ at infinity. Both of these place the infinite line at the x-axis.

To send ip2z to origin (For a complete list of global functions, see Appendix 2):

```
In[49]:= ip2z[ip_, x_, y_] := Module[{p, l, A, B},
           If[Length[ip] ≠ 3, Print["not infinite point"];
             Abort[], If[Abs[ip[[3]]] > 1.*^-6, Print["not infinite point"];
               Abort[], p = Chop[ip, 1.*^-6]]];
           B = {{1.414213562373095`, 0.`, 0.`}, {0.`, 0.`, 1.414213562373095`},
               {0.`, -0.7071067811865475`, -0.7071067811865475`}};
           If[Abs[p[[1]]] < 1.*^-6, Return[B]];
           ral = line[p, {0, 0}, x, y];
           B.klRotation[l, x, x, y]]
```

Send line *l* to the infinite line; that is, delete it from the affine plane:

```
In[50]:= rrit1 = {-0.7812879011987071`, -0.3168196688992504`, 0.9191639531555853`};
         rrit2 = 1.4009431979825675`;
         iTransform[l_, x_, y_] := Module[{ph},
             ph = N[{Coefficient[l, x], Coefficient[l, y], l /. Thread[{x, y} → {0, 0}]}];
             ph = ph / Norm[ph];
             {rrit1, {0, 0, rrit2}, ph}];
```

Programming note: iTransform uses random numbers rrit1, rrit2. On this page, they are set to specific random numbers. In Appendix 2, they are set once per initialization. Thus using this function out of this specific context may give different answers.

As examples of these two transformations, consider the hyperbola $y = \frac{1}{x}$ and the lines $l = x - y,\ m = x + y$:

```
In[53]:= ContourPlot[{x y == 1, x - y == 0, x + y == 0}, {x, -5, 5}, {y, -5, 5}]
```

Out[53]=
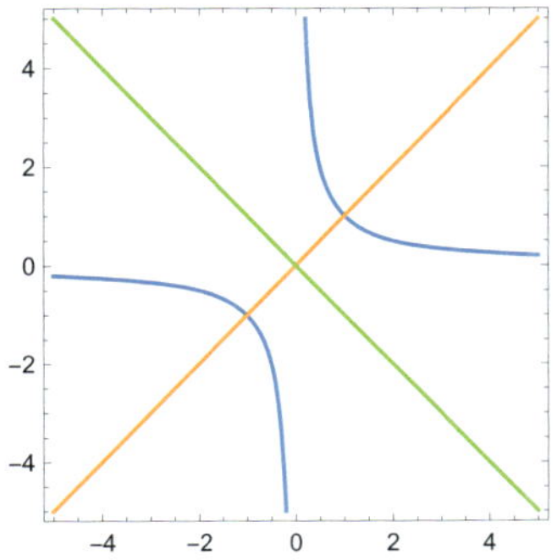


```
In[54]:= A = ip2z[{0, 1, 0}, x, y]

Out[54]= {{1.41421, 0., 0.}, {0., 0., 1.41421}, {0., -0.707107, -0.707107}}
```

In[55]:=
```
B1 = iTransform[x - y, x, y]
```

Out[55]=
```
{{-0.781288, -0.31682, 0.919164}, {0, 0, 1.40094}, {0.707107, -0.707107, 0.}}
```

In[56]:=
```
B2 = iTransform[x + y, x, y]
```

Out[56]=
```
{{-0.781288, -0.31682, 0.919164}, {0, 0, 1.40094}, {0.707107, 0.707107, 0.}}
```

In[57]:=
```
f = x y - 1;
```

In[58]:=
```
g = FLT[f, N[A], x, y]
```

Out[58]= $-1.\,x - 0.5\,x\,y - 0.5\,y^2$

In[59]:=
```
h1 = FLT[f, B1, x, y]
```

Out[59]= $-0.410548 + 0.54473\,x + 0.829297\,x^2 - 0.357399\,y - 1.08821\,x\,y - 0.152528\,y^2$

In[60]:=
```
h2 = FLT[f, B2, x, y]
```

Out[60]= $-2.29478 - 7.1986\,x - 4.63541\,x^2 + 4.72303\,y + 6.08262\,x\,y - 2.50493\,y^2$

In[61]:=
```
{ContourPlot[{g == 0, y == 0}, {x, -5, 5}, {y, -5, 5}, ContourStyle → {Blue, Dashed},
   ImageSize → Small], ContourPlot[{h1 == 0, y == 0}, {x, -2, 2},
   {y, -2, 2}, ContourStyle → {Blue, Dashed}, ImageSize → Small],
 ContourPlot[{h2 == 0, y == 0}, {x, -2, 2}, {y, -2, 2},
   ContourStyle → {Blue, Dashed}, ImageSize → Small]}
```

Out[61]=
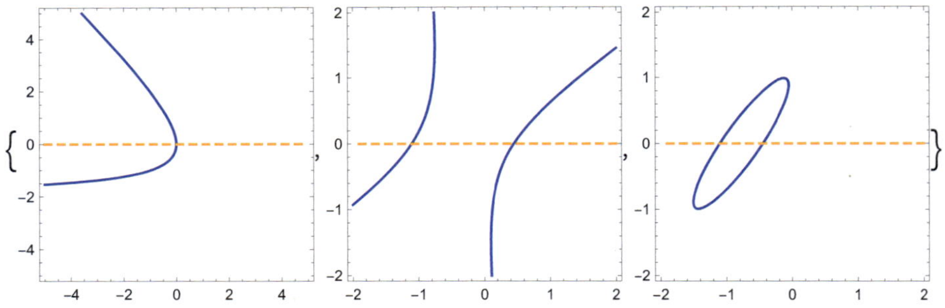

In these examples, the dashed line is the original infinite line. In the first case, ip2z shows the intersection of the hyperbola with the infinite line at {0, 1, 0}. The infinite line is now at the origin, but the intersection with {1, 0, 0} remains infinite under the transformation and the transformed equation is actually a parabola. In the second of the two cases, both of the original infinite points are shown, but we get a hyperbola (or ellipse, depending on whether the removed line intersects the original hyperbola in two or zero points). It was Johann Kepler who first noted in 1604 that the parabola, hyperbola and ellipse were the same except for the number of infinite points. We will say more on conics and rational function curves in Chapter 7.

6.4 Applications

I will give three important applications of FLTs here. In the next chapter, I will give applications to classical mathematics.

Projective Singular Points

In Chapter 5, we saw that it was important to find all complex projective singular points in order to decide if a curve is nonsingular. To properly describe the projective graph of a curve, it is necessary to have at least all real projective singular points. The method of Chapter 5 required checking for singular points in all three specializations. Using FLTs, we will see that it is usually enough to check once. The strategy is to pick a random affine line and remove it from the projective plane. We get the image of this specialization by using FLT. We then use the procedure below for finding all complex affine singular points. This procedure has a fourth parameter, eps, which gives a tolerance. If you think the curve has no singular points, use a very loose tolerance such as eps = .01 to be sure you get all the singular points available. If the result is the empty set { }, you can be fairly confident that the curve is nonsingular. However, you may get some false singular points, but by decreasing the tolerance, the false ones will be eliminated. Also note that singular points that appear on one run of the algorithm but not others are probably false, but be careful with infinite points that are only defined up to a constant.

Note that the last third of this procedure is devoted to displaying the points nicely, encouraging infinite singular points to display as infinite and real points to display as real.

For a complete list of global functions, see Appendix 2:

```
In[62]:= complexProjectiveSingularPoints[h_, u_, v_, eps_] :=
  Module[{f, w, A, a, b, sol1, sol2, solz, sing, sing1, sing2},
   A = {RandomReal[{-1, 1}, 3], {0, 0, RandomReal[{1, 2}]}, RandomReal[{-1, 1}, 3]};
   f = FLT[h, A, u, v];
   {a, b} = RandomComplex[{-1 - I, 1 + I}, 2];
   sol1 = {u, v} /. NSolve[{f, a D[f, u] + b D[f, v]}];
   {a, b} = RandomComplex[{-1 - I, 1 + I}, 2];
   sol2 = {u, v} /. NSolve[{f, a D[f, u] + b D[f, v]}];
   solz = Join[sol1, sol2];
   sing =
    Flatten[Take[Select[Tally[solz, Norm[#1 - #2] < eps &], #[[2]] > 1 &], All, 1], 1];
   sing1 = Table[{sing[[j, 1]], If[Abs[sing[[j, 2]]] > eps, sing[[j, 2]], 0]},
     {j, Length[sing]}];
   sing2 = flti[#, Inverse[A]] & /@ sing1;
   sing2 = Table[Which[Length[p] == 2, p, Abs[p[[1]]] > eps, p / p[[1]],
      Abs[p[[2]]] > eps, p / p[[2]], True, {0, 0, p[[3]]}], {p, sing2}];
   Chop[sing2, eps]];
```

Here is an example:

```
In[63]:= fs = -1 - x y + x^3 y;

In[64]:= sing = complexProjectiveSingularPoints[fs, x, y, .1]

Out[64]= {{0.662658, -2.69047}, {0, 1., 0}}
```

Here, each time you run this, you will probably get a different answer. You will often get two finite real answers that are near turning points that do not look singular:

```
In[65]:= ContourPlot[fs == 0, {x, -5, 5}, {y, -5, 5}]
```

Out[65]=

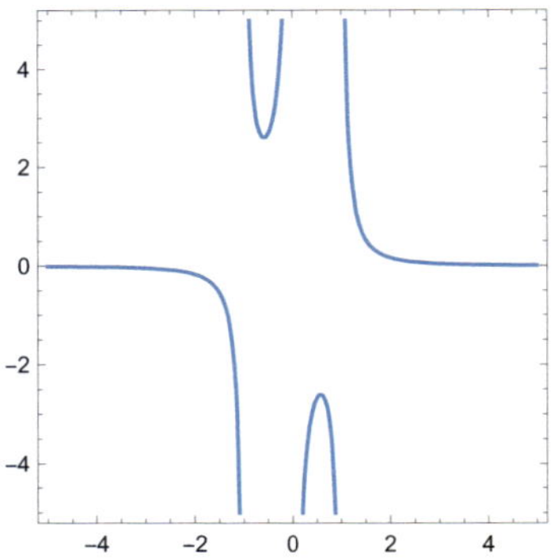


```
In[66]:= sing = complexProjectiveSingularPoints[f, x, y, .0001]

Out[66]= {}
```

Our next examples involve perturbing curve g slightly. In the affine real case, we saw that a small perturbation of a curve with a singular point often made the singular point appear to vanish. We now explore the complex projective case.

The second run with tighter tolerance indicates that the infinite point {0, 1, 0} is probably the only singular point:

```
In[67]:= g = -1 + x^2 - 2 x^3 - y + x^2 y;
         h = -0.9975` + x^2 - 2 x^3 - 0.9925` y + x^2 y + 0.0075` y^2 + 0.0025` y^3;

In[69]:= singg = complexProjectiveSingularPoints[g, x, y, .01]
         singh = complexProjectiveSingularPoints[h, x, y, .01]

Out[69]= {{0, 1., 0}}

Out[70]= {}

In[71]:= singg = complexProjectiveSingularPoints[g, x, y, .000001]
         singh = complexProjectiveSingularPoints[h, x, y, .000001]

Out[71]= {{0, 1., 0}}

Out[72]= {}
```

Here, g has an infinite singular point, but h, which is g with a small perturbation, has no singular points, real or complex.

Projective Complex Intersection Points

The same technique works for finding all the complex projective intersection points of two curves.

In this case, NSolve does the work of making the output look nice so we do not need the post-processing (For a complete list of global functions, see Appendix 2):

```
In[73]:= complexProjectiveIntersectionpoints[f_, g_, x_, y_] := Module[{f1, g1, A, sol, sol2},
    A = {RandomReal[{-1, 1}, 3], {0, 0, RandomReal[{1, 2}]}, RandomReal[{-1, 1}, 3]};
    f1 = FLT[f, A, x, y];
    g1 = FLT[g, A, x, y];
    sol = {x, y} /. NSolve[{f1, g1}];
    flti[#, Inverse[A]] &/@sol];
```

A complicated example comes from Gauss curves:

```
In[74]:= f = -4 - 20 x + x^2 + x^3 + x^5 - y^2 - 3 x y^2 - 10 x^3 y^2 + 5 x y^4;
    g = 16 x - 8 x^3 + x^5 + 24 x y^2 - 10 x^3 y^2 + 5 x y^4;
```

```
In[76]:= ContourPlot[{f == 0, g == 0}, {x, -4, 4}, {y, -4, 4}]
```

Out[76]=

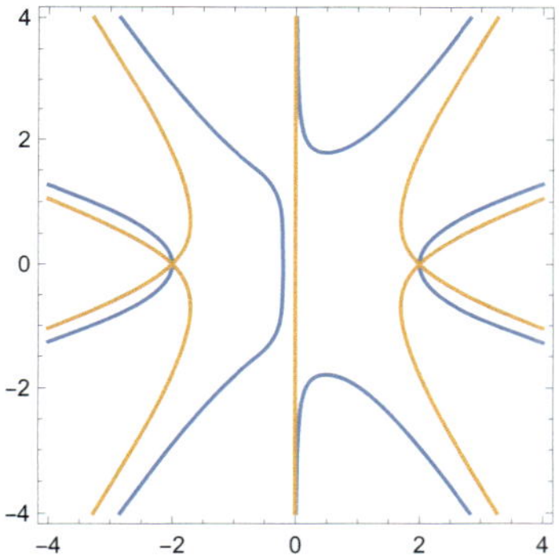


```
In[77]:= L = Chop[complexProjectiveIntersectionpoints[f, g, x, y], 1.*^-6]
    Length[L]
```

... NSolve: Scale of possible solution value $-2. \times 10^{-13}$ for variable y is too small. Must set this value to zero in order to avoid generation of spurious solutions. This may cause the loss of some solutions.

```
Out[77]= {{2., 0}, {2., 0}, {1.02181, 0. + 1.02676 ⅈ}, {1.02181, 0. - 1.02676 ⅈ}, {-0.225946, 0. + 0.894521 ⅈ},
    {-0.225946, 0. - 0.894521 ⅈ}, {0, 0. + 2. ⅈ}, {0, 0. + 2. ⅈ}, {0, 0. - 2. ⅈ}, {0, 0. - 2. ⅈ},
    {-0.962529, 0. + 0.970831 ⅈ}, {-0.962529, 0. - 0.970831 ⅈ}, {-2., 0}, {-2., 0},
    {-3.66266, -1.19007, 0}, {-3.66266, -1.19007, 0}, {-1.19859, 0.389444, 0},
    {-1.19859, 0.389444, 0}, {-0.573887, 0.789887, 0}, {-0.573887, 0.789887, 0},
    {0, 1.15776, 0}, {0, 1.15776, 0}, {0, 1.15776, 0}, {1.5744, 2.16698, 0}, {1.5744, 2.16698, 0}}
```

```
Out[78]= 25
```

We may want to do additional post-processing:

```
In[79]:= Li = Tally[Select[L, Length[#] == 3 &], Norm[#1 - #2] < .01 &]
         La = Tally[Select[L, Length[#] == 2 &], Norm[#1 - #2] < .01 &]

Out[79]= {{{20.7055, 6.72763, 0}, 2}, {{2.37853, -0.772833, 0}, 2},
          {{0.97795, -1.34603, 0}, 2}, {{0, -1.74627, 0}, 3}, {{-1.80563, -2.48523, 0}, 2}}

Out[80]= {{{2., 0}, 2}, {{1.02181, 0. - 1.02676 i}, 1},
          {{1.02181, 0. + 1.02676 i}, 1}, {{-0.225946, 0. - 0.894521 i}, 1},
          {{-0.225946, 0. + 0.894521 i}, 1}, {{-0.962529, 0. - 0.970831 i}, 1},
          {{-0.962529, 0. + 0.970831 i}, 1}, {{0, 0. - 2. i}, 2}, {{0, 0. + 2. i}, 2}, {{-2., 0}, 2}}
```

We conclude there are, as Bézout tells us, 25 intersection points counted by multiplicity. These include five real infinite points, one $\{0, 1, 0\}$ of multiplicity 3 and four of multiplicity 2. There are two real affine intersection points $\{\pm 2, 0\}$ of multiplicity 2, two complex affine multiple two-intersection points $\{0, \pm 2\, i\}$ and three pairs of complex conjugate affine simple intersection points.

Note that finding intersection points is more stable than finding singular points. complexProjectiveIntersectionpoints does not require a tolerance, and usually, running it once is enough. It is better to use this second function to find singular points that come from intersections.

Constructions with Curves

One way to make interesting new curves from old curves is to add or subtract infinite points by removing an affine line ℓ using iTransform. The new infinite points correspond to the affine points on ℓ while infinite points not on ℓ become affine points.

Newton observed that there were cubic curves with four affine components. We saw this with Newton hyperbolas. Another way to construct such a curve is to start with a common cubic such as

$$f = y^2 - x^3 + x;$$

and remove a line ℓ intersecting the infinite component in two chosen real points. We can see, as Bézout determined, it will also intersect in an additional point that must also be real and in the infinite (versus the finite) component.

Choose some parameters:

```
In[81]:= l = 10 x - y - 20

Out[81]= -20 + 10 x - y
```

In[82]:=
```
A = iTransform[l, x, y]
h = FLT[f, A, x, y]
{ContourPlot[{f == 0, l == 0}, {x, -4, 4}, {y, -4, 4}],
  ContourPlot[h == 0, {x, -8, 8}, {y, -8, 8}]}
```

Out[82]= {{−0.781288, −0.31682, 0.919164}, {0, 0, 1.40094}, {0.446767, −0.0446767, −0.893534}}

Out[83]= $2334.57 + 6575.53\,x + 5771.27\,x^2 + 1708.5\,x^3 - 13.4489\,x^4 - 50.9921\,x^5 + 3130.82\,y + 9202.62\,x\,y + 7680.04\,x^2\,y + 2214.69\,x^3\,y + 158.703\,x^4\,y + 923.941\,y^2 + 3710.16\,x\,y^2 + 2710.06\,x^2\,y^2 + 500.496\,x^3\,y^2 - 268.81\,y^3 + 344.416\,x\,y^3 + 238.519\,x^2\,y^3 - 162.882\,y^4 - 29.7743\,x\,y^4 - 23.6031\,y^5$

Out[84]=
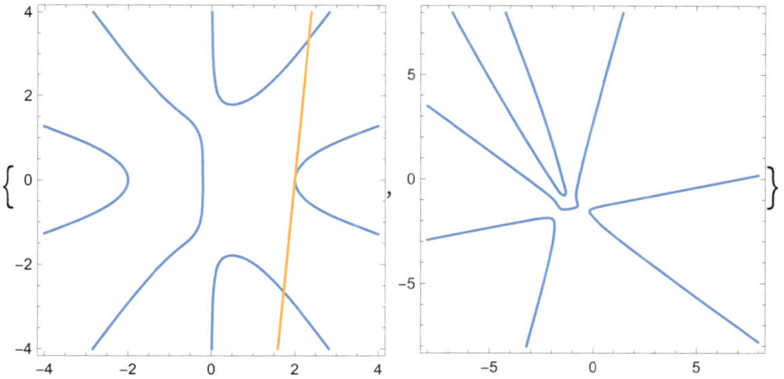

Programming note: Since iTransform *is partially random, the curve will come out differently when you run this example. The curve will still have four affine components, but you may have to adjust the scale in order to be able to see them.*

Projective real curves of even degree can often be represented by affine curves with no infinite points. We will see in Chapter 9 that this is not true of odd degree curves. This is easy to check if you can find a line that does not intersect the curve in a real projective point.

We illustrate this with Gauss curves. Let:

In[85]:=
```
g = gaussCurve[z^6 - 6 z + 6, x, y, z]
```

Out[85]= $6 - 6\,x + x^6 - 15\,x^4\,y^2 + 15\,x^2\,y^4 - y^6$

We see below that this curve does not meet the *x*-axis in an affine or infinite point:

In[86]:=
```
B = iTransform[y, x, y]
```

Out[86]= {{−0.781288, −0.31682, 0.919164}, {0, 0, 1.40094}, {0., 1., 0.}}

In[87]:=
```
g1 = FLT[g, B, x, y];
{ContourPlot[{g == 0, y == 0}, {x, -6, 6}, {y, -6, 6}],
  ContourPlot[g1 == 0, {x, -6, 6}, {y, -10, 10}]}
```

Out[88]=

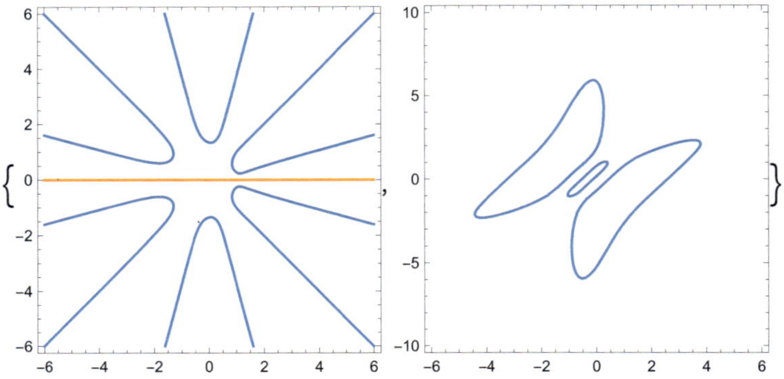

In[89]:=
```
h = gaussCurve[(z^2-1) (z^2-4) (z^2-9), x, y, z]
```

Out[89]= $-36+49x^2-14x^4+x^6-49y^2+84x^2y^2-15x^4y^2-14y^4+15x^2y^4-y^6$

This curve does not meet the *y*-axis:

In[90]:=
```
A = iTransform[x, x, y]
```

Out[90]= {{-0.781288, -0.31682, 0.919164}, {0, 0, 1.40094}, {1., 0., 0.}}

In[91]:=
```
h1 = FLT[h, A, x, y];
{ContourPlot[{h == 0, x == 0}, {x, -6, 6}, {y, -6, 6}],
  ContourPlot[h1 == 0, {x, -4, 4}, {y, -3, 3}]}
```

Out[92]=

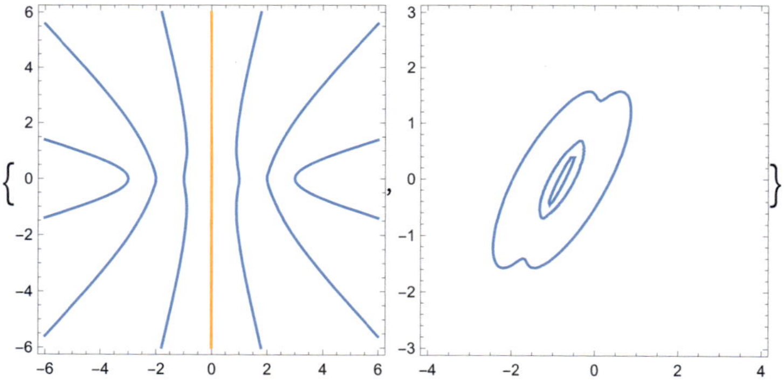

In Chapter 9, we call the innermost oval a depth 2 oval since it is contained in the interior of a curve component, which itself is contained in the interior of a curve component.

6.5 Summary of Fractional Linear Transformations

The single idea of transforming curves has several different facets. We can give the transformation as a function or use a matrix representation. Since we can represent a curve as an equation, bivariate polynomial or a point set, we can transform either. Further, we can decide if we want to transform just affine points or if we want to include infinite points. In the latter case, we can still work primarily in the affine plane or move to the projective plane.

As a function on the affine plane, a fractional linear transformation is given as

$$F(x,\ y) = \left(\frac{a_{1,1}\ x + a_{1,2}\ y + a_{1,3}}{d_1\ x + d_2\ y + d_3},\ \frac{a_{2,1}\ x + a_{2,2}\ y + a_{2,3}}{d_1\ x + d_2\ y + d_3}\right).$$

The domain is the plane subset where the denominator is not zero, which is either the plane or the plane with a single line removed. This may be represented in matrix form:

$$A = \begin{pmatrix} a_{1,1} & a_{1,2} & a_{1,3} \\ a_{2,1} & a_{2,2} & a_{2,3} \\ d_1 & d_2 & d_3 \end{pmatrix}$$

$$F : \{x,\ y\} \mapsto A \begin{pmatrix} x \\ y \\ 1 \end{pmatrix} = \{u,\ v,\ w\} \mapsto \left\{\frac{u}{w},\ \frac{v}{w}\right\}$$

This is consistent with transformations in the Wolfram Language. See the documentation pages for the Wolfram Language functions TransformationMatrix and Transformation for more information.

Our matrices agree, in the affine case, with TransformationMatrix, and one can generally represent the transformation that works on points using either flt or the Transformation function.

For example, consider a random 3 × 3 matrix and a random real point:

```
In[93]:= A = {{-2, 2, 7}, {-1, -6, 0}, {3, 3, 8}};
         p = {3.4552977152086983`, -3.719873903204607`};
         flt[p, A]
         TransformationFunction[A][p]

Out[95]= {-1.01999, 2.61771}

Out[96]= {-1.01999, 2.61771}
```

The difference is that flt can detect when a point is not in the domain whereas TransformationFunction returns an error message:

```
In[97]:= q = {-3 , 1/3};
         flt[q, A]
         TransformationFunction[A][q]

Out[98]= {}

   ... Power: Infinite expression 1/0 encountered.

Out[99]= {ComplexInfinity, ComplexInfinity}
```

An important feature of this code is that we can transform it on the equation level. For this, in the affine case, we use FLT:

```
In[100]:= FLT[y-x^2, A, x, y]
          FLT[y-x^2, N[A], x, y]
```

Out[100]= $-\frac{38}{961}+\frac{577\,x}{6727}-\frac{312\,x^2}{6727}-\frac{146\,y}{6727}+\frac{3\,x\,y}{6727}-\frac{67\,y^2}{6727}$

Out[101]= $-0.0395421+0.0857737\,x-0.0463803\,x^2-0.0217036\,y+0.000445964\,x\,y-0.00995986\,y^2$

If we want to include infinite points as exceptions, we can use flti:

```
In[102]:= q1 = flti[q, A]
          flti[q1, Inverse[A]]
```

Out[102]= $\left\{\frac{41}{3}, 1, 0\right\}$

Out[103]= $\left\{-3, \frac{1}{3}\right\}$

If we want to work completely projectively, we can use flth on the point level and FLTh on the homogeneous polynomial level:

```
In[104]:= flth[Append[p, 1], A]
          flth[Append[q, 1], A]

Out[104]= {-7.35034, 18.8639, 7.20627}
```

Out[105]= $\left\{\frac{41}{3}, 1, 0\right\}$

```
In[106]:= FLTh[y z-x^2, A, x, y, z]
```

Out[106]= $-\frac{312\,x^2}{6727}+\frac{3\,x\,y}{6727}-\frac{67\,y^2}{6727}+\frac{577\,x\,z}{6727}-\frac{146\,y\,z}{6727}-\frac{38\,z^2}{961}$

Finally, in Chapter 7, we will be discussing how to transform an affine point in the symbolic form using flts:

```
In[107]:= flts[{Cos[t], Sin[t]}, A]
```

Out[107]= $\left\{\frac{7-2\,\text{Cos}[t]+2\,\text{Sin}[t]}{8+3\,\text{Cos}[t]+3\,\text{Sin}[t]}, \frac{-\text{Cos}[t]-6\,\text{Sin}[t]}{8+3\,\text{Cos}[t]+3\,\text{Sin}[t]}\right\}$

More to Explore

For more information on linear fractional transformation, check out Wolfram MathWorld (wolfr.am/sV04LrrN)

To see other samples from the symmetry group, try looking at the page about the Klein bottle (wolfr.am/sV01dMoQ)

To see an excellent 3D version of a Klein bottle, try I. C. Antonio's Demonstration (wolfr.am/vmYgN0jE)

7 | Asymptotes, Conics, and Cubics

In this chapter, we apply our fractional linear transformations to some classical algebraic geometry. A major goal is to parameterize these low-degree curves. In the last section, we give Arthur Caley's proof that a curve of genus 0 is rational by giving the parameterization using simple cases.

7.1 Asymptotes and Rational Function Curves

We have touched on the subject of asymptotes briefly in the past, but in this section, we will carefully define them and give a procedure to calculate them.

Asymptotes are an affine concept but are associated with infinite points. If the infinite point is regular, there is a single asymptote, which is the tangent line to the curve at the infinite point. However, if the infinite point is singular, then there may be several asymptotes.

The standard example of the latter case is a rational function curve with several vertical asymptotes.

Consider the rational function $y = \frac{1}{(x^2-4)\,x}$, which gives the algebraic curve:

In[1]:=
```
f = y (x^2 - 4) x - 1;
```

In[2]:=
```
ContourPlot[{f == 0, x == 0, x == 2, x == -2, y == 0}, {x, -3, 3}, {y, -6, 6},
  ContourStyle → {Black, Directive[Orange, Dashed], Directive[Orange, Dashed],
    Directive[Orange, Dashed], Directive[Green, Dashed]}]
```

Out[2]=

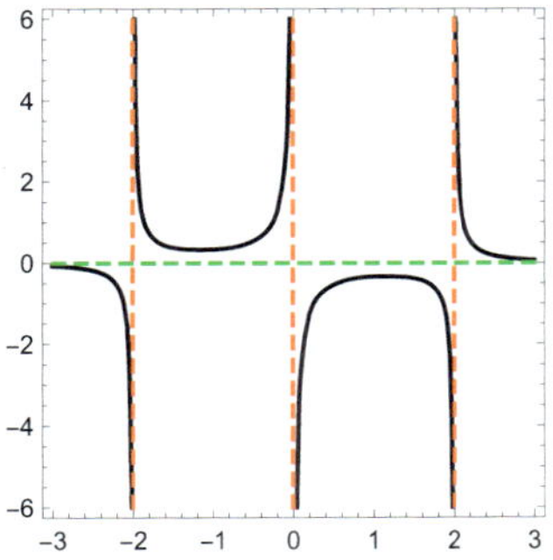

By elementary algebra considerations, we have three vertical asymptotes associated with the infinite point {0, 1, 0} and one horizontal asymptote associated with the infinite point {1, 0, 0}.

We inspect the infinite points:

```
In[3]:= A = ip2z[{1, 0, 0}, x, y];
        B = ip2z[{0, 1, 0}, x, y];
        f1 = FLT[f, A, x, y]
        f2 = FLT[f, B, x, y]
```

Out[5]= $2.\,x + 3.\,x\,y - 0.5\,x\,y^2 - 0.75\,x\,y^3 - 0.25\,y^4$

Out[6]= $-0.5\,x^3 - 0.25\,x^3\,y + 2.\,x\,y^2 + 1.\,x\,y^3 - 0.25\,y^4$

```
In[7]:= {ContourPlot[{f1 == 0, y == 0}, {x, -1, 1}, {y, -1, 1}, ContourStyle → {Blue, Dashed}],
         ContourPlot[{f2 == 0, y == 0}, {x, -1, 1}, {y, -1, 1}, ContourStyle → {Blue, Dashed}]}
```

Out[7]=

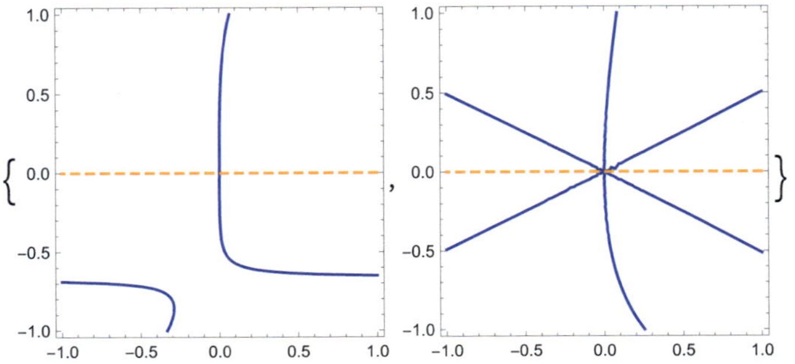

In these plots, the dashed line represents the affine infinite line and the point {0, 0} represents the original infinite point. In the left plot, we see that there is a regular point with what appears to be the vertical tangent line $x = 0$.

The corresponding asymptote is calculated by:

```
In[8]:= FLT[x, Inverse[A], x, y]
```

Out[8]= 1.41421 y

```
In[9]:= FLT[x, Inverse[ip2z[{1, 0, 0}, x, y]], x, y]
```

Out[9]= 1.41421 y

It is otherwise known by the affine line $y = 0$. The right plot shows an ordinary singularity at {0, 0}, which represents the original infinite point {0, 1, 0}.

Multiplying by 4, we see:

```
In[10]:= f2min = -2 x^3 + 8 x y^2;
         Factor[f2min]
```

Out[11]= $-2\,x\,(x - 2\,y)\,(x + 2\,y)$

Giving the asymptotes:

```
In[12]:= FLT[x, Inverse[B], x, y]

Out[12]= 1.41421 x

In[13]:= FLT[x−2 y, Inverse[B], x, y]

Out[13]= −2.82843 + 1.41421 x

In[14]:= FLT[x+2 y, Inverse[B], x, y]

Out[14]= 2.82843 + 1.41421 x
```

These are otherwise given by $x = 0$, $x = 2$, $x = -2$, respectively.

We recall from Chapter 3 (For a complete list of global functions, see Appendix 2):

```
In[15]:= form[f_, k_, x_, y_] :=
           FromCoefficientRules[Select[CoefficientRules[f, {x, y}], Total[#[[1]]] == k &], {x, y}];
         MaxForm[f_, x_, y_] := Module[{A, Bd},
              A = <|CoefficientRules[f, {x, y}]|>;
              d = Max[Table[Total[k], {k, Keys[A]}]];
              Form[f, d, x, y]];
         pointMinForm[f_, p_, x_, y_] := Module[{A, h, j, tol},
              tol = 1.*^-9;
              h = Chop[Expand[f /. Thread[{x, y} → {x, y} + p]], tol];
              A = <| CoefficientRules[h, {x, y}] |>;
              j = Min[Table[Total[k], {k, Keys[A]}]];
              form[h, j, x, y]];
```

Now we can find asymptotes. This simple procedure assumes the point p is an infinite point of the curve f. The user may wish to add checks on these assumptions. Recall that the graph of a form will be a (possibly empty) union of lines through the origin. We intersect this union with a random line that does not go through the origin. The equation of each line is a factor of the form.

In Chapter 3, we defined a function finding infinite real points using a random real line. If working with a live notebook, the following code should be initialized now.

Note that this code differs slightly from that in Appendix 2 as the random line is set for this chapter:

```
In[18]:= rnl7 = −1 + 0.6916206531506335` x − 0.6706463898598312` y

Out[18]= −1 + 0.691621 x − 0.670646 y

In[19]:= infiniteRealPoints[f_, x_, y_] := Module[{fm, sol},
              fm = maxForm[f, x, y];
              sol = NSolve[{fm, rnl7}, {x, y}, Reals];
              If[Length[sol] == 0, Return[{}]];
              sol = DeleteDuplicates[{x, y} /. sol, Norm[#1 − #2] < .001 &];
              Table[Append[sol[[i]], 0], {i, Length[sol]}]
           ];
```

We next define our function to find asymptotes using the same fixed "random line" for this chapter:

```
In[20]:= asymptotes[f_, p_, x_, y_] := Module[{g, gm, A, sol, k},
           A = ip2z[p, x, y];
           g = FLT[f, A, x, y];
           gm = pointMinForm[g, {0, 0}, x, y];
           sol = DeleteDuplicates[
               {x, y} /. NSolve[{gm, rnl7}, {x, y}, Reals], Norm[#1 - #2] < .001 &];
           k = Length[sol];
           If[k == 0, Return[{}]];
           Table[FLT[sol[[i, 2]] x - sol[[i, 1]] y, Inverse[A], x, y], {i, k}]
         ];
```

We first look at some rational function curves.

Let $y = \frac{2x^4+x-1}{x^3-x^2-2x}$ (For a complete list of global functions, see Appendix 2):

```
In[21]:= f = y x (x + 1) (x - 2) - (2 x ^ 4 - 1);
         inf = infiniteRealPoints[f, x, y]

Out[22]= {{-1.53924, -3.07848, 0}, {0., -1.4911, 0}}
```

```
In[23]:= asy1 = asymptotes[f, inf[[1]], x, y]
         asy2 = asymptotes[f, inf[[2]], x, y]

Out[23]= {24.3055 + 24.3055 x - 12.1528 y}

Out[24]= {-3.96919 + 1.9846 x, -2.10873 x, -1.03813 - 1.03813 x}
```

```
In[25]:= a1 = Expand[asy1[[1]] / Coefficient[asy1[[1]], y]]
         a2 = Expand[asy2[[1]] / Coefficient[asy2[[1]], x]]
         a3 = Expand[asy2[[2]] / Coefficient[asy2[[2]], x]]
         a4 = Expand[asy2[[3]] / Coefficient[asy2[[3]], x]]

Out[25]= -2. - 2. x + 1. y

Out[26]= -2. + 1. x

Out[27]= 1. x

Out[28]= 1. + 1. x
```

```
In[29]:= ContourPlot[{f == 0, a1 == 0, a2 * a3 * a4 == 0}, {x, -5, 8},
           {y, -10, 20}, ContourStyle → {Black, Dashed, Dashed}]
```

Out[29]=

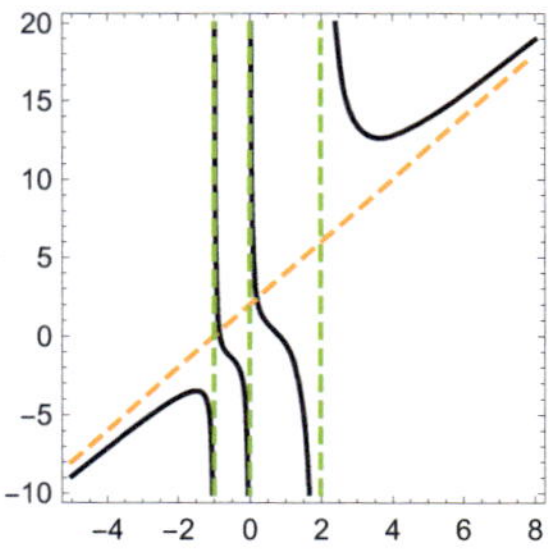

Using a different rational function curve, $y = \frac{x^4}{x^2+1}$:

In[30]:=
```
g = y (x^2 + 1) - x^4;
inf = infiniteRealPoints[g, x, y]
```

Out[31]= {{0, -1.4911, 0}}

In[32]:=
```
asy = asymptotes[g, inf[[1]], x, y]
```

Out[32]= {-2.04478}

In Chapter 1, we mentioned the line $c = 0$ was the infinite line. So we have the one infinite point {0, 1, 0} with infinite asymptote.

Before leaving rational function curves, we should note that from the rational function point of view, the function is continuous and differentiable for all points (except when the denominator is zero, in which case there are no affine crossing points), but making the rational function into an algebraic curve by multiplying by the denominator can, in some cases, introduce a vertical component that meets other components at an affine point. For example, a curve from the family of curves known as *the Trident of Newton* can have an affine singularity of this type.

The extra line is obtained by converting a rational function curve to an algebraic curve:

In[33]:=
```
f = 3 x - 3 x^2 + 2 x^3 - x y
{Plot[(3 x - 3 x^2 + 2 x^3)/x, {x, -4, 4}, PlotRange → {0, 8}, Axes → False,
    Frame → True, AspectRatio → 1], ContourPlot[f == 0, {x, -4, 4}, {y, 0, 8}]}
```

Out[33]= $3x - 3x^2 + 2x^3 - xy$

Out[34]=

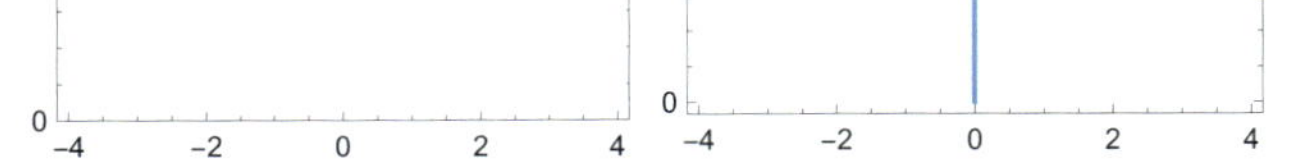

There is a singularity at {0, 3} as well as an infinite singular point at {0, 1, 0}.

Here is a different, interesting example: a 4th degree curve with two infinite singular points:

In[35]:=
```
h = 0.9` - 2 x^2 + x^4 - 2 y^2 - 2 x^2 y^2 + y^4;
```

In[36]:=
```
inf = infiniteRealPoints[h, x, y]
```

Out[36]= {{47.6775, 47.6775, 0}, {0.73407, -0.73407, 0}}

```
In[37]:= asy1 = asymptotes[h, inf[[1]], x, y]
         asy2 = asymptotes[h, inf[[2]], x, y]

Out[37]= {5.50671 + 5.50671 x - 5.50671 y, -0.862295 + 0.862295 x - 0.862295 y}

Out[38]= {5.50671 - 5.50671 x - 5.50671 y, -0.862295 - 0.862295 x - 0.862295 y}

In[39]:= {ContourPlot[{h == 0, asy1[[1]] == 0, asy1[[2]] == 0},
           {x, -3, 3}, {y, -3, 3}, ContourStyle → {Thick, Dashed, Dashed}],
          ContourPlot[{h == 0, asy2[[1]] == 0, asy2[[2]] == 0}, {x, -3, 3},
           {y, -3, 3}, ContourStyle → {Thick, Dashed, Dashed}]}
```

Out[40]=

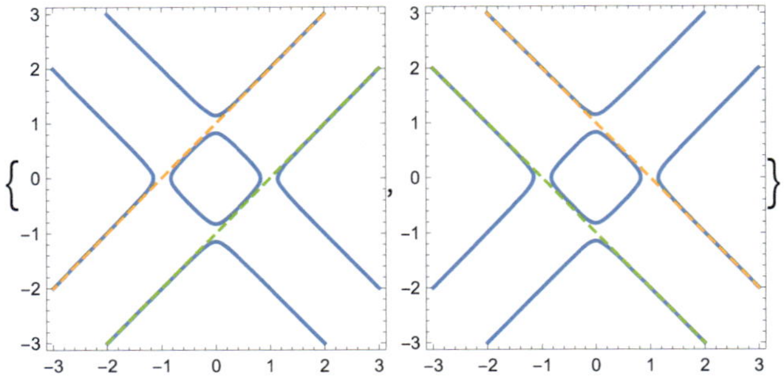

Finally, we remind the reader to not confuse rational function curves with rational curves. A rational curve is a curve that has a rational parameterization. Of course, a rational function curve $y\,g[x] - f[x]$ (where f, g are polynomials) has parameterization $\phi[t] = \left\{t, \frac{f[t]}{g[t]}\right\}$, so it is a rational curve, but the converse is not true; we will see in the next section that all conics have a rational parameterization. In the last section of this chapter, we will discuss a theorem of Caley, which shows that irreducible curves with a maximum number of singularities (all double points) are rational curves.

7.2 Euclidean Geometry of Conics

In elementary algebra, real affine nonsingular conics appear in the form

$\frac{x^2}{a^2} + \frac{y^2}{b^2} = 1$ for an ellipse,

$\frac{x^2}{a^2} - \frac{y^2}{b^2} = 1$ for a hyperbola or

$y = ax^2$ for a parabola.

In this section, we see that every nonsingular conic (exact or numeric) is equivalent to a conic in one of these forms. In an advanced course for mathematics teachers in the early '90s, I attempted an exact example by hand, which ended up taking an entire class period. Or see the recent book by Thomas Garrity (2013), where in Section 1.2, no actual example was even given—neither in the text nor as an exercise. In precalculus or calculus, it is often shown how conics in the previous forms can be described in terms of foci and directrix. I have not seen a description of how to do this for a general conic, but since we're using FLT, we can get the actual transformation as well as the final form, and by taking the inverse, we can find the foci and directrix.

We start with a conic $f = a_1 x^2 + a_2 x y + a_3 y^2 + a_4 x + a_5 y + a_6$ where the coefficients could be exact rational numbers or numerical machine numbers. We could check (see Chapter 1 or Appendix 2 for the code) complex projective plane, where in this particular case, Factor could be used. If the checks in the previous sentence fail, the conic is the union of two (possibly nondistinct) lines and we are done. The next step is to find the infinite real points. If there are none, the curve is an ellipse. If there is a single infinite point (or infinite points differing by a constant), it is a parabola. Finally, if there are two projectively nonequivalent infinite points, it is a hyperbola. We will discuss these three cases separately.

Parabola

Let the infinite point be $\{u, v, 0\}$. Then $vx - uy = 0$ is the equation of the line through the infinite point and origin. Using A1 = klRotation[vx – uy, x, x, y], the line $b x - a y$ is rotated to the y-axis and the resulting conic is in the form $g = ax^2 + bx + c$. Use

```
tangentRealPoints[f_, l_, x_, y_] :=
  With[{J = Det[D[{f, l}, {{x, y}}]]}, {x, y} /. NSolve[{f, J}, {x, y}, Reals]]
```

with tangentRealPoints[g, y, x, y] to find the vertex p and A2 = PQTranslation[p, {0, 0}] to move it to the origin. To get the equation in the form $b y = ax^2$, divide by b; note that this will not change the transformation.

The single FLT transformation A = A1.A2 is the transformation from the original conic to the conic $y = a x^2$. Letting $q = \frac{1}{4a}$, this parabola has focus $\{0, q\}$ and directrix $y = -q$. Applying the FLT given by Inverse[A] gives the focus and directrix for the original conic.

For example, find the standard form, focus and directrix:

```
In[41]:= f = 3.3244920269981044` -5.643417116045773` x+0.0836822330181507` x^2+
           4.251598647968187` y-0.7266575251177807` x y+1.5774888520713963` y^2

Out[41]= 3.32449 - 5.64342 x + 0.0836822 x^2 + 4.2516 y - 0.726658 x y + 1.57749 y^2
```

```
In[42]:= ipts = infiniteRealPoints[f, x, y]

Out[42]= {{1.86165, 0.428778, 0}}
```

This is a parabola because there is only one infinite point:

```
In[43]:= ipt = ipts[[1]];
         k = ipt.{-y, x, 1}

Out[44]= 0.428778 x - 1.86165 y
```

```
In[45]:= A1 = klRotation[k, x, x, y]

Out[45]= {{0.224445, -0.974487, 0.}, {0.974487, 0.224445, 0.}, {0., 0., 1.}}
```

```
In[46]:= g = FLT[f, A1, x, y]

Out[46]= 3.32449 - 5.40976 x + 1.66117 x^2 - 4.54519 y
```

In[47]:=
```
p = tangentRealPoints[g, y, x, y][[1]]
```
Out[47]= {1.6283, −0.237583}

In[48]:=
```
A2 = pqTranslation[p, {0, 0}];
h1 = FLT[g, A2, x, y]
```
Out[49]= $1.66117\, x^2 - 4.54519\, y$

Here is the transformation:

In[50]:=
```
A = A2.A1
```
Out[50]= {{0.224445, −0.974487, −1.6283}, {0.974487, 0.224445, 0.237583}, {0., 0., 1.}}

In[51]:=
```
h = Expand[FLT[f, A, x, y]/-Coefficient[h1, y]]
```
Out[51]= $0.365479\, x^2 - 1.\, y$

Now let:

In[52]:=
```
q = 1/4/Coefficient[h, x^2]
```
Out[52]= 0.684033

Here is the focus:

In[53]:=
```
F = flt[{0, q}, Inverse[A]]
```
Out[53]= {0.800523, −1.48655}

Here is the directrix:

In[54]:=
```
d = FLT[y+q, Inverse[A], x, y]
```
Out[54]= $0.921617 + 0.974487\, x + 0.224445\, y$

Finally, here is what our original parabola with the focus and directrix looks like:

In[55]:=
```
plot1 := ContourPlot[{f == 0, d == 0}, {x, -4, 6}, {y, -6, 4}];
plot2 := ListPlot[{F}, PlotStyle -> {Red}, PlotMarkers -> {Automatic, Medium}];
Show[plot1, plot2]
```
Out[55]=

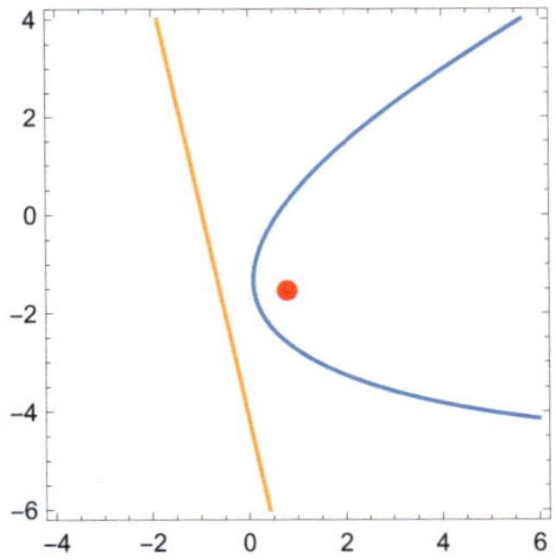

Ellipse

We first put the ellipse in the standard form above; then use the classical formulas to find the eccentricity, a focus and the directrix; and then transform it back.

To find the standard form, we first find the center by intersecting two diagonals, that is lines connecting points with parallel tangents. After transforming the center to {0, 0}, the default critical points of Chapter 3 give the closest and farthest points from the origin and hence the major and minor axes, which we can then rotate to the *x* and *y* axes, respectively, to get the standard form.

We would normally use a randomly generated example, but we are giving this example so you can reproduce our results:

```
In[58]:= f = 2 x^2 + 2 x y + 3 y^2 - 4 x - 2 y - 3;
         ContourPlot[f == 0, {x, -2, 4}, {y, -4, 2}]
```

Out[58]=

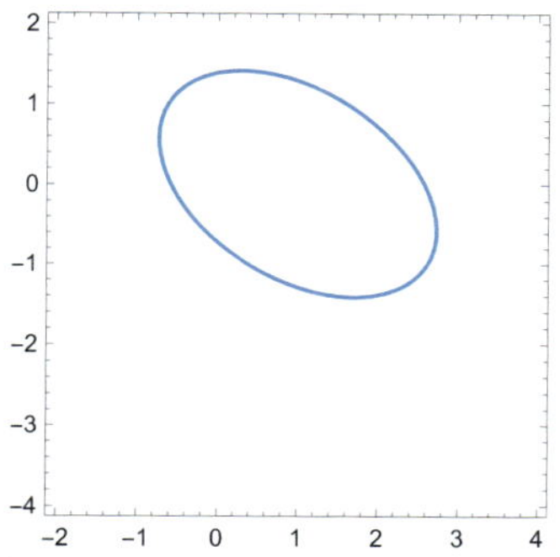


```
In[60]:= tp1 = tangentRealPoints[f, x, x, y]
Out[60]= {{2.73205, -0.57735}, {-0.732051, 0.57735}}

In[61]:= tp2 = tangentRealPoints[f, y, x, y]
Out[61]= {{0.292893, 1.41421}, {1.70711, -1.41421}}
```

We use the line function from Chapter 1 to find the major and minor axes.

As a reminder, make sure you run Appendix 2 before viewing this chapter if you are reading the notebook version of this book:

```
In[62]:= l1 = line[#1, #2, x, y] &@@tp1
         l2 = line[#1, #2, x, y] &@@tp2
Out[62]= -1.70838 + 1.70838 x + 5.12514 y
Out[63]= 1.17809 - 1.17809 x - 0.589043 y
```

Here is the center of our ellipse:

```
In[64]:= p = {x, y} /. NSolve[{l1, l2}][[1]]
```

Out[64]= $\{1., 1.89335 \times 10^{-17}\}$

We move the center to the origin by a translation:

```
In[65]:= B1 = pqTranslation[p, {0, 0}]
```

Out[65]= $\{\{1, 0, -1.\}, \{0, 1, -1.89335\times 10^{-17}\}, \{0, 0, 1\}\}$

```
In[66]:= f1 = FLT[f, B1, x, y]
```

Out[66]= $-5.+2.\,x^2+2.\,x\,y+3.\,y^2$

```
In[67]:= pqTranslation[{1.`, 0.`}, {0, 0}]
```

Out[67]= {{1, 0, −1.}, {0, 1, 0.}, {0, 0, 1}}

```
In[68]:= FLT[-3-4x+2x^2-2y+2xy+3y^2, pqTranslation[{1.`, 0.`}, {0, 0}], x, y]
```

Out[68]= $-5.+2.\,x^2+2.\,x\,y+3.\,y^2$

We need to rotate this ellipse to put the major axis on the *x*-axis.

We again use the critical points but pick the largest pair:

```
In[69]:= cps = SortBy[criticalPoints[f1, x, y], Norm[#] &]
```

Out[69]= {{0.618034, 1.}, {−0.618034, −1.}, {−1.61803, 1.}, {1.61803, −1.}}

The major axis of f1 = 0 is:

```
In[70]:= l3 = Chop[line[#1, #2, x, y] &@@ Take[cps, -2]]
```

Out[70]= 1.90069 x + 3.07538 y

Now rotate this to a horizontal line:

```
In[71]:= B2 = klRotation[l3, y, x, y]
         f2 = FLT[f1, B2, x, y]
         B = B2.B1;
         FLT[f, B, x, y]
```

Out[71]= {{0.850651, −0.525731, 0.}, {0.525731, 0.850651, 0.}, {0., 0., 1.}}

Out[72]= $-5.+1.38197\,x^2+3.61803\,y^2$

Out[74]= $-5.+1.38197\,x^2+3.61803\,y^2$

Which is equivalent to:

```
In[75]:= g = Expand[f2/f2[[1]]]
```

Out[75]= $1.-0.276393\,x^2-0.723607\,y^2$

Which, in turn, is of the form $\frac{x^2}{a^2}+\frac{y^2}{b^2}=1$ where:

```
In[76]:= a = 1/Sqrt[-Coefficient[g, x^2]]
         b = 1/Sqrt[-Coefficient[g, y^2]]
```

Out[76]= 1.90211

Out[77]= 1.17557

We note that $a > b$. If that is not the case, then we rotated incorrectly.

Now from the classical formula, the curve has focus:

In[78]:=
```
F1 = {Sqrt[a^2 − b^2], 0}
```
Out[78]= {1.49535, 0}

And eccentricity:

In[79]:=
```
e = Sqrt[a^2 − b^2]/a
```
Out[79]= 0.786151

And directrix:

In[80]:=
```
d1 = x − a/e
```
Out[80]= −2.41953 + x

Here is the transformed conic with its focus and directrix:

In[81]:=
```
Show[ContourPlot[{f2 == 0, d1 == 0}, {x, −3, 3}, {y, −3, 3}],
  ListPlot[{F1}, PlotMarkers → Automatic]]
```
Out[81]=

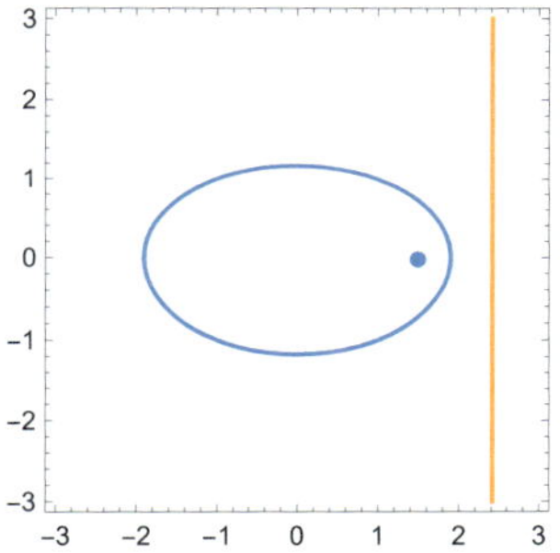

Applying the inverse FLT, we have the focus and directrix of the original ellipse:

In[82]:=
```
F = flt[F1, Inverse[B]]
d = FLT[d1, Inverse[B], x, y]
```
Out[82]= {2.27202, −0.786151}

Out[83]= −3.27018 + 0.850651 x − 0.525731 y

Here is the plot:

In[84]:=
```
plot3 := ContourPlot[{f == 0, d == 0}, {x, −2, 4}, {y, −3.5, 2.5}];
plot4 := ListPlot[{F}, PlotStyle −> {Red}, PlotMarkers −> {Automatic, Medium}];
Show[plot3, plot4]
```
Out[84]=

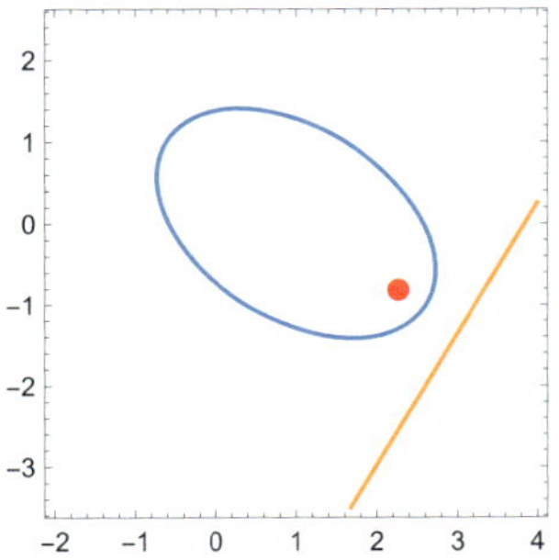

Hyperbola

We can proceed in much the same way as the ellipse, but the center will be the intersection of the two asymptotes. There will be only two real default critical points, but rotating the line through these to the x-axis ($y = 0$) will give the desired standard form. The focus of $\frac{x^2}{a} - \frac{y^2}{b} = 1$ is $\{c, 0\}$, the eccentricity is $e = \frac{c}{a}$ and the directrix is $x = \frac{a}{e} = \frac{a^2}{c}$ where c = Sqrt[a² + b²].

Do the example as an exercise with this suggested hyperbola:

```
In[87]:= h = 1 - 2 x - 3 y - x^2 + x y + 2 y^2;
```

Hint: this parabola transforms to

$$\frac{x^2}{1.90211^2} - \frac{y^2}{1.17557^2} = 1,$$

where the focus is $\{0.558859, 0\}$ and the directrix is $x = 0.26509$.

Transforming it back, the original curve has focus {-0.00391255, 0.79937} and directrix 0.987087 x - 0.160182 y = -0.425676:

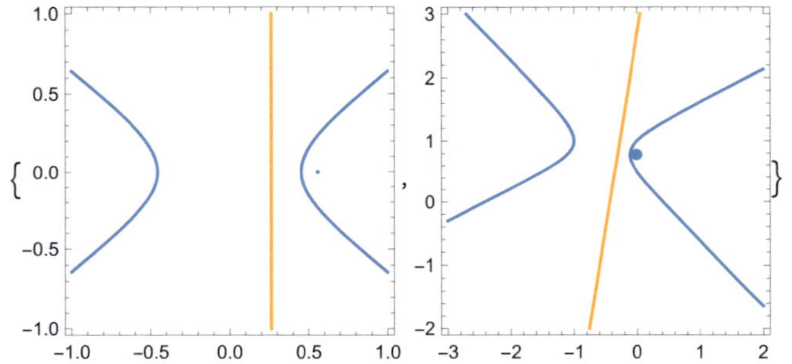

A converse to this problem is to start with the foci and find the conic. For the ellipse and hyperbola, the symmetry shows that there are actually two foci. The line containing them is the *major axis* and the conic is symmetric about this line. It turns out that the foci and one point on the conic determines the conic. Our goal is to find the equation from this information and the knowledge of whether this conic should be an ellipse or hyperbola. We should note that in the classical situation, the extra point q is where the *semilatus rectum* meets the ellipse or hyperbola, that is, a point q on the conic where the closest point on the major axis to q is a focus.

The additional point cannot be completely random. To be sure to get a conic with real coefficients, we restrict ourselves to *good* points. Given our desired foci p_1, p_2, a point q will be called *good* if it is not on the major axis but the closest point on the major axis l to q, call it q_c, is between points p_1, p_2 and no farther from l than half the distance from p_1, p_2. This latter quantity is known as the *linear eccentricity* c.

Our technique is to transform our data p_1, p_2, q to standard position $\{0, c\}$, $\{0, -c\}$, q_1 using an FLT with matrix M. We then solve a nonlinear system to find a, b where, in the ellipse case, $c^2 = a^2 - b^2$ and q_1 satisfies the equation $\frac{x^2}{a^2} + \frac{y^2}{b^2} = 1$. In the hyperbola case, we use the equations $c^2 = a^2 + b^2$ and $-\frac{x^2}{a^2} + \frac{y^2}{b^2} = 1$. We then apply the FLT with matrix M^{-1} to the latter equation to get our desired equation.

The details are similar to the earlier calculations in this section, but now we have a function to do this automatically. In this function, the default case is the ellipse case, which we will consider in the next chapter, an option conic → "hyperbola" will give the hyperbola case. The function will check to make sure for the given p_1, p_2 that q is good.

As an example, we find the ellipse with foci {1, 3}, {3, 5} and q = {2, 2}:

```
In[91]:= p1 = {1, 3};
         p2 = {3, 5};
         q = {2, 5};
         f = conicWithFoci[p1, p2, q, x, y]
```

```
Out[94]= 9.11146 + 0.944272 x + 1. x^2 - 5.52786 y - 1.23607 x y + 1. y^2
```

```
In[95]:= ContourPlot[{f == 0, line[p1, p2, x, y] == 0}, {x, 0, 4},
           {y, 2, 6}, Epilog → {PointSize[Large], Red, Point[{p1, p2, q}]}]
```

Out[95]=

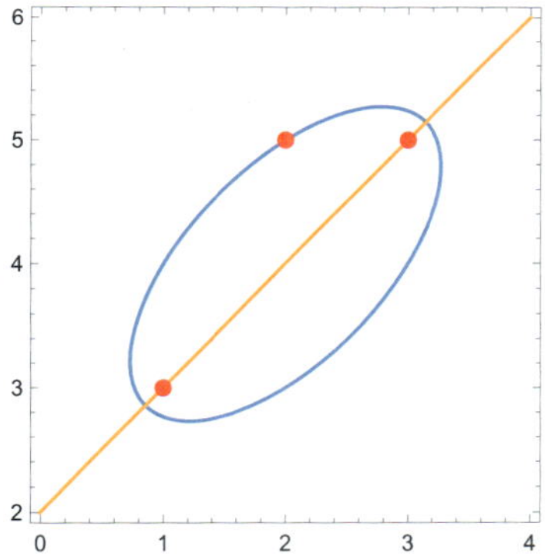

The focus is a Euclidean concept only. When using Euclidean transformations for plotting conics, it is important that the x-range is the same as the y-range. Otherwise, plotting introduces an affine, non-Euclidean transformation that distorts the ellipse or hyperbola. For example, doing the same plot with a different scale makes the major axis not look like the major axis!

You can also adjust for this distortion using AspectRatio:

```
In[96]:= {ContourPlot[{f == 0, line[p1, p2, x, y] == 0}, {x, .7, 3.5},
           {y, 1, 7}, Epilog → {PointSize[Large], Red, Point[{p1, p2, q}]}],
         ContourPlot[{f == 0, line[p1, p2, x, y] == 0}, {x, .7, 3.5}, {y, 1, 7},
           Epilog → {PointSize[Large], Red, Point[{p1, p2, q}]}, AspectRatio → 2.1]}
```

Out[96]=

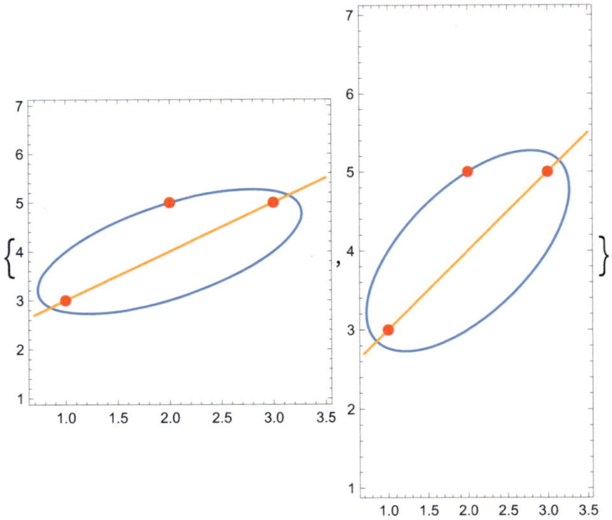

7.3 Projective Conics and Parameterization

All nonsingular projective conics are equivalent under FLT. We can use this to parameterize any nonsingular conic, either by a rational or trigonometric parameterization.

A key to this, both here and in the coming sections, is the cTransform[f, p, x, y] function.

This takes a nonsingular point (finite or infinite) and places it at the infinite point {0, 1, 0} with the infinite tangent line at {0, 1, 0} (For a complete list of global functions, see Appendix 2):

```
In[97]:= cTransform[f_, p_, x_, y_] := Module[{fh, ph, nh, t, cs, A, d},
            d = tDeg[f, x, y];
            fh = Expand[t^d*(f /. Thread[{x, y} → {x/t, y/t}])];
            ph = If[Length[p] == 2, N[Append[p, 1]], N[p]];
            ph = ph/Norm[ph];
            nh = {D[fh, x], D[fh, y], D[fh, t]} /. Thread[{x, y, t} → ph];
            nh = nh/Norm[nh];
            cs = Cross[nh, ph];
            cs = cs/Norm[cs];
            A = {cs, ph, nh}];
```

To find an FLT taking a nonsingular conic $f = 0$ to the parabola $y = x^2$, first find a real point p on the conic and use A1 = CTransform[f, p, x, y]. The result of this transform will be a parabola with vertical axis $f1 = a\,x^2 + b\,x + c - d\,y$.

Next, find the vertex q using q = tangentrealpoints[f1, y, x, y][[1]], which will transform by A2 = pqTranslation[p, {0, 0}] to f 2 = a x^2 + b y. We now use A3 = homothety[r, 1], r = ± Sqrt[Abs[a / b]] to get $a\,x^2 - a\,y$, which is equivalent to $x^2 - y$.

We illustrate this using the hyperbola of the previous example:

```
In[98]:= h = 1-2 x-3 y-x^2+x y +2 y^2;
         cps = criticalPoints[h, x, y]
Out[99]= {{-1.06504, 0.829443}, {0.233947, 0.202123}}

In[100]:= p = cps[[1]]
Out[100]= {-1.06504, 0.829443}

In[101]:= A1 = cTransform[h, p, x, y]
          h1 = FLT[h, A1, x, y]
Out[101]= {{-0.614438, -0.788965, -1.13716 × 10^-15},
            {-0.633966, 0.493726, 0.59525}, {0.469631, -0.365744, 0.803541}}
Out[102]= 0.647831 + 4.47491 x + 1.35217 x^2 + 1.21618 y

In[103]:= tp = tangentRealPoints[h1, y, x, y]
Out[103]= {{-1.65471, 2.51156}}

In[104]:= A2 = pqTranslation[tp[[1]], {0, 0}]
Out[104]= {{1, 0, 1.65471}, {0, 1, -2.51156}, {0, 0, 1}}

In[105]:= f2 = FLT[h, A2.A1, x, y]
Out[105]= 1.35217 x^2 + 1.21618 y

In[106]:= r = Sqrt[Coefficient[f2, x^2]/Coefficient[f2, y]]
Out[106]= 1.05443

In[107]:= A3 = homothety[r, -1]
Out[107]= {{1.05443, 0, 0}, {0, -1, 0}, {0, 0, 1}}

In[108]:= f3 = FLT[h, A3.A2.A1, x, y]
Out[108]= 1.21618 x^2 - 1.21618 y
```

Which is the desired form $y - x^2$ up to constant multiple, with FLT given by:

```
In[109]:= A = A3.A2.A1
Out[109]= {{0.171521, -1.47004, 1.402}, {1.81347, -1.41231, 1.42289}, {0.469631, -0.365744, 0.803541}}
```

Rational Parameterization

Once we have an FLT given by matrix A, which takes our curve to the parabola $y - x^2$, we can parameterize the latter by F t = {t, t^2}, and apply the FLT given by Inverse[A] to get our rational parameterization.

A technical detail to note is that the function flt expects a pair of real numbers rather than variables.

We use the variant flts (For a complete description of the various functional linear transformation functions, see Section 6.5):

```
In[110]:= flts[{x_, y_}, A_] := Module[{u, v, w},
             {u, v, w} = A.{x, y, 1};
             {u/w, v/w}];
```

Taking the previous example, we then do:

```
In[111]:= F = flts[{t, t^2}, Inverse[A]]
```

Out[111]= $\left\{\frac{-0.105892-0.582723\,t+0.633966\,t^2}{2.29854-1.04857\times10^{-15}\,t-0.59525\,t^2}, \frac{2.17979-0.748242\,t-0.493726\,t^2}{2.29854-1.04857\times10^{-15}\,t-0.59525\,t^2}\right\}$

The parametric plot is:

```
In[112]:= ParametricPlot[F, {t, -10, 10}, Frame → True, Axes → False]
```

Out[112]=

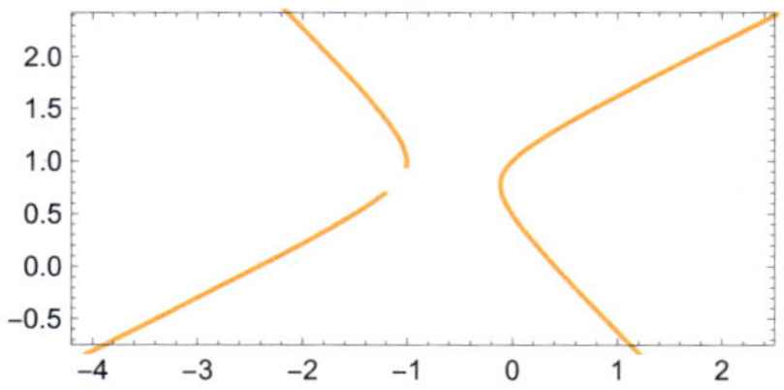

This has a gap near the point on the curve:

```
In[113]:= Limit[F, t → Infinity]
```

Out[113]= {-1.06504, 0.829443}

It is interesting to note that we have a converse.

If a parameterized curve has a parameterization of the form

$$F = \left\{\frac{a_{1,1}\,t + a_{1,2}\,t^2 + a_{1,3}}{a_{3,1}\,t + a_{3,2}\,t^2\,a_{3,3}}, \frac{a_{2,1}\,t + a_{2,2}\,t^2 + a_{2,3}}{a_{3,1}\,t + a_{3,2}\,t^2\,a_{3,3}}\right\},$$

then letting $t = x$, $t^2 = y$, we see this parameterization is in the form of an FLT with matrix

$$A = \begin{pmatrix} a_{11} & a_{12} & a_{13} \\ a_{21} & a_{22} & a_{23} \\ a_{31} & a_{32} & a_{33} \end{pmatrix}.$$

This curve can be obtained in the form FLT[x² – y, A, x, y] and is generally a conic.

Consider the curve parameterized by:

In[114]:=
```
F = {3 t - t ^ 2 + 1, -2 t + t ^ 2 - 2} / (1 + t + t ^ 2)
```

Out[114]= $\left\{\frac{1+3t-t^2}{1+t+t^2}, \frac{-2-2t+t^2}{1+t+t^2}\right\}$

Let A be given by:

In[115]:=
```
A = N[{{3, -1, 1}, {-2, 1, -2}, {1, 1, 1}}];
```

Then:

In[116]:=
```
g = FLT[x ^ 2 - y, A, x, y]
```

Out[116]= $-0.0833333 + 0.5\,x + 0.25\,x^2 + 0.5\,y + 0.5\,x\,y + 0.333333\,y^2$

In[117]:=
```
P1 := ParametricPlot[F, {t, -10, 10}, PlotStyle → Black];
P2 := ContourPlot[g == 0, {x, -4, 3}, {y, -3, 3}];
```

In[119]:=
```
{P1, P2, Show[P2, P1]}
```

Out[119]=

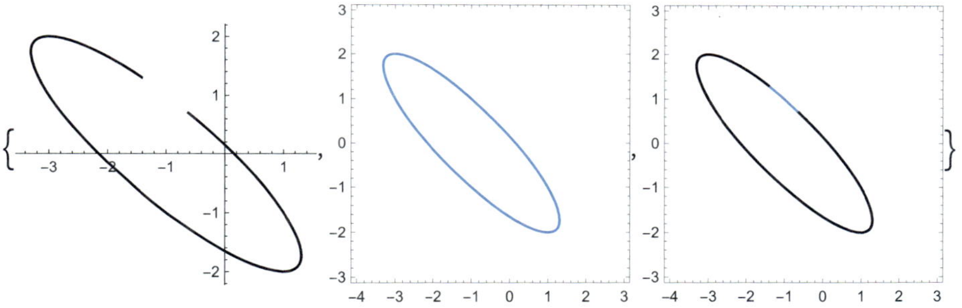

Trigonometric Parameterization

An FLT that takes the parabola $y = x^2$ to the circle $x^2 + y^2 = 1$ is given by:

```
P2C = pqTranslation[{0, -1}, {0, 0}].ip2z[{0, 1, 0}, x, y];
```

Therefore, start with any nonsingular conic and follow the transformation to $y - x^2$ by P2C. We get an FLT transformation from the conic to the standard unit circle. The unit circle can be parameterized by F = {Cos[t], Sin[t]}. Applying the inverse transformation parameterizes our original conic by cosine and sine.

Using our example hyperbola from above:

In[121]:=
```
h = 1 - 2 x - x² - 3 y + x y + 2 y²;
```

We saw that when:

In[122]:=
```
A = {{0.17152143109999518`, -1.4700447703390642`, 1.4019961464680566`},
     {1.8134718990760177`, -1.4123120357497518`, 1.422889116329615`},
     {0.4696313544127659`, -0.3657437507250866`, 0.8035408513311424`}};
```

Then:

In[123]:= **FLT[h, P2C.A, x, y]**

Out[123]= −0.608092 + 0.608092 x^2 + 0.608092 y^2

This is equivalent to $x^2 + y^2 - 1$. The following gives our parameterization:

In[124]:= **F = flts[{Cos[t], Sin[t]}, Inverse[P2C.A]]**

Out[124]= $\left\{\frac{-0.373405 - 0.412047\,\text{Cos}[t] - 0.523158\,\text{Sin}[t]}{-1.20441 - 1.01959\times 10^{-15}\,\text{Cos}[t] + 2.04622\,\text{Sin}[t]}, \frac{-1.19223 - 0.529087\,\text{Cos}[t] + 1.89046\,\text{Sin}[t]}{-1.20441 - 1.01959\times 10^{-15}\,\text{Cos}[t] + 2.04622\,\text{Sin}[t]}\right\}$

In[125]:= **ParametricPlot[F, {t, 0, 2 Pi}]**

Out[125]=
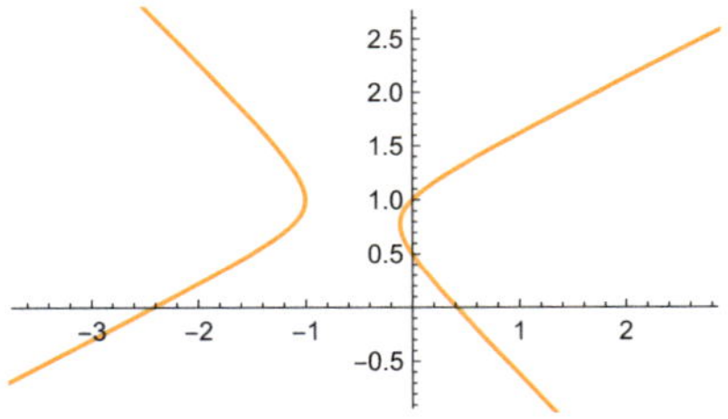

7.4 Inflection Points

A point p is an *inflection point* of an algebraic curve if the intersection multiplicity at p of the curve and its tangent line is 3 or higher. Inflection points are important in elementary mathematics for understanding graphs of single variable polynomial functions, which can be considered a special case of an algebraic curve. Lines and conics do not have inflection points. We will see in the next section that inflection points play an important role in the theory of cubic curves. In general, however, there may be many inflection points, but they are hard to compute and do not play a significant role in real curve theory.

Since we are mostly interested in inflection points of conics, the most efficient method of finding inflection points is the classical one. Intersect the curve with its Hessian curve. The Hessian curve is best viewed as a projective curve, even if we are only interested in real inflection points, so the first step is to homogenize the equation of the curve $f = 0$ using the following (from Section 5.2):

```
fh = homog[f, x, y, z]
```

Next, we calculate the Hessian matrix:

$$\begin{pmatrix} \partial_{x,x}\text{fh} & \partial_{x,y}\text{fh} & \partial_{x,z}\text{fh} \\ \partial_{y,x}\text{fh} & \partial_{y,y}\text{fh} & \partial_{y,z}\text{fh} \\ \partial_{z,x}\text{fh} & \partial_{z,y}\text{fh} & \partial_{z,z}\text{fh} \end{pmatrix}$$

Classical textbooks, e.g. (Fischer, 2001), often give shortcuts for hand calculation since fh is a homogeneous polynomial, but with Mathematica, we can simply use D[fh, {{x, y, z}, 2 to get the Hessian matrix. We then do a rare use of the determinant in numerical math to get the Hessian curve.

For example:

```
In[126]:= f = x^3 - 5 x^2 + 7 x - y^2
```

Out[126]= $7x - 5x^2 + x^3 - y^2$

```
In[127]:= fh = homog[f, x, y, z]
```

Out[127]= $x^3 - 5x^2 z - y^2 z + 7 x z^2$

```
In[128]:= H = Det[D[fh, {{x, y, z}, 2}]]
```

Out[128]= $8(-3xy^2 + 4x^2 z + 5y^2 z - 35 x z^2 + 49 z^3)$

To get the real affine inflection points, set $z = 1$ and solve with f:

```
In[129]:= inflH = {x, y} /. NSolve[{f, H, z - 1}, {x, y, z}, Reals]
```

Out[129]= {{1.72929, 1.52452}, {1.72929, −1.52452}}

```
In[130]:= ha = H /. {z → 1}
```

Out[130]= $8(49 - 35x + 4x^2 + 5y^2 - 3xy^2)$

```
In[131]:= Show[ContourPlot[{f == 0, ha == 0}, {x, -1, 4}, {y, -4, 4}],
  ListPlot[inflH, PlotStyle → Red, PlotMarkers → Automatic]]
```

Out[131]=

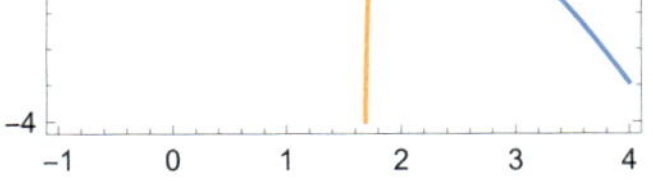

Note that in this case, the Hessian is of degree 3. More generally, the degree of the Hessian curve is $3(d-2)$ where d is the degree of the original curve, so there can be as many as $3d(d-2)$ inflection points for a curve of degree d, but as in this example, many are complex and some may be infinite.

Here is the code for the general affine inflection point procedure (For a complete list of global functions, see Appendix 2):

```
In[132]:= affineInflectionPoints[f_, x_, y_] := Module[{H, fh, d, z},
    fh = Homog[f, x, y, z];
    H = Det[D[fh, {{x, y, z}, 2}]];
    {x, y} /. NSolve[{fh, H, z-1}, {x, y, z}, Reals]
  ];
```

To find all projective inflection points, we could use the trick from Chapter 6 for finding projective singular points.

But since we are already working projectively, we can modify the above by replacing $z-1$ with the random line $l-1$ in the NSolve step:

```
In[133]:= rinfl = 8.390338886809685` x+1.572094833417954` y-1.8964087979890598`;
allInflectionPoints[f_, x_, y_] := Module[{H, fh, d, z, l, ips, sol},
    fh = homog[f, x, y, z];
    H = Det[D[fh, {{x, y, z}, 2}]];
    l = homog[rinfl, x, y, z];
    sol = NSolve[{fh, H, l-1}, {x, y, z}, Reals];
    If[Length[sol] == 0, Return[{}]];
    ips = {x, y, z} /. sol;
    ips = DeleteDuplicates[ips, Norm[#1-#2] < 1.*^-10 &];
    Which[Abs[#[[3]]] > 1.*^-9, Take[#/#[[3]], 2],
        Abs[#[[2]]] > 1.*^-9, #/#[[2]], True, #/#[[1]]] &/@ips
  ];
```

As an example:

```
In[135]:= f = 1-4x^3+2y+y^2;
allInflectionPoints[f, x, y]

Out[136]= {{0., 1., 0.}, {0., -1.}}
```

7.5 Cubics

As mentioned in Chapter 1, most discussions of cubic curves are curves of the form

$$g = u[x] - y^2 \text{ where } u = u[x] = x^3 + r x^2 + s x + t$$

and r, s, t are integers. In this book, and with the Wolfram Language, we can use machine numbers as well.

With standard cubics, topologically and geometrically, the curve will be one of the following:

Description	Type	Zeros of u[x]	Example	Picture
one oval	nonsingular	one real, 2 complex	$y^2 = x^3 + x$	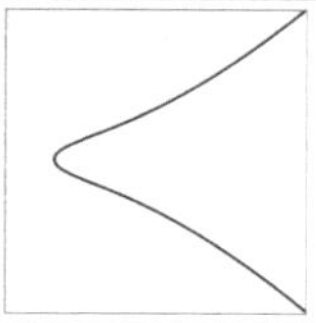
two ovals	nonsingular	a < b < c real	$y^2 = x^3 - x$	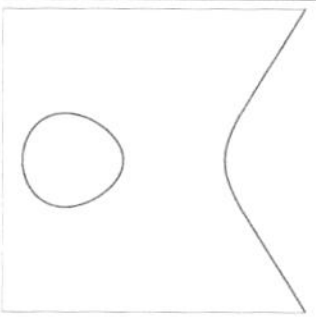
oval and isolated point	singular	a = b < c real	$y^2 = x^3 - x^2$	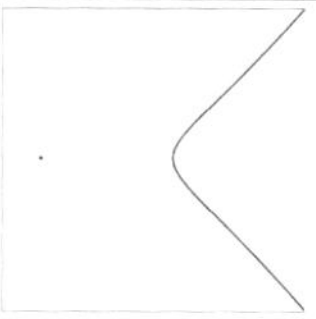
node	singular	a < b = c real	$y^2 = x^3 + x^2$	
cusp	singular	a = b = c real	$y^2 = x^3$	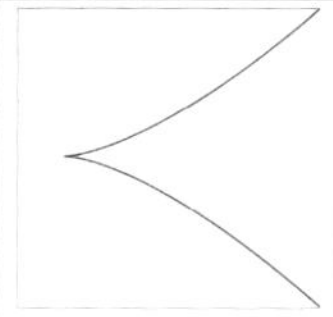

Reducible cubics are either the union of three lines or are the union of a line and a nonsingular conic. If the coefficients are integers, use Factor to get this reduction; otherwise, find points and use aCurve for the equation. The main result of this section is to show that every irreducible exact or numerical cubic can be reduced to the form $u[x] - y^2$ shown previously or to the even simpler Weierstrass normal form $y^2 = x^3 + s\,x + t$.

This is a classical result, but it is usually described theoretically only or using examples that have been purposely chosen so that there are obvious inflection points with rational coefficients, e.g. (Garrity, 2013).

We start with an example:

```
In[138]:= f = newtonHyperbola[836, x, y]
```

Out[138]= $1 + x + 0.2\,x^2 - 0.008\,x^3 - y + x\,y + 0.2\,x^2\,y + 0.2\,y^2 - 0.2\,x\,y^2 - 0.008\,y^3$

Our first step is to calculate real inflection points, including infinite ones:

```
In[139]:= ips = allInflectionPoints[f, x, y]

Out[139]= {{-0.12874, 22.0683}, {6.57711, 12.6084}, {15.8884, -0.527148}}
```

We can pick any of these points. If we had an infinite one, we would give preference to that one. Here we just pick the first.

Our first step is to do a cTransform:

```
In[140]:= B1 = cTransform[f, ips[[1]], x, y]
          f1 = FLT[f, B1, x, y]

Out[140]= {{-0.576197, 0.0336427, -0.816618},
            {-0.00582762, 0.998958, 0.0452666}, {-0.81729, -0.0308415, 0.575401}}

Out[141]= 0.0220105 - 0.293535 x + 1.09218 x^2 - 0.985939 x^3 - 0.673757 y + 1.39562 x y + 0.189345 y^2
```

This is what we expect: the xy^2, $x^2\ y$ and y^3 terms are missing. If any were present at this step, they would probably have very small coefficients, which we can use Chop to shorten.

Next, we find a line:

```
In[142]:= k = Expand[Coefficient[f1, y]/Coefficient[f1, y^2]/2 + y]

Out[142]= -1.77918 + 3.68539 x + y
```

Apply a shear:

```
In[143]:= B2 = {{1, 0, 0}, {Coefficient[k, x], 1, k[[1]]}, {0, 0, 1}}
          f2 = FLT[f1, B2, x, y]

Out[143]= {{1, 0, 0}, {3.68539, 1, -1.77918}, {0, 0, 1}}

Out[144]= -0.577356 + 2.18952 x - 1.47952 x^2 - 0.985939 x^3 + 0.189345 y^2
```

This is already up to a constant in the form $y^2 = u[x]$, so we could stop here if our only interest was in parameterizing the curve with an FLT. The next two steps are fairly easy, so we continue next with a homothety.

This makes the coefficients of x^3 and y^2 equal, so we can divide by that coefficient to get 1:

```
In[145]:= B3 = homothety[CubeRoot[-(Coefficient[f2, x^3]/Coefficient[f2, y^2])], 1]
          f3 = FLT[f2, B3, x, y]

Out[145]= {{1.73327, 0, 0}, {0, 1, 0}, {0, 0, 1}}

Out[146]= -0.577356 + 1.26323 x - 0.492481 x^2 - 0.189345 x^3 + 0.189345 y^2
```

```
In[147]:= f3 = -Expand[f3/Coefficient[f3, y^2]]

Out[147]= 3.04923 - 6.6716 x + 2.60097 x^2 + 1. x^3 - 1. y^2
```

Getting rid of the x^2 term can be done with a linear change-of-variables trick that goes back to Gerolamo Cardano in 1539.

This is the final step:

```
In[148]:= B4 = {{1, 0, Coefficient[f3, x^2]/3}, {0, 1, 0}, {0, 0, 1}}
          f4 = FLT[f3, B4, x, y]

Out[148]= {{1, 0, 0.86699}, {0, 1, 0}, {0, 0, 1}}

Out[149]= 10.1368 - 8.92661 x + 1. x^3 - 1. y^2

In[150]:= {ContourPlot[f == 0, {x, -25, 20}, {y, -20, 20}], ContourPlot[f4 == 0, {x, -4, 5}, {y, -10, 10}]}
```

Out[150]=

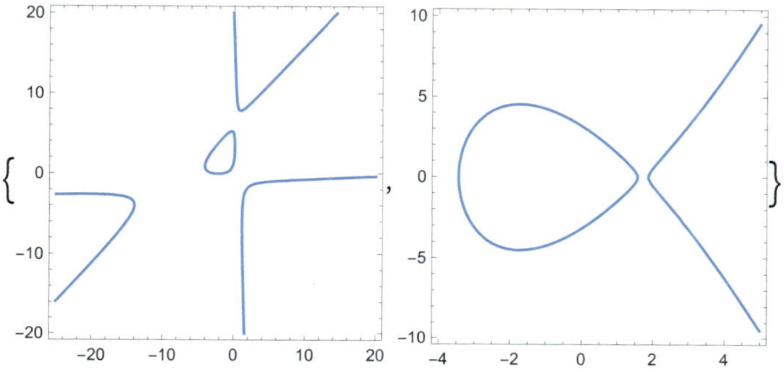

Here is the FLT connecting them:

```
In[151]:= B = B4.B3.B2.B1

Out[151]= {{-1.70729, 0.0315726, -0.91655},
            {-0.675234, 1.17782, -3.98803}, {-0.81729, -0.0308415, 0.575401}}
```

To parameterize the original, we have two parts. Use flts:

```
In[152]:= F1 =
            flts[{x, Sqrt[10.136815015201705` - 8.926609369242879` x + 1.0000000000000002` x^3]},
             Inverse[B]]
```

Out[152]= $\left\{\frac{-0.550184 - 0.320043\,x - 0.00582762\sqrt{10.1368 - 8.92661\,x + 1.\,x^3}}{1.14786 - 0.567393\,x + 0.0452666\sqrt{10.1368 - 8.92661\,x + 1.\,x^3}},\ \frac{3.57118 - 2.10464\,x + 0.998958\sqrt{10.1368 - 8.92661\,x + 1.\,x^3}}{1.14786 - 0.567393\,x + 0.0452666\sqrt{10.1368 - 8.92661\,x + 1.\,x^3}}\right\}$

And (not expanded but with a sign change before the radical in the numerator):

```
In[153]:= F2 = flts[{x, -Sqrt[10.136815015201705` -
                   8.926609369242879` x + 1.0000000000000002` x^3]}, Inverse[B]];
```

```
In[154]:= ParametricPlot[{F1, F2}, {x, -3.44, 40}, PlotRange → 20, Frame → True, Axes → False]
```

Out[154]=

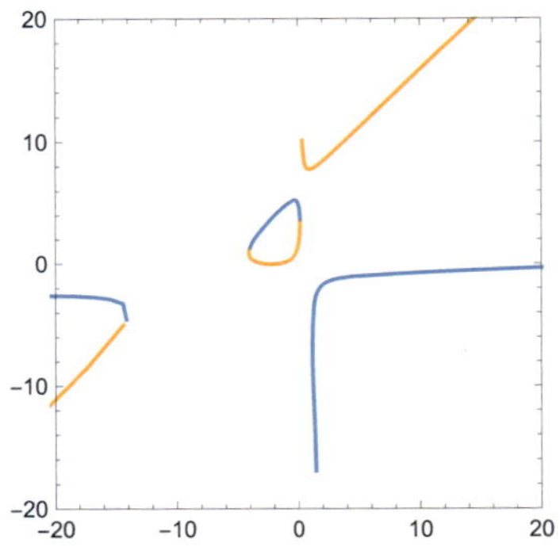

We note that in the interval $-3.44 \leq x \leq 40$, there is an interval where the cubic under the radical is negative and a point where the denominator is zero. However, ParametricPlot takes care of this for us.

As noted previously, we could have also used f2 and B = B2.B1.

Since there are no other choices, we can put these steps together in an almost black-box algorithm. This is difficult with numerical algorithms since unexpected numerical errors may crop up.

We attempt to catch the common problem here:

```
In[155]:= weierstrassNormalForm[f_, ip_, x_, y_] :=
  Module[{B1, B2, B3, B4, B, f1, f2, f3, f4, j, k, c4, c6, cy2},
    B1 = cTransform[f, ip, x, y];
    f1 = Chop[FLT[f, B1, x, y], 1.*^-9];
    cy2 = Coefficient[f1, y^2];
    If[Length[cy2] > 0, Print["Sorry, this example requires special handling"];
      Abort[]];
    k = Expand[Coefficient[f1, y]/cy2/2+y];
    B2 = {{1, 0, 0}, {Coefficient[k, x], 1, k[[1]]}, {0, 0, 1}};
    f2 = FLT[f1, B2, x, y];
    B3 = homothety[CubeRoot[-(Coefficient[f2, x^3]/Coefficient[f2, y^2])], 1];
    f3 = FLT[f2, B3, x, y];
    f3 = -Expand[f3/Coefficient[f3, y^2]];
    B4 = {{1, 0, Coefficient[f3, x^2]/3}, {0, 1, 0}, {0, 0, 1}};
    f4 = FLT[f3, B4, x, y];
    B = B4.B3.B2.B1;
    c4 = Coefficient[f4, x]/27;
    c6 = f4[[1]]/54;
    j = 1728 c4^3/(c4^3-c6^2);
    {f4, B, j}]
```

In this implementation, we calculate the inflection points separately and choose one. We also return not only the normal form and FLT B but a number called the j-invariant (which we explore below). The user could change the last line and only get the normal form, if preferred.

In general, for a nonsingular cubic in Weierstrass form $y^2 = x^3 + a\,x + b$, the j-invariant is

$$j - 1728\,\frac{4\,a^3}{4\,a^3 - 27\,b^2}.$$

Note that we use different inflection points:

```
In[156]:= weierstrassNormalForm[f, ips[[1]], x, y]

Out[156]= {10.1368 - 8.92661 x + 1. x^3 - 1. y^2, {{-1.70729, 0.0315726, -0.91655},
           {-0.675234, 1.17782, -3.98803}, {-0.81729, -0.0308415, 0.575401}}, 874.894}

In[157]:= weierstrassNormalForm[f, ips[[2]], x, y]

Out[157]= {174.903 - 59.605 x + 1. x^3 - 1. y^2, {{0.9287, -0.816791, 4.19022},
           {1.10238, 0.570604, -0.189017}, {0.141862, -0.151586, 0.978211}}, 874.894}

In[158]:= weierstrassNormalForm[f, ips[[3]], x, y]

Out[158]= {0.928686 - 1.81411 x + 1. x^3 - 1. y^2, {{0.0847194, 0.751728, -0.949785},
           {0.866318, -0.344688, 1.98246}, {-0.025385, 0.659709, 0.751093}}, 874.894}
```

We get different normal forms and FLT, but the j-invariant is always 874.894. That is one reason we call it an *invariant*; it does not depend on the inflection point chosen.

Suppose we transform $f = 0$ by an arbitrary, or random, FLT:

```
In[160]:= A = {{0.07485986157215319`, 0.9693047762451474`, 0.24950067263711073`},
              {0.6094077133922`, -0.4539225635800057`, -0.5247878943188331`},
              {-0.02373700112965471`, 0.3027242271117272`, -0.048647756985034984`}};
          g = FLT[f, A, x, y]

Out[160]= -472.146 + 405.534 x - 123.155 x^2 + 12.9945 x^3 - 48.5671 y +
           14.5051 x y - 0.651304 x^2 y + 0.19564 y^2 - 0.587121 x y^2 - 0.0026431 y^3

In[161]:= ipg = allInflectionPoints[g, x, y]

Out[161]= {{-1.58521, -16.0531}, {3.58884, -0.620103}, {3.26009, -1.60068}}

In[162]:= weierstrassNormalForm[g, ipg[[2]], x, y]

Out[162]= {8746.52 - 809.05 x + 1. x^3 - 1. y^2, {{4.39651, -0.40465, -16.0293},
           {1.86759, 1.42205, -2.04384}, {0.262081, -0.0382526, -0.964287}}, 874.894}
```

We still get the same j-invariant. Therefore, cubics that differ by a fractional linear transformation have the same j-invariant. The converse is partly true. Suppose we have two cubics in Weierstrass form with the same j-invariant:

$$C_1 : y^2 = x^3 + a_1\,x + b_1$$
$$C_2 : y^2 = x^3 + a_2\,x + b_2$$

Assuming b_1 is not zero, I claim that the FLT homothety[d₁, d₂] where

$$d_1 = \sqrt[3]{[b_2/b_1]}$$
$$d_2 = -{d_1}^{3/2}$$

takes cubic C_1 to cubic C_2. I leave the easy case $b_1 = 0$, $a_1 \neq 0$ to the reader.

Now if B_1, B_2 are FLT taking cubics f to C_1 and g to C_2, then the FLT Inverse[B₂]. homothety[d₁, d₂]. B_1 takes f to g.

One technical detail is that if b_1, b_2 are of opposite signs, then d_1 is negative, so d_2 is imaginary. Thus, the claim that having the same j-invariant implies the cubics are FLT equivalent works only over the complex numbers. In particular, the formulas for a, b real, $b \neq 0$,

$$C_1 : y^2 = x^3 + ax + b$$
$$C_2 : y^2 = x^3 + ax - b,$$

have the same j-invariant but are not equivalent via a real FLT because they are equivalent by homothety[−1, −ⅈ].

It is also true that given any real number j (other than 0), there is a nonsingular cubic with that j-invariant. One can pick j and b to be arbitrary real numbers and use NSolve to solve for a in the single variable polynomial equation $j(4\,a^3 - 27\,b^2) - 6912\,a^3 = 0$. Because it is of degree 3 in a, one zero must be real.

7.6 Caley's Theorem

For the curves in this chapter, a recurring problem is to find a parameterization. Although the techniques are somewhat different, we end this chapter with Caley's theorem from 1865, which says that an irreducible curve of any degree with a maximal number of singularities, each of which is a double point, is a rational curve, that is, the curve has a rational parameterization. This is a special case of a more general theorem in modern algebraic geometry that says a genus 0 curve is a rational curve. The interesting thing about Caley's proof is that it is constructive. Using the Wolfram Language, we can actually find this parameterization, following the argument in (Popescu-Pampu, 2016).

First, we say what we mean by a *double point*. It is a point p on a curve $f = 0$ such that the intersection multiplicity with a random line through this point is 2. Since a regular point on a curve locally behaves like its tangent line, this also means that the intersection multiplicity of any curve g that intersects f at p at a regular point of g has intersection multiplicity 2 with f at p.

Some examples are shown:

```
In[163]:= {Labeled[ContourPlot[y^2-x^3-x^2 == 0, {x, -1, 1}, {y, -1, 1}], "node"],
  Labeled[ContourPlot[y^2-x^3 == 0, {x, -1, 1}, {y, -1, 1}], "cusp"],
  Labeled[ContourPlot[(y-x^2) (y+x^4) == 0, {x, -1, 1}, {y, -1, 1}], "common tangent"],
  Labeled[ContourPlot[x y (x-y) == 0, {x, -1, 1}, {y, -1, 1}], "triple point"]}
```

Out[163]=

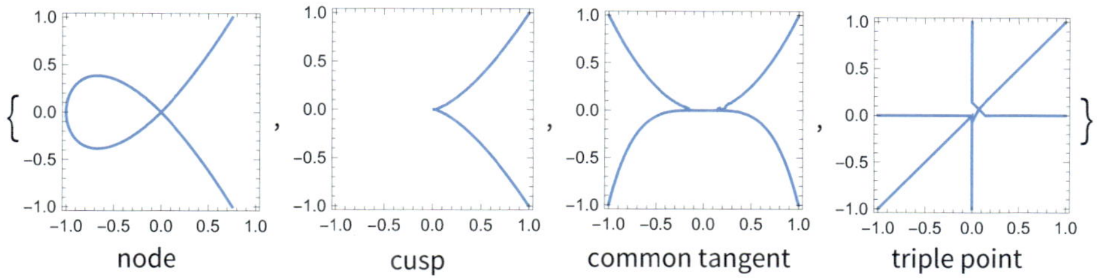

Pick a random line:

```
In[164]:= l = RandomReal[{-1, 1}, 2].{x, y}
Out[164]= 0.278835 x - 0.609392 y
```

```
In[165]:= {IntersectionMultiplicity[y^2-x^3-x^2, l, {0, 0}],
  IntersectionMultiplicity[y^2-x^3, l, {0, 0}],
  IntersectionMultiplicity[(y-x^2) (y+x^4), l, {0, 0}],
  IntersectionMultiplicity[x y (x-y), l, {0, 0}]}
```

```
Out[165]= {IntersectionMultiplicity[-x^2 - x^3 + y^2, 0.278835 x - 0.609392 y, {0, 0}],
  IntersectionMultiplicity[-x^3 + y^2, 0.278835 x - 0.609392 y, {0, 0}],
  IntersectionMultiplicity[(-x^2 + y) (x^4 + y), 0.278835 x - 0.609392 y, {0, 0}],
  IntersectionMultiplicity[x (x - y) y, 0.278835 x - 0.609392 y, {0, 0}]}
```

In particular, nodes and cusps are double points.

Caley's theorem is that if an irreducible plane curve of degree d has $\frac{(d-1)(d-2)}{2}$ double points, then it can be parameterized by rational parameters.

We know that an irreducible plane curve cannot have more than $\frac{(d-1)(d-2)}{2}$ singularities. A curve with this many singularities has genus 0.

Let $f = 0$ be our irreducible curve of degree d with $\frac{(d-1)(d-2)}{2}$ double points. We start our proof/algorithm by forming a set S consisting of the singularities and $2d - 3$ additional random points on f. These should not be too bunched up.

Our first serendipity is that the number of points in S is as follows:

```
In[166]:= Expand[(d-1) (d-2)/2+2 d-3]
Out[166]= 1.71194 - 2.35645 x + 0.361803 x^2 + 1.45637 y - 0.447214 x y + 0.138197 y^2
```

This can be written as $\frac{(d+1)d}{2} - 2$, which is exactly one less than the number of points needed (see the end of Chapter 2) for interpolating a unique bivariate polynomial of degree $d - 1$. We now choose (randomly but with some care) two points: q_1, q_2 not on f, and, in practice, not too near f. Then the sets $\mathcal{S} \cup \{q_1\}$ and $\mathcal{S} \cup \{q_2\}$ will produce two curves g_1, g_2 of degree $d-1$ by interpolation. The randomness should be enough to require $g_1 \neq g_2$. Nonetheless, both curves should contain all the points in $\mathcal{S}$, but because they contain q_1 or q_2 and are of smaller degree, they are also distinct from f.

For each real number, let t, $g_t = q_1 + t\, q_2$. This is, again, a curve of degree $d-1$ containing the points of $\mathcal{S}$.

Our second serendipity is that by applying Bézout, each g_t intersects f in $d(d-1)$ points counting intersection multiplicities.

But then each double point counts twice so $\mathcal{S}$ accounts for the resulting output's points:

```
In[167]:= Expand[2 ((d-1) (d-2)/2)+2 d-3]
```

```
Out[167]= 12.9642-6.41421 x+0.723607 x^2+3.9642 y-0.894427 x y+0.276393 y^2
```

Note that this is exactly one less than the Bézout total. The intersection of f, g_t contains exactly one point not in $\mathcal{S}$, so we can name this point $\Phi[t]$, but since t is a real variable, a small change in t gives a small perturbation in $g_t = q_1 + t\, q_2$, so the solution set will only have a small change, but $\mathcal{S}$ does not change, so instead, $\Phi[t]$ has a small change. Thus, $\Phi[t]$ will be a continuous function giving a continuous parameterization of at least part of the curve $f = 0$.

Before proceeding, let's look at a famous example, the hypocycloid that has degree 4:

```
In[168]:= f = Expand[3 (x^2+y^2)^2+8 x (-x^2+3 y^2)+6 (x^2+ y^2)-1]
```

```
Out[168]= -1+6 x^2-8 x^3+3 x^4+6 y^2+24 x y^2+6 x^2 y^2+3 y^4
```

```
In[169]:= cp = criticalPoints[f, x, y]
```

```
Out[169]= {{-0.5, -0.866025}, {1., 0.}, {0.166667, 0.288675},
            {-0.333333, 0.}, {-0.5, 0.866025}, {0.166667, -0.288675}}
```

Checking each critical point, we see there are three double points of this 4th degree irreducible curve:

```
In[170]:= S = cp[[{1, 2, 5}]]
```

```
Out[170]= {{-0.5, -0.866025}, {1., 0.}, {-0.5, 0.866025}}
```

An example for the first:

```
In[171]:= rl = RandomReal[{-1, 1}, 2].{x-S[[1, 1]], y-S[[1, 2]]};
          intersectionMultiplicity[f, rl, S[[1]], 1.*^-10]
```

```
Out[172]= 2
```

This shows the first singular point is a double point. The reader with Mathematica can check points 2 and 3. Our hypotheses are satisfied.

We need 2 d − 3 = 5 more points. We can take the other three critical points and perhaps two points near {1, 0}:

```
In[173]:= S = Join[S, cp[[{3, 4, 6}]], {x, y} /. NSolve[{x - .5, f}, {x, y}, Reals]]

Out[173]= {{-0.5, -0.866025}, {1., 0.}, {-0.5, 0.866025}, {0.166667, 0.288675},
  {-0.333333, 0.}, {0.166667, -0.288675}, {0.5, -0.126437}, {0.5, 0.126437}}
```

We also need two points near the curve. We should use random points, but for expository purposes, we will use the following, which were generated randomly:

```
In[174]:= S1 = {{1.1354546845775721`, -0.3327801966326215`},
     {-0.27241452246256326`, 1.1514296886703472`}};
```

We now construct our two cubic curves:

```
In[175]:= g1 = aCurve[Append[S, S1[[1]]], x, y]
  g2 = aCurve[Append[S, S1[[2]]], x, y]
```

Out[175]= $0.0372925 + 0.0167626\,x - 0.227522\,x^2 + 0.173467\,x^3 - 0.218895\,y + 0.875579\,x\,y - 0.951857\,x^2\,y - 0.294573\,y^2 - 0.721553\,x\,y^2 + 1.19287\,y^3$

Out[176]= $0.116539 + 0.052383\,x - 0.711006\,x^2 + 0.542084\,x^3 - 0.0663244\,y + 0.265297\,x\,y - 0.288409\,x^2\,y - 0.920538\,y^2 - 2.25485\,x\,y^2 + 0.361434\,y^3$

```
In[177]:= {ContourPlot[{f == 0, g1 == 0}, {x, -1.2, 1.2}, {y, -1.2, 1.2}],
  ContourPlot[{f == 0, g2 == 0}, {x, -1.2, 1.2}, {y, -1.2, 1.2}]}
```

Out[177]=
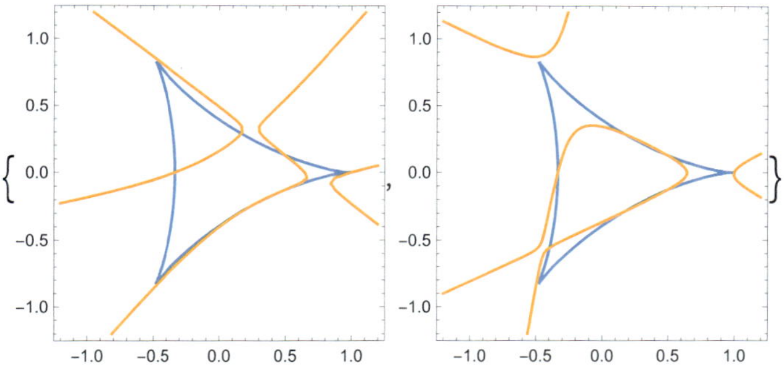

Combining them, we get g[t] and an example:

```
In[178]:= g := Expand[g1 + # g2] &;

In[179]:= g[.5]
```

Out[179]= $0.0955619 + 0.0429541\,x - 0.583025\,x^2 + 0.444509\,x^3 - 0.252057\,y + 1.00823\,x\,y - 1.09606\,x^2\,y - 0.754842\,y^2 - 1.84898\,x\,y^2 + 1.37358\,y^3$

We check the claim that g[.5] will intersect f in exactly one more point than S:

```
In[180]:= S
          sol = DeleteDuplicates[{x, y} /. NSolve[{f, g[.5]}], Norm[#1 - #2] < 1.*^-9 &]

Out[180]= {{-0.5, -0.866025}, {1., 0.}, {-0.5, 0.866025}, {0.166667, 0.288675},
           {-0.333333, 0.}, {0.166667, -0.288675}, {0.5, -0.126437}, {0.5, 0.126437}}

Out[181]= {{-0.5, -0.866025}, {-0.280452, -0.615736}, {1., 0.}, {-0.5, 0.866025}, {0.5, 0.126437},
           {0.166667, 0.288675}, {-0.333333, 0.}, {0.5, -0.126437}, {0.166667, -0.288675}}

In[182]:= Show[ContourPlot[{f == 0, g[.5] == 0}, {x, -1.2, 1.2}, {y, -1.2, 1.2}],
            ListPlot[{sol[[-1]]}, PlotMarkers -> Automatic]]
```

Out[182]=

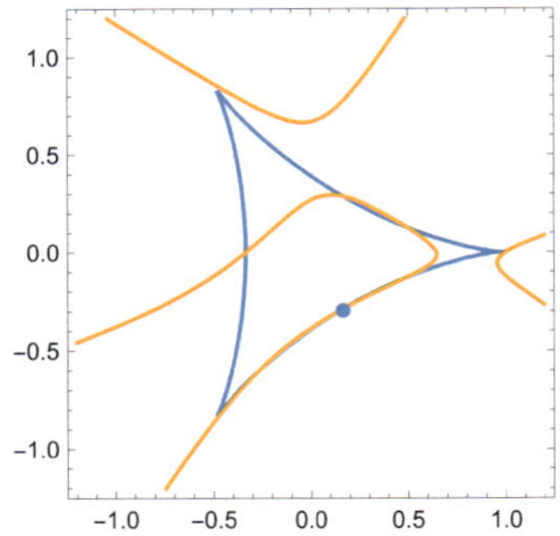

We can parameterize the hypocycloid by doing the following (note that because we must use numerical points, the selection process is more complicated):

```
In[183]:= φ[t_, S_] := Module[{sol},
              sol = {x, y} /. NSolve[{g[t], f}];
              Select[sol, Total[Boole[Table[Norm[p - #] < 1.*^-10, {p, S}]]] == 0 &][[1]]];

In[184]:= L = Table[φ[t, S], {t, 0, 2, .1}];

In[185]:= Show[ContourPlot[f == 0, {x, -2, 2}, {y, -2, 2}, ContourStyle → Cyan],
            ListLinePlot[L, PlotStyle → Black]]
```

Out[185]=

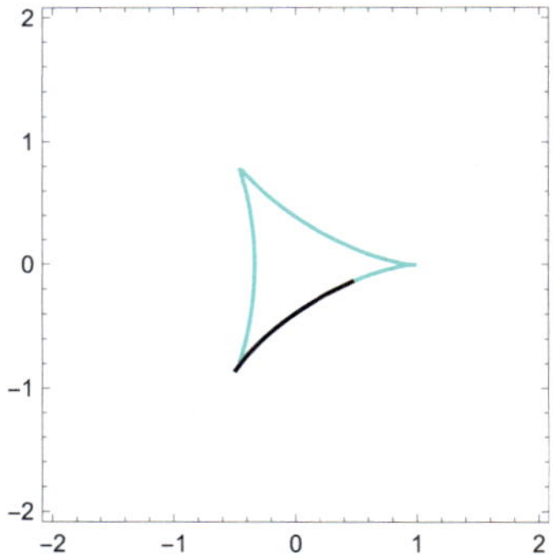

This example shows we have a parameterization of the hypocycloid, but we need to show that ϕ is a rational function of t. To do so, we must introduce the classical resultant.

Prior to the twenty-first century, solving nonlinear polynomial systems was, in general, an ad-hoc process, but there was an established method for 2×2 systems using the resultant. The reader who is unfamiliar may look this up in many references in the theory of equations, algebra or algebraic geometry. The resultant is implemented in Mathematica and, in short, if f and g are bivariate polynomials in x and y, then Resultant[f, g, y] is a polynomial in the one variable x whose zeros are all the x-coordinates of solutions of the system. By substituting these back in f and g and solving for y, one can get all the solutions to the system. Typically, the resultant is the product of degree(f) and degree(g), which almost proves Bézout's theorem except for the problem of infinite solutions.

The key to Caley's method is the well-known fact in elementary algebra: for a monic polynomial p in the one variable x of degree d, the coefficient of x^{d-1} is the sum of the zeros. More generally, the sum of the zeros is –Coefficient[p, x^(d – 1)]/Coefficient[p, x^d] after dividing by the leading coefficient. However, as we have seen above, all but one zero of the intersection of g_t and f are known, so all but one zero of the resultants Resultant[f, g_t x] and Resultant[f, g_t, y] are known. Finding the last one just requires an addition and a subtraction. As we see in the example, however, the coefficients are functions of t, so we get a rational expression in t. This is our rational parameterization!

Continuing with our example, we first note that our sum of known points in S gives:

```
In[186]:= TSx = -Total[Flatten[Take[S, All, 1]]]
          TSy = -Total[Flatten[Take[S, All, -1]]]

Out[186]= -1.

Out[187]= 8.32667 × 10^-15
```

We calculate our resultants, suppressing the very large display. Note that we need a little programming trick:

```
In[188]:= Rx = Resultant[f, SetPrecision[g[t], 20], y];
          Ry = Resultant[f, SetPrecision[g[t], 20], x];
```

The display version of our parametric equations are:

```
In[190]:= ϕx = TSx - Coefficient[Rx, x^11]/Coefficient[Rx, x^12];
          ϕy = TSy - Coefficient[Ry, y^11]/Coefficient[Ry, y^12];
          dϕx = N[Together[ϕx], 6]
          dϕy = N[Together[ϕy], 6]
```

Out[192]= $$-\frac{0.394628\left(-0.506011+1.02712\,t+3.78094\,t^2+2.66877\,t^3+1.\,t^4\right)}{0.42908+1.23842\,t+2.20367\,t^2+1.89059\,t^3+1.\,t^4}$$

Out[193]= $$-\frac{0.557736\left(0.108147+1.04665\,t+3.47556\,t^2+4.26037\,t^3+1.\,t^4\right)}{0.42908+1.23842\,t+2.20367\,t^2+1.89059\,t^3+1.\,t^4}$$

The limiting point in both directions as $t \to \pm \infty$ is:

```
In[194]:= lx = {dϕx[[1]], dϕy[[1]]}

Out[194]= {-0.394628, -0.557736}
```

```
In[195]:= Show[ContourPlot[f == 0, {x, -2, 2}, {y, -2, 2}, ContourStyle → Cyan],
  ParametricPlot[{ϕx, ϕy}, {t, -10, 10}, PlotStyle → Black],
  ListPlot[{lx}, PlotMarkers -> Automatic]]
```

Out[195]=

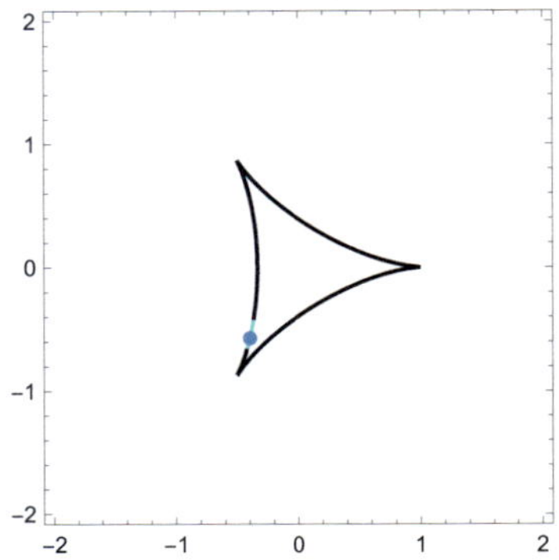

More to Explore

We already showed the Wolfram|Alpha page for elliptic curves in Chapter 1. We saw that by allowing real points and coefficients, the theory of elliptic curves is fairly simple. However, if we insist that coefficients are integers and points have rational coordinates, then this is one of the hardest problems in mathematics.

Take a look at the Wolfram|Alpha page for conic sections (wolfr.am/wPJz0Q2b)

To get some flavor of this that is not written at a too-advanced level, see this PDF (wolfr.am/tgwKoH2P)

For more information on the trident of Netwon, try Wolfram|Alpha (wolfr.am/tgwSor7D)

For more information on the Hessian matrix, check out Wolfram MathWorld (wolfr.am/tgwVXXJw)

For some more information on conic sections, look at this Wolfram|Alpha entry (wolfr.am/tgx2INDs)

To play with some conic sections and their polar equations, try this Demonstration by A. Brown (wolfr.am/tgx7yAVy)

Have fun experimenting with I. Hafner and M. Razpet's Zindler curves Demonstration (wolfr.am/vuzcLF2x)

8 | Global Plots

In Chapter 4, we discussed plotting in circular disks around the origin. In this chapter, we discuss plotting the entire curve, including infinite points. We will discuss several methods illustrating curves globally.

8.1 Plotting in Elliptical Regions

We will simply use an FLT to change from a circular disk to an elliptical region. If we want a new affine plotting region, we will use the function from Chapter 6 pqTranslation, possibly followed by a Homothety or klRotation in the affine case. We will illustrate this by examples.

Here is an interesting curve:

```
In[1]:= f = newtonHyperbola[1 050 752, x, y]
```

```
Out[1]= 1 + x + 0.2 x^2 + 0.008 x^3 + 0.000064 x^4 + 1.024 × 10^-7 x^5 + y + x y - 0.2 x^2 y -
        0.008 x^3 y + 0.000064 x^4 y + 0.2 y^2 + 0.2 x y^2 + 0.2 x^2 y^2 + 0.008 x^3 y^2 + 0.008 y^3 +
        0.008 x y^3 + 0.008 x^2 y^3 + 0.000064 y^4 + 0.000064 x y^4 - 1.024 × 10^-7 y^5
```

Since we don't expect singularities, we can find its critical points by:

```
In[2]:= cp = criticalPoints[f, x, y]
```

```
Out[2]= {{3865.95, -14.9506}, {-1.46007, 259.696}, {-0.0584974, -77.3069}, {-28.7503, 2.16897},
        {-25.0575, 0.201784}, {-12.8489, -12.7587}, {-0.254255, -26.1364}, {0.230938, -25.9504},
        {0.0000290324, -4.90484}, {-1.15024, 0.306826}, {2.22015 × 10^-7, -1.34015}}
```

We will worry about the larger ones in the top rows later, but most of the action lies in the disk of radius 35:

```
In[3]:= gaussPlot[f, 35, cp]
```

Out[3]=

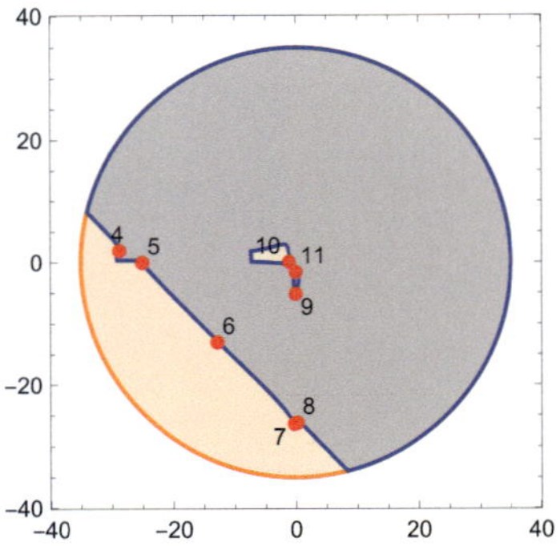

We need to inspect further the region around 4, 5; around 7, 8; and also around 9, 10, 11. To this end, we first modify gaussPlot to accept an association:

```
In[4]:= gaussPlotA[f_, r_, as_] :=
          Module[{plta, pltb, pltc}, plta := RegionPlot[x^2+y^2 < r^2, {x, −1.1 r, 1.1 r},
               {y, −1.1 r, 1.1 r}, PlotStyle → LightOrange, BoundaryStyle → Orange];
            pltb := RegionPlot[{f > 0&&x^2+y^2 < r^2}, {x, −r, r},
               {y, −r, r}, BoundaryStyle → Blue];
            pltc := Graphics[{{PointSize[Large], Yellow, Point[Values[as]]},
                 Table[Text[k, as[k]], {k, Keys[as]}]}];
            Show[plta, pltb, pltc]];
```

Then:

```
In[5]:= A1 = pqTranslation[{0, 0}, {−26, 0}];
        f1 = FLT[f, Inverse[A1], x, y];
        gp1 = gPoints[f1, 5, x, y];
        gp1A = <| Table[20+i → gp1[[i]], {i, 9}]|>;
        A2 = pqTranslation[{0, 0}, {0, −26}];
        f2 = FLT[f, Inverse[A2], x, y];
        gp2 = gPoints[f2, 2, x, y];
        gp2A = <| Table[30+i → gp2[[i]], {i, 3}]|>;
        gp3 = gPoints[f, 7, x, y];
        gp3A = <| Table[40+i → gp3[[i]], {i, 5}]|>;
        {gaussPlotA[f1, 5, gp1A], gaussPlotA[f2, 2, gp2A], gaussPlotA[f, 7, gp3A]}
```

Out[15]=

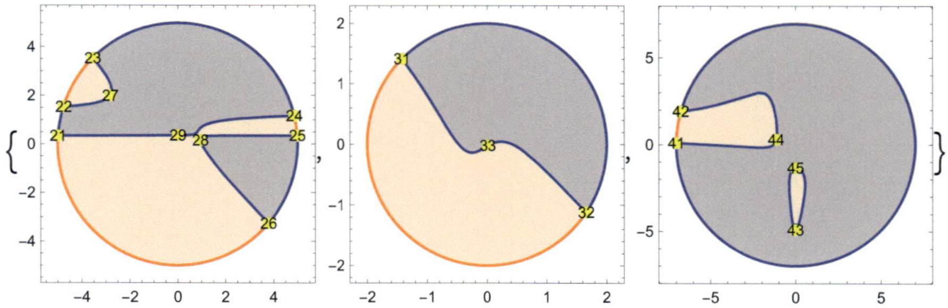

We can now use our path-finding algorithms from Chapter 4 to trace the various paths between points in the previous plots:

```
In[16]:= L1 = plotPath[f1, {1 → 2, 2 → 3}, {gp1A[22], gp1A[27], gp1A[23]}, x, y, .4]
         K1 = flt[#, A1] &/@L1

Out[16]= {{-4.75784, 1.53719}, {-4.35893, 1.56578}, {-3.96106, 1.6053},
          {-3.56589, 1.66355}, {-3.1805, 1.75939}, {-2.81361, 1.99067}, {-2.81361, 1.99067},
          {-2.75679, 2.28787}, {-2.89469, 2.64301}, {-3.10904, 2.97814}, {-3.54097, 3.53009}}

Out[17]= {{-30.7578, 1.53719}, {-30.3589, 1.56578}, {-29.9611, 1.6053},
          {-29.5659, 1.66355}, {-29.1805, 1.75939}, {-28.8136, 1.99067}, {-28.8136, 1.99067},
          {-28.7568, 2.28787}, {-28.8947, 2.64301}, {-29.109, 2.97814}, {-29.541, 3.53009}}
```

Likewise:

```
In[18]:= L2 = plotPath[f1, {1 → 2, 2 → 3}, {gp1A[24], gp1A[29], gp1A[21]}, x, y, .4];
         K2 = flt[#, A1] &/@L2;
         L3a = pathFinder[f1, gp1A[26], gp1A[28], .4, x, y];
         L3b = pathFinder[f1, gp1A[28], gp1A[25], .2, x, y];
         L3 = Join[L3a, L3b];
         K3 = flt[#, A1] &/@L3;
         L4 = plotPath[f2, {1 → 2, 2 → 3}, {gp2A[32], gp2A[33], gp2A[31]}, x, y, .3];
         K4 = flt[#, A2] &/@L4;
         L5 = plotPath[f, {1 → 2, 2 → 3}, {gp3A[41], gp3A[44], gp3A[42]}, x, y, .6];
         L6 = plotPath[f, {1 → 2, 2 → 1}, {gp3A[43], gp3A[45]}, x, y, .3];
```

Now let:

```
In[28]:= cpA = <| Table[i → cp[[i]], {i, 11}] |>;
         gp1B = Map[flt[#, A1] &, gp1A];
         gp2B = Map[flt[#, A2] &, gp2A];
```

So all points of interest so far are given by the following:

```
In[31]:= allpoints = Join[cpA, gp1B, gp2B, gp3A]

Out[31]= <| 1 → {3865.95, -14.9506}, 2 → {-1.46007, 259.696}, 3 → {-0.0584974, -77.3069},
          4 → {-28.7503, 2.16897}, 5 → {-25.0575, 0.201784}, 6 → {-12.8489, -12.7587},
          7 → {-0.254255, -26.1364}, 8 → {0.230938, -25.9504}, 9 → {0.0000290324, -4.90484},
          10 → {-1.15024, 0.306826}, 11 → {2.22015×10^-7, -1.34015},
          21 → {-30.9882, 0.342605}, 22 → {-30.7578, 1.53719}, 23 → {-29.541, 3.53009},
          24 → {-21.1373, 1.16382}, 25 → {-21.012, 0.345628}, 26 → {-22.2096, -3.26078},
          27 → {-28.8136, 1.99067}, 28 → {-25.0537, 0.159529}, 29 → {-26.004, 0.361991},
          31 → {-1.42134, -24.593}, 32 → {1.6584, -27.1179}, 33 → {0.0112772, -26.0198},
          41 → {-6.99931, 0.0979789}, 42 → {-6.73099, 1.92193}, 43 → {0.0000290324, -4.90484},
          44 → {-1.15024, 0.306826}, 45 → {2.22015×10^-7, -1.34015} |>
```

For the left and center plot, we applied `flt[#, A]&/@L` to the list `L` obtained where `A` is `A1` or `A2`.

We then have the following plots from the gaussPlot picking index set:

```
In[32]:= S1 = {23, 22, 21, 24, 26, 25, 31, 32, 41, 42, 43, 45};
```

```
In[33]:= Show[ContourPlot[f == 0, {x, -32, 10}, {y, -30, 10}, ContourStyle → Gray],
           Graphics[{{Blue, Thick, Line[K1]}, {Blue, Thick, Line[K2]}, {Blue, Thick, Line[K3]},
              {Red, Thick, Line[K4]}, {Green, Thick, Line[L5]}, {Green, Thick, Line[L6]},
              {Table[{PointSize[Large], Yellow, Point[allpoints[j]]}, {j, S1}]},
              {Table[Text[j, allpoints[j]], {j, S1}]}}]]
```

Out[33]=

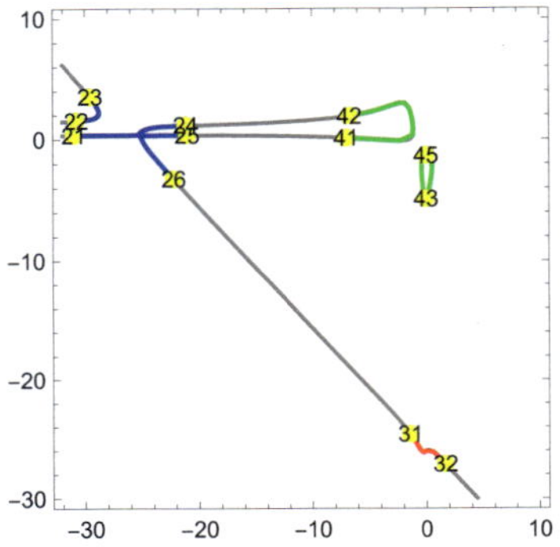

We note that the part of the blue curve starting at point 21 and continuing through 42, 41, 25, 26 to 32 is one continuous curve, so we can finish our decomposition in this region by tracing the relatively straight paths from 24 to 42, 41 to 25 and 26 to 31.

Here we use pathFinder and allpoints:

```
In[34]:= L7 = pathFinder[f, allpoints[42], allpoints[24], .5, x, y];
         L8 = pathFinder[f, allpoints[25], allpoints[41], .5, x, y];
         L9 = pathFinder[f, allpoints[31], allpoints[26], 1.0, x, y];
```

```
In[37]:= Show[ContourPlot[f == 0, {x, -32, 10}, {y, -30, 10}, ContourStyle → Gray],
           Graphics[{{Blue, Thick, Line[K1]}, {Red, Thick, Line[K2]}, {Red, Thick, Line[K3]},
              {Red, Thick, Line[K4]}, {Red, Thick, Line[L5]}, {Green, Thick, Line[L6]},
              {Red, Thick, Line[L7]}, {Red, Thick, Line[L8]}, {Red, Thick, Line[L9]},
              {Table[{PointSize[Large], Yellow, Point[allpoints[j]]}, {j, S1}]},
              {Table[Text[j, allpoints[j]], {j, S1}]}}]]
```

Out[37]=

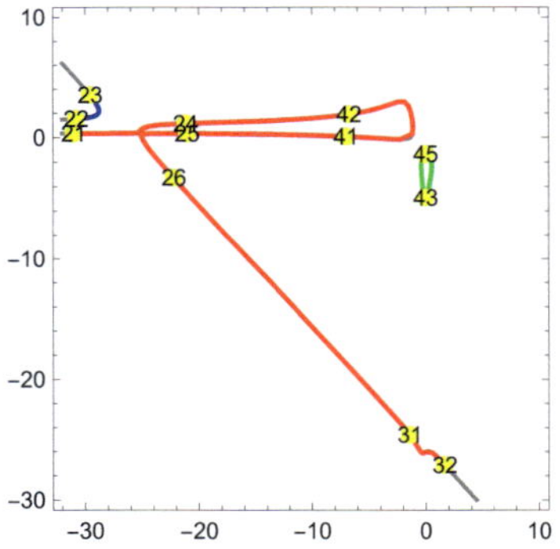

Finally, we note that the green loop in the previous plots is another topological component. We will see in the next section that in the projective plane, the blue and red components above are actually in the same topological component.

Adding the infinite points to our list at positions 12 to 16 of interesting points of the running example Newton Hyperbola 1050752 of the first part of this chapter, we get:

```
In[38]:= KeySelect[allpoints, 1 ≤ # < 20 &]

Out[38]= <|1 → {3865.95, −14.9506}, 2 → {−1.46007, 259.696}, 3 → {−0.0584974, −77.3069},
  4 → {−28.7503, 2.16897}, 5 → {−25.0575, 0.201784}, 6 → {−12.8489, −12.7587},
  7 → {−0.254255, −26.1364}, 8 → {0.230938, −25.9504}, 9 → {0.0000290324, −4.90484},
  10 → {−1.15024, 0.306826}, 11 → {2.22015 × 10^-7, −1.34015}|>
```

8.2 Handling Infinite Points

To continue with our main example, we must deal with the part of the curve that is far away from the origin {0, 0}. This is best done by changing the origin to infinite points using the FLT transformation ip2z. The best approach is to start with a circle of a small radius around the origin and use the inverse transformation to get a hyperbolic region, which will transform back to the circle.

The first step is to find the infinite points, recalling that each infinite point returned is generally approached at both ends of each asymptote:

```
In[39]:= ipts = InfiniteRealPoints[f, x, y]

Out[39]= {{−0.999998, 0.00220813, 0.}, {−0.999983, 0.00584369, 0.},
  {0.00136632, 0.999999, 0.}, {−0.00944406, 0.999955, 0.}, {−0.707098, 0.707116, 0.}}
```

We will start with the infinite point {−.707098, .707116, 0}. We will focus on the other pairs of seemingly close infinite points later:

```
In[40]:= circ = x^2 + y^2 − .25;
  A3 = ip2z[ipts[[5]], x, y]
  icirc3 = FLT[circ, Inverse[A3], x, y]
  pts1 = {x, y} /. NSolve[{f, icirc3}, {x, y}, Reals]
  g3 = FLT[f, A3, x, y]
  fpts1 = flt[#, A3] &/@pts1

Out[45]= {{−1.00001, −0.999987, 0.}, {0., 0., 1.41421}, {−0.499993, 0.500007, −0.707107}}

Out[46]= 1.875 − 0.176774 x + 0.937528 x^2 + 0.176779 y + 2.125 x y + 0.937472 y^2

Out[47]= {{−64.9326, 39.2747}, {37.7813, −63.3901}}

Out[48]= −0.007808 x − 1.02904 × 10^-6 x^2 + 0.003968 x^3 + 5.41377 × 10^-8 x^4 − 0.000504 x^5 +
  0.141512 y − 0.00429488 x y − 0.070576 x^2 y + 0.00113772 x^3 y + 0.0088445 x^4 y +
  0.0830235 y^2 − 0.00673687 x y^2 − 0.0205753 x^2 y^2 − 0.00142281 x^3 y^2 − 0.299864 y^3 −
  0.00742899 x y^3 + 0.1311 x^2 y^3 − 0.291369 y^4 − 0.252079 x y^4 + 0.107591 y^5

Out[49]= {{0.499242, 0.0275158}, {−0.499239, −0.0275713}}
```

```
In[50]:= {Show[gaussPlot[icirc3, 100, pts1],
    ContourPlot[f == 0, {x, -100, 100}, {y, -100, 100}, ContourStyle → Brown]],
  Show[gaussPlot[circ, 1, fpts1], ContourPlot[{g3 == 0, y == 0}, {x, -1, 1},
     {y, -1, 1}, ContourStyle → {Brown, Directive[Orange, Dashed]}]]}
```

Out[50]=

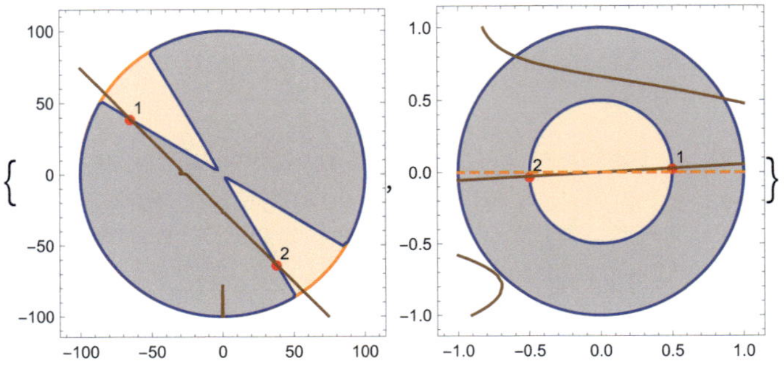

The left-hand picture shows the intersection of the curve $f = 0$ with the affine interior of icirc. The right-hand picture shows the transform of icirc to the unit disk, with the transform $g = 0$ of $f = 0$. A nice feature of the transformations ip2z is that the infinite line appears as the dashed x-axis. The component of the curve that goes through the infinite point ipts[[5]], which is the origin on the right, is the almost straight line in the right-hand plot.

Now we look at the first pair of infinite points. Since these are close together, we take a smaller circle of radius $r = .2$:

```
In[51]:= circ2 = x^2 + y^2 - .04;
  p = (ipts[[1]] + ipts[[2]]) / 2
  A4 = ip2z[p, x, y]
  icirc2 = FLT[circ2, Inverse[A4], x, y]
  pts2 = {x, y} /. NSolve[{f, icirc2}, {x, y}, Reals]
  g1 = FLT[f, A4, x, y]
  fpts2 = flt[#, A4] &/@ pts2
```

Out[57]= {-0.99999, 0.00402591, 0.}

Out[58]= {{-0.00569351, -1.4142, 0.}, {0., 0., 1.41421}, {-0.707101, 0.00284675, -0.707107}}

Out[59]= $1.98 - 0.0399997\,x - 0.0199673\,x^2 + 0.000161037\,y + 0.0162646\,x\,y + 1.99997\,y^2$

Out[60]= {{-28.9431, 2.72694}, {-23.476, -1.91751}, {-16.7574, 1.28589}, {-11.1648, 0.227686}}

Out[61]= $1.47732\times10^{-7} + 7.56527\times10^{-8}\,x - 0.011177\,x^2 - 0.00572367\,x^3 - 0.000045536\,x^4 - 5.04587\times10^{-8}\,x^5 + 0.000281644\,y - 0.0135879\,x\,y + 0.124506\,x^2\,y - 0.00232664\,x^3\,y - 5.18837\times10^{-6}\,x^4\,y - 0.0118938\,y^2 + 0.122101\,x\,y^2 + 0.0627849\,x^2\,y^2 - 0.00128835\,x^3\,y^2 + 0.120312\,y^3 + 0.485834\,x\,y^3 + 0.034936\,x^2\,y^3 - 0.222754\,y^4 + 0.0332064\,x\,y^4 + 0.0338173\,y^5$

Out[62]= {{-0.186765, 0.0715466}, {0.179099, 0.089015},
  {-0.154598, 0.126884}, {-0.0359519, 0.196742}}

In[63]:=
```
{Show[gaussPlot[icirc2, 50, pts2],
    ContourPlot[f == 0, {x, −50, 50}, {y, −50, 50}, ContourStyle → Brown]],
  Show[gaussPlot[circ2, .2, fpts2], ContourPlot[{g1 == 0, y == 0}, {x, −.2, .2},
    {y, −.2, .2}, ContourStyle → {Brown, Directive[Orange, Dashed]}]]}
```

Out[63]=

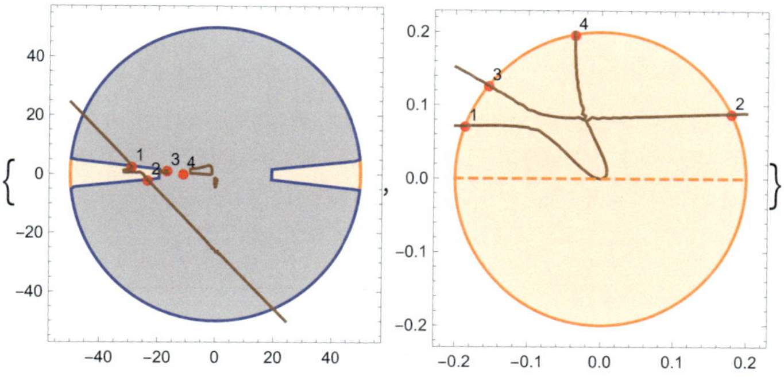

At first glance, the right-hand plot looks like the curve is actually tangential to the infinite line at p, and perhaps the two different infinite points of the pair is a numerical artifact. This is very possible with rational function curves. However, the Newton hyperbolas (while technically exact) behave more like numerical curves, so an actual tangency at the infinite line is very unlikely.

The seemingly singular point in this plot is the close encounter we saw before. We need to zoom in the picture farther:

In[64]:=
```
Show[ContourPlot[{g1 == 0, y == 0},
    {x, −.01, .01}, {y, −.01, .005}, ContourStyle → {Blue, Dashed}],
  ListPlot[{Labeled[c, "image cp[[1]]"]}, PlotStyle → Red, PlotMarkers → ■]]
```

Out[64]=

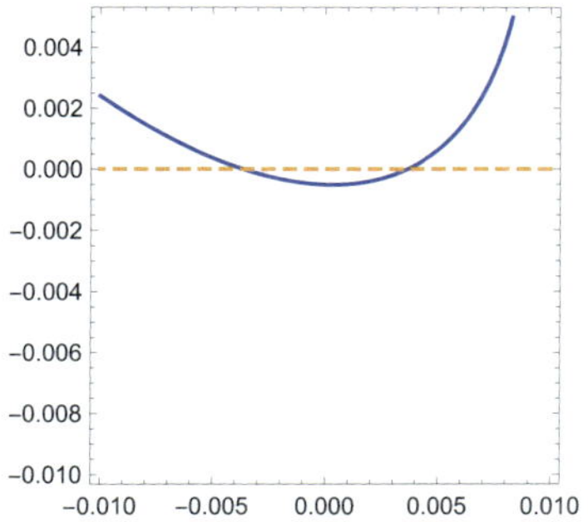

There are two close (but still distinct) infinite points here. The part of the curve above the dashed infinite line in this plot will appear in the part of the original plane where $x > 0$. In fact, this explains the first critical point $\{3865.95, -14.9506\}$, which will be the closest point to the origin on this affine curve component.

Putting together what we have so far, we see that we have one large loop that goes through three infinite points. Why this happens will be explained at the end of this chapter.

We next look at the final pair of critical points near {0, 1, 0} —ipts[[3]] and ipts[[4]]:

```
In[65]:= circ3 = x^2 + y^2 - .01;
         A5 = ip2z[{0, 1, 0}, x, y];
         icirc5 = FLT[circ3, Inverse[A5], x, y];
         g2 = FLT[f, A5, x, y]
         fpts3 = flt[#, A5] &/@{cp[[2]], cp[[3]]}
```

Out[70]= $5.79262\times10^{-7}+0.000181019\,x-0.0113137\,x^2+0.00565685\,x^3-0.0000226274\,x^4+1.81019\times10^{-8}\,x^5+0.000182467\,y-0.0109517\,x\,y+0.124451\,x^2\,y+0.00848528\,x^3\,y-0.0109502\,y^2+0.124722\,x\,y^2+0.203647\,x^2\,y^2+0.00424264\,x^3\,y^2+0.124723\,y^3-0.220527\,x\,y^3+0.104652\,x^2\,y^3-0.220527\,y^4+0.0339524\,x\,y^4+0.0339525\,y^5$

```
Out[71]= {{0.0112013, -0.00767177}, {-0.00153322, 0.02621}}
```

```
In[72]:= {Show[gaussPlot[icirc5, 300, {cp[[2]], cp[[3]]}],
            ContourPlot[f == 0, {x, -300, 300}, {y, -300, 300}, ContourStyle → Brown]],
          Show[gaussPlot[circ3, .05, Append[ fpts3, {0, 0}]], ContourPlot[{g2 == 0, y == 0},
            {x, -.05, .05}, {y, -.05, .05}, ContourStyle → {Brown, Directive[Orange, Dashed]}]]}
```

Out[72]=

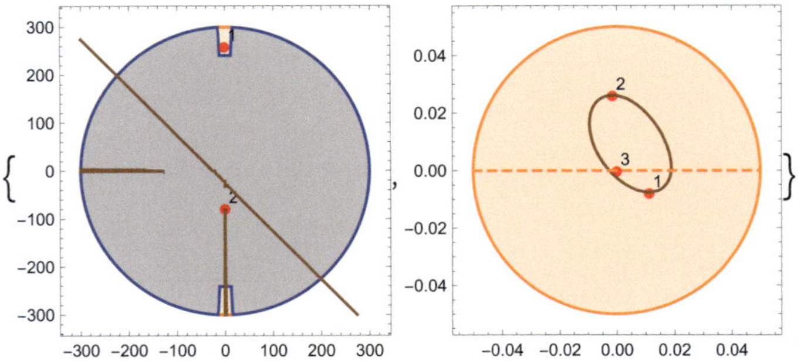

These plots show a third topological component, which goes through critical points 2 and 3, labeled as 1 and 2 on both plots and loops around the infinite point {0, 1, 0}, labeled as 3 in the right-hand plot.

We will put all this together in the next section.

8.3 The Hemisphere

It is well known that the plane is topologically equivalent to the inside of an open (without boundaries) disk of any radius $r > 0$ about the origin. Including the boundary, circle $x^2 + y^2 = r^2$ compactifies the circle. We can consider these new points to be the infinite points, which creates a model of the plane with infinite points (which works well with the Gauss decomposition of Chapter 4).

By picking the positive real constant r and using the transformation, we take the point p to the following:

```
In[73]:= ghsMap[p_, r_] := If[Length[p] == 2, N[6*r/Pi*ArcTan[Norm[p]/6/r*Pi]]*Normalize[p],
            3*r*N[Normalize[Take[p, 2]]]];
```

This transformation will distort the disk of radius *r* a relatively small amount but points of norm larger than *r* increasingly. Infinite points will be on the circle of radius 3 *r*, so the entire plane, including infinite points, will be transformed onto the closed disk of radius 3 *r*.

For example, if *r* = 10, then:

```
In[74]:= L = {ghsMap[{2, 3}, 10], ghsMap[{20, 30}, 10],
           ghsMap[{200, 345}, 10], ghsMap[{2, 3, 0}, 10], ghsMap[{-2, -3, 0}, 10]}
Out[74]= {{1.97674, 2.9651}, {11.4804, 17.2206},
          {14.5875, 25.1635}, {16.641, 24.9615}, {-16.641, -24.9615}}

In[75]:= Norm /@ L
Out[75]= {3.56361, 20.6966, 29.086, 30., 30.}
```

No matter how large the coordinates of a point are, the norm of the image and (in particular) the coordinates will be less than or equal to 30. As in Chapters 1 and 5, multiplying an infinite point (third coordinate 0) by a positive real number does not affect the result, but multiplying by a negative number multiplies the coordinates by – 1.

Circles centered around the origin remain circles around the origin, but other circles may be distorted:

```
In[76]:= CC = Table[ghsMap[r {Sin[t], Cos[t]}, 10], {r, {2, 10, 50, 250, 1250, 6250}}, {t, 0, 2 Pi, .1}];
         CO = Table[ghsMap[{20 Sin[t] + 10, 20 Cos[t] - 22}, 10], {t, 0, 2 Pi, .1}];
         {Show[ContourPlot[x^2 + y^2 == 910, {x, -31, 31}, {y, -31, 31}, ContourStyle → Red],
            ListLinePlot[CC, PlotStyle → Cyan]],
          Show[ContourPlot[x^2 + y^2 == 910, {x, -31, 31}, {y, -31, 31}, ContourStyle → Red],
            ListLinePlot[CO, PlotStyle → Cyan]]}
```

Out[78]=

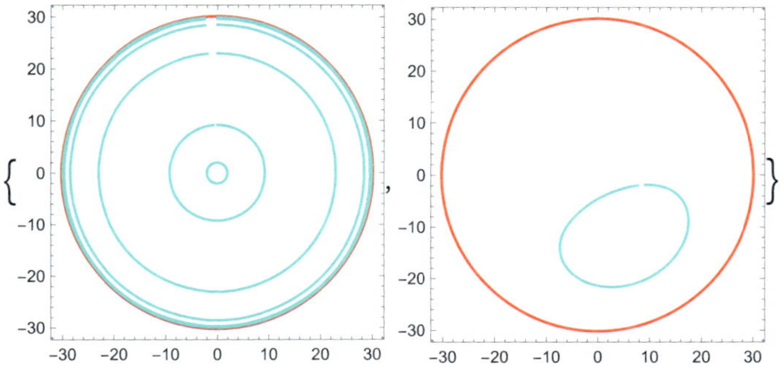

Lines through the origin remain unchanged. However, lines that don't go through the origin are arcs meeting the parallel line through the origin at the common infinite points by both ends of the line.

Perhaps this is why this model is called the hemisphere:

```
In[79]:= LC = Table[ghsMap[{x, 3 x + a}, 10],
            {a, {0, -20, 20, -50, 50, -100, 100, -250, 250}}, {x, -250, 250, .5}];
         Show[ContourPlot[{x^2 + y^2 == 900}, {x, -30, 30}, {y, -30, 30}, ContourStyle → Red],
          ListLinePlot[LC, PlotStyle → Cyan, PlotRange → All]]
```

Out[80]=

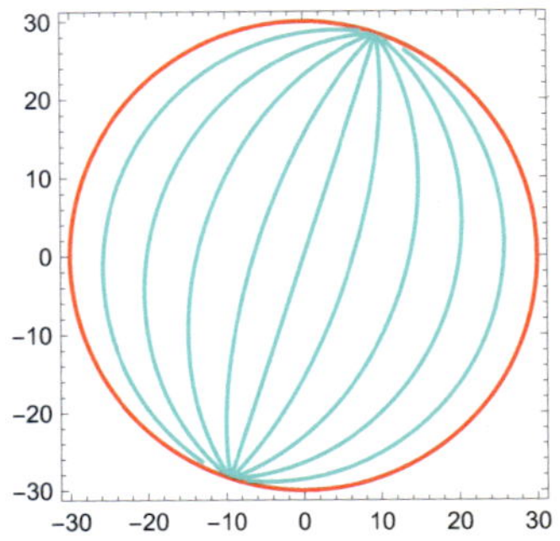

In order to plot an algebraic curve in the hemisphere, we need to break down the curve into piecewise linear parameterized curves (as we did in Chapter 4 and earlier in this chapter). Then we can combine ghsMap with the parameterizations to draw parameterized curves. It's suggested that you first find parameterizations on the closed disk of radius r (as in Chapter 4) then look at the regions around the infinite points and find the paths from the infinite points to the boundary points on the disk of radius r.

Each infinite point (in the sense of Chapter 3) gives two antipodal infinite points in the Gauss hemisphere. If there are no segments of the interior that connect to a given hemisphere infinite point, then that point is no longer considered an infinite point. An isolated infinite point would be a possible exception.

We will complete the main example of this section later. To begin, we will do some easier examples, starting with function curves. Recall the function gPoints from Chapter 4 or Appendix 2.

Our first example is the quadratic function $y = x^2 - 8$ with $r = 10$:

```
In[81]:= f = x^2 - 8 - y;
         gp = gPoints[f, 10, x, y]
         infiniteRealPoints[f, x, y]

Out[82]= {{-4.13578, 9.10469}, {4.13578, 9.10469}, {0., -8.}, {-2.73861, -0.5}, {2.73861, -0.5}}

Out[83]= {{0, 1.13564, 0}}
```

Using gPoints, it is easy to find one list giving a path from $gp[[1]]$ to $gp[[2]]$. The one infinite point is equivalent to $\{0, 1, 0\}$.

We can now calculate:

In[84]:= **A = ip2z[{0, 1, 0}, x, y]**

Out[84]= {{1.41421, 0., 0.}, {0., 0., 1.41421}, {0., −0.707107, −0.707107}}

In[85]:= **g = FLT[f, A, x, y]**

Out[85]= $0.5\,x^2 + 1.\,y - 3.5\,y^2$

In[86]:= **b1 = flt[gp[[1]], A]**

Out[86]= {0.818587, −0.197928}

In[87]:= **b2 = flt[gp[[2]], A]**

Out[87]= {−0.818587, −0.197928}

It helps to look at the picture:

In[88]:= **{gaussPlot[f, 10, gp],**
Show[gaussPlot[g, 1, {b1, b2}], ListLinePlot[{{−1, 0}, {1, 0}}, PlotStyle → Dashed]]}

Out[88]= 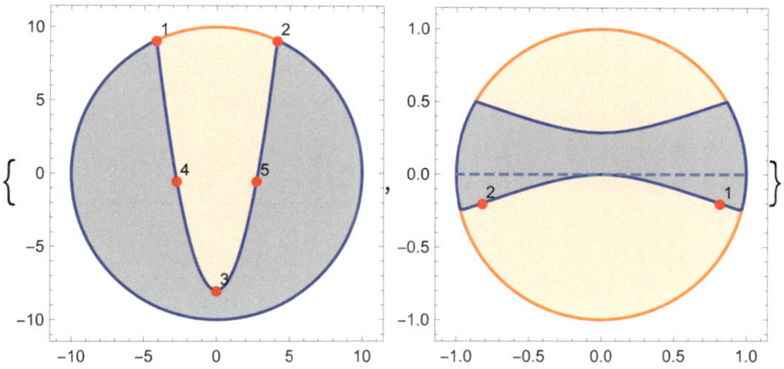

We see that the curve seems tangential to the (dashed) infinite line. Here we know that this does happen. We can trace this curve from b1, to {0, 0}, to b2. In this example, we use a step size of .05.

A small step size around an infinite point is important because after transformation, the distance between points becomes large:

In[89]:= **L1a = pathFinder[g, b1, {0, 0}, .05, x, y]**

Out[89]= {{0.818587, −0.197928}, {0.771267, −0.181777}, {0.72389, −0.165797}, {0.676445, −0.150019}, {0.628923, −0.134477}, {0.581309, −0.119216}, {0.53359, −0.104291}, {0.485747, −0.0897698}, {0.437758, −0.0757388}, {0.3896, −0.0623066}, {0.341244, −0.0496098}, {0.292662, −0.0378195}, {0.243827, −0.0271464}, {0.194717, −0.0178431}, {0.14533, −0.0101966}, {0.0956885, −0.00450705}, {0, 0}}

In[90]:= **L1b = pathFinder[g, {0, 0}, b2, .05, x, y]**

```
Out[90]= {{0, 0}, {-0.0499385, -0.00124153}, {-0.0997637, -0.00489261},
  {-0.149385, -0.0107533}, {-0.198749, -0.0185467},
  {-0.247835, -0.0279725}, {-0.296648, -0.0387458}, {-0.34521, -0.0506176},
  {-0.393548, -0.0633804}, {-0.441692, -0.0768663}, {-0.489667, -0.0909411},
  {-0.5375, -0.105498}, {-0.58521, -0.120454}, {-0.632815, -0.135739},
  {-0.680331, -0.151302}, {-0.72777, -0.167098}, {-0.818587, -0.197928}}
```

We transform it back to the affine plane:

In[91]:= **H1a = Reverse[flti[#, Inverse[A]] &/@L1a]**

```
Out[91]= {{0., -1.41421, 0}, {-21.2309, 442.749}, {-14.2529, 195.144},
  {-10.9128, 111.088}, {-8.9819, 72.6745}, {-7.7384, 51.8828}, {-6.87856, 39.3146},
  {-6.25295, 31.0993}, {-5.77984, 25.4065}, {-5.41103, 21.2792},
  {-5.11636, 18.1771}, {-4.87609, 15.7762}, {-4.67679, 13.8724}, {-4.50906, 12.3316},
  {-4.36611, 11.0629}, {-4.24294, 10.0025}, {-4.13578, 9.10469}}
```

In[92]:= **H1b = Reverse[flti[#, Inverse[A]] &/@L1b]**

```
Out[92]= {{4.13578, 9.10469}, {4.35534, 10.969}, {4.49651, 12.2186},
  {4.66199, 13.7341}, {4.85839, 15.6039}, {5.09486, 17.9576},
  {5.38445, 20.9923}, {5.74623, 25.0192}, {6.20931, 30.5555},
  {6.81997, 38.512}, {7.65627, 50.6185}, {8.85996, 70.4989}, {10.7162, 106.836},
  {13.8921, 184.99}, {20.3907, 407.779}, {40.2233, 1609.91}, {0., -1.41421, 0}}
```

We see that the infinite point has the wrong sign, which we fix:

```
In[93]:= H1a[[1]] = {0, 1, 0};
         H1b[[-1]] = {0, 1, 0};
```

We now get the finite part. The long output is suppressed, except for endpoints:

```
In[95]:= H2 = plotPath[f, {1 → 4, 4 → 3, 3 → 5, 5 → 2}, gp, x, y, .5];
         H2[[1]]

Out[96]= {-4.13578, 9.10469}
```

```
In[97]:= H2[[-1]]

Out[97]= {4.13578, 9.10469}
```

Altogether, we have the following (again with the output suppressed):

In[98]:= **H = Join[H1a, H2, H1b];**

Now we apply ghsMap (again with the output suppressed):

In[99]:= **GH = ghsMap[#, 10] &/@H;**

We do the same to the other infinite path from {0, 0} to b2. We also apply these maps to the finite path from b2 to b1, getting the (hidden) paths K and K6.

We finish with the picture in the Gauss hemisphere:

```
In[100]:= Show[ContourPlot[{x^2+y^2 == 900, x^2+y^2 == 100}, {x, −30, 30},
            {y, −30, 30}, ContourStyle → {Red, Pink}], ListLinePlot[GH, PlotStyle → Black]]
```

Out[100]=
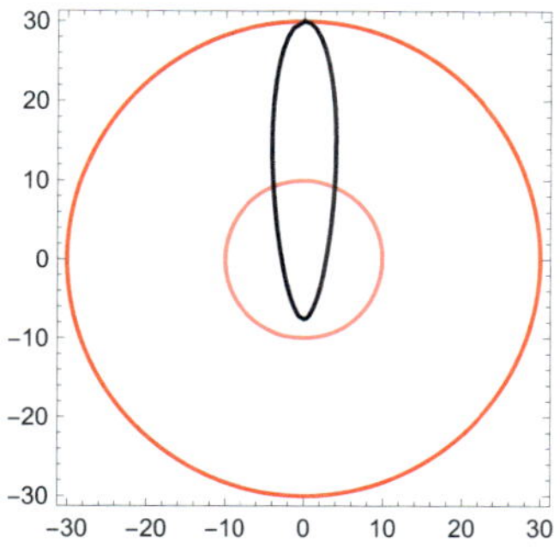

This picture was anticipated by Kepler, who was the first to think about infinite points on conics in 1604. From our perspective, he was wrong in characterizing a parabola as a conic with one infinite point in that this infinite point actually has multiplicity 2. In fact, Bézout's theorem says that every conic has two (possibly complex) infinite points counting multiplicity, but Kepler was correct in noticing that "being a parabola" is a positional concept rather than a geometrical one.

In the Gauss hemisphere plots, we do the inner disk, showing approximately what we would get with finite plotting (as explained in Chapter 4). The outside shows the rest of the plot out to infinity.

For our next few examples, we will again suppress many calculations. Consider the cubic function curve $y = .2\,x^3 - 5\,x$. Again, the only infinite point is $\{0, 1, 0\}$.

If we plot around this point, we get the plot of g = FLT[f, A, x, y]:

```
In[101]:= f2 = .2 x^3−5 x−y;
          A = ip2z[{0, 1, 0}, x, y]
          gp2 = gPoints[f2, 10, x, y]
          g2 = FLT[f2, A, x, y]

Out[102]= {{1.41421, 0., 0.}, {0., 0., 1.41421}, {0., −0.707107, −0.707107}}

Out[103]= {{−3.16228, 9.48683}, {−2.78434, 9.60455}, {5.67868, 8.23119},
            {3.16228, −9.48683}, {2.78434, −9.60455}, {−5.67868, −8.23119},
            {4.94813, −0.510621}, {−4.94813, 0.510621}, {0., 0.}}
```

Out[104]= $0.0707107\,x^3 + 0.707107\,y^2 - 1.76777\,x\,y^2 + 0.353553\,y^3$

We plot this, with points 7, 8, 9, and 10 being in the only relative positions in the middle plot:

In[105]:=
```
Row[{gaussPlot[f2, 12, gp2], Show[gaussPlot[g2, 2, flti[#, A] &/@gp2],
    ListPlot[{Labeled[{0, 0}, 10], Labeled[{1.94, .45}, 7], Labeled[{1.94, -.45}, 8],
        Labeled[{-1.94, .45}, 8], Labeled[{-1.94, -.45}, 7], Labeled[{.77, 1.84}, 9]}]
      ImageSize → Small], Show[RegionPlot[g2 > 0, {x, -1, 1}, {y, -.3, .3}],
    ListLinePlot[{{-1, 0}, {1, 0}}, PlotStyle → Directive[Dashed, Orange]],
    ListPlot[{Labeled[{0, 0}, 10]}]]}]
```

Out[105]=

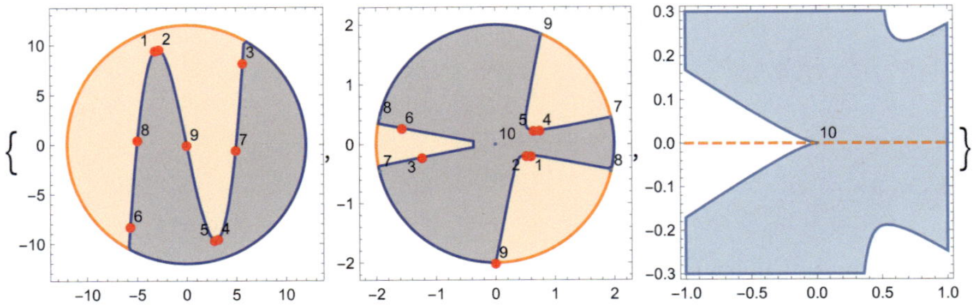

The right-hand plot above gives a detail in the middle that has a singularity at {0, 0}—the infinite point. A differentiable function curve has no affine singularities but (as noted in Chapter 5) has genus 0. Since nonsingular cubics have genus 1, g must (by Max Noether's formula) have a singularity. This is true for all function curves of degree 3 or more: there must be an infinite singularity.

Careful tracing (not shown) produces the following plot in the hemisphere:

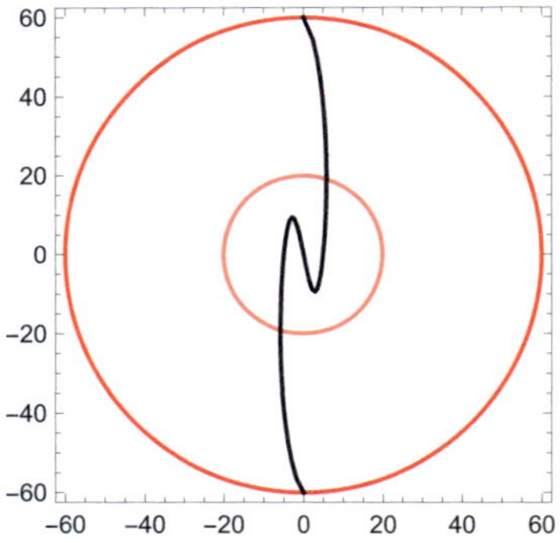

Consider this example of a rational function curve, $y = \frac{x^4 - 10}{x^3 - 2x^2}$:

In[106]:=
```
Plot[(x^4 - 10)/(x^3 - 2 x^2), {x, -15, 15}, Exclusions → {x == 2}]
```

Out[106]=

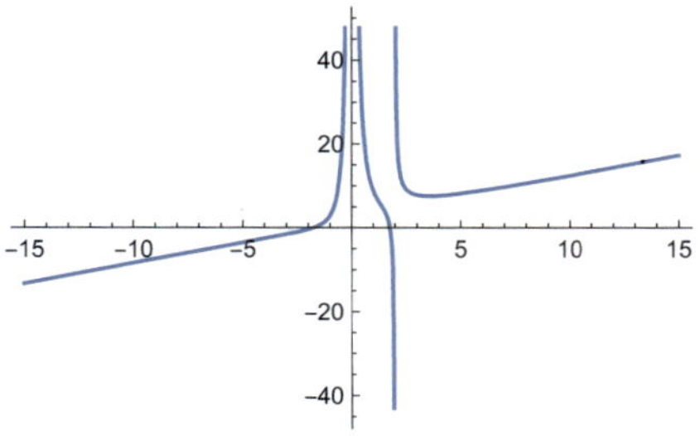

We plot $g = -x^4 + (x^3 - 2\,x^2)\,y + 10$ on the Gauss hemisphere. This shows that there are two (projective) infinite points: {0, 1, 0} and {1, 1, 0}. The latter is a regular point and, by now, handling it should be routine.

The other is a nonordinary singularity, which we look at:

In[107]:=
```
g = Expand[y (x^3 - 2 x^2) - x^4 + 10]
A = ip2z[{0, 1, 0}, x, y];
h = FLT[g, A, x, y]
bp = {x, y} /. NSolve[{h, x^2 + y^2 - .25}, {x, y}, Reals]
gaussPlot[h, .5, Append[bp, {0, 0}]]
```

Out[107]= $10 - x^4 - 2\,x^2\,y + x^3\,y$

Out[109]= $-0.5\,x^3 - 0.25\,x^4 + 1.\,x^2\,y - 0.25\,x^3\,y + 0.5\,x^2\,y^2 + 2.5\,y^4$

Out[110]= {{−0.400998, −0.298665}, {0.442834, 0.23216}, {−0.463049, −0.188641}, {0.328713, −0.37676}}

Out[111]=
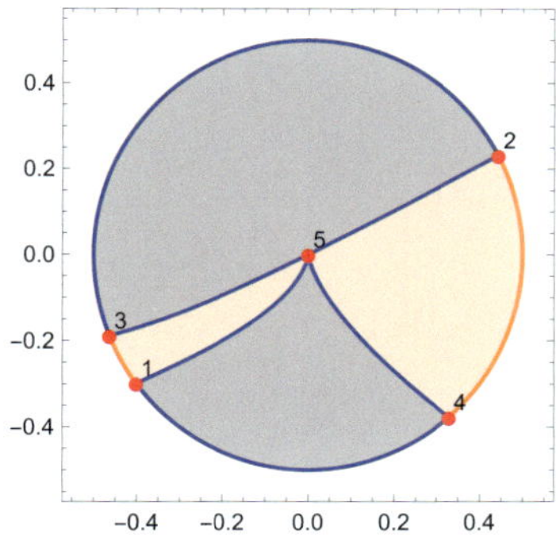

For each path into the origin (point 5), we first find the path then convert to affine coordinates and then to hemisphere coordinates:

In[112]:=
```
L1 = pathFinderT[h, bp[[1]], {0, 0}, .05, x, y]
```

Out[112]= {{−0.400998, −0.298665}, {−0.355576, −0.277763}, {−0.310304, −0.256538}, {−0.265359, −0.234626}, {−0.220978, −0.211586}, {−0.177478, −0.186912}, {−0.135298, −0.160026}, {−0.0950856, −0.130249}, {−0.0578805, −0.0967398}, {−0.0255688, −0.058384}, {0, 0}}

In[113]:=
```
J1 = flti[#, Inverse[A]] &/@L1
```

Out[113]= {{1.34264, 5.69647}, {1.28014, 6.20038}, {1.20958, 6.79611}, {1.13099, 7.52422}, {1.04439, 8.45244}, {0.949525, 9.70021}, {0.84547, 11.4979}, {0.730031, 14.3552}, {0.598311, 19.674}, {0.437942, 33.256}, {0., −1.41421, 0}}

In[114]:=
```
K1 = ghsMap[#, 15] &/@J1
```

Out[114]= {{1.32441, 5.61915}, {1.25989, 6.1023}, {1.18696, 6.66898}, {1.10547, 7.35443}, {1.01516, 8.21585}, {0.91524, 9.34997}, {0.803807, 10.9313}, {0.67663, 13.3052}, {0.524227, 17.2379}, {0.324317, 24.6277}, {0., −45.}}

```
In[115]:= L2 = pathFinderT[h, bp[[2]], {0, 0}, .05, x, y];
          J2 = flti[#, Inverse[A]] &/@L2;
          K2 = ghsMap[#, 15] &/@J2;
          L3 = pathFinderT[-h, bp[[3]], {0, 0}, .05, x, y];
          J3 = flti[#, Inverse[A]] &/@L3;
          K3 = ghsMap[#, 15] &/@J3;
          L4 = PathFinderT[-h, bp[[4]], {0, 0}, .05, x, y];
          J4 = flti[#, Inverse[A]] &/@L4;
          K4 = ghsMap[#, 15] &/@J4;
```

The infinite point at the end of each segment may have the wrong sign:

```
In[124]:= K1[[-1]] = {0, 45}; K2[[-1]] = {0, -45}; K3[[-1]] = {0, 45}; K4[[-1]] = {0, 45};
```

```
In[125]:= Show[ContourPlot[{x^2+y^2 == 2025, x^2+y^2 == 225}, {x, -45, 45},
            {y, -45, 45}, ContourStyle → {Red, Pink}], ListLinePlot[{K1, K2, K3, K4}]]
```

Out[125]=

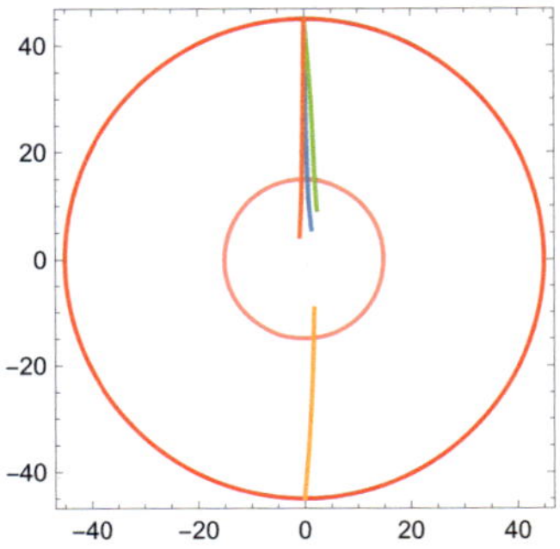

Now we can easily connect these segments inside the disk of radius 15 to get the final picture:

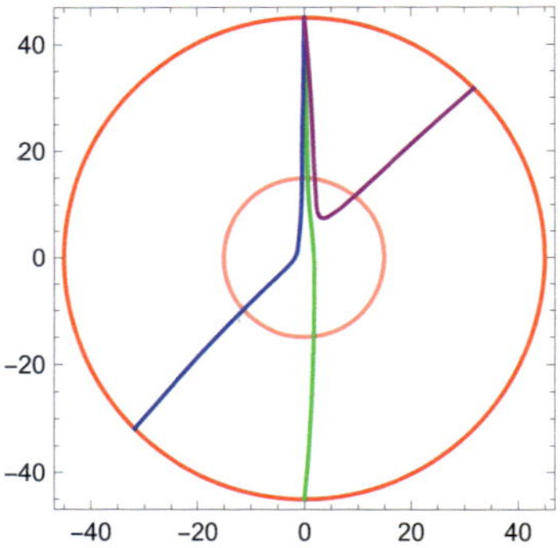

We now go back to our main example for this section.

In Sections 8.1 and 8.2, we looked around all the critical and infinite points and traced the curve inside the disk of radius 30. All that remains is to trace near the infinite points.

Then, as in the previous example, convert to affine coordinates then to hemisphere coordinates and put them all together. Our result is the following. In this plot, the affine topological components of the curve have different colors. However, the blue, orange, and tiny black components (the latter on the far right side) meet up at infinite points, as do the magenta and purple components.

There are actually three projective components, each a continuous loop. A detail plot shows that there is no singularity in the blue component.

We will look at this again in the next section:

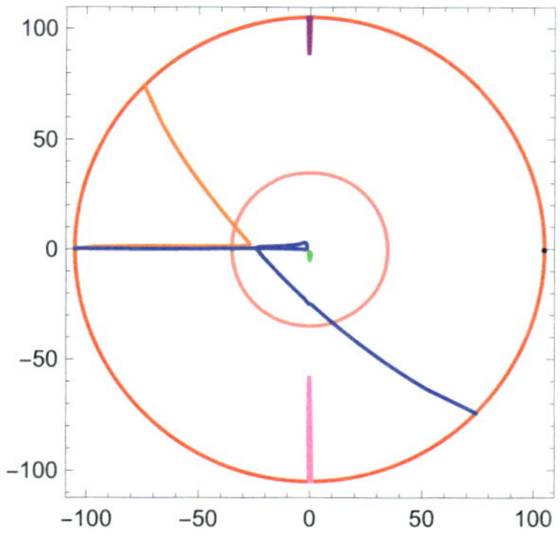

A nice feature of the hemisphere model is, like the plain affine model, it supports positive and negative regions. The best way to see this is to note that our ghsMap transform has an inverse.

This inverse will return infinite points as affine points with a very large norm.

Leaving the function in this form will ensure that the function works with symbolic q:

```
In[126]:= inverseGHSmap[q_, r_] := Tan[Norm[q]/6/r*Pi]*6*r/Pi*Normalize[q];
```

```
In[127]:= inverseGHSmap[{3, 4}, 10]
```

Out[127]= $\left\{\frac{36\left(2-\sqrt{3}\right)}{\pi}, \frac{48\left(2-\sqrt{3}\right)}{\pi}\right\}$

```
In[128]:= p = inverseGHSmap[ghsMap[{3, 4, 0}, 10], 10]
```

Out[128]= $\{1.87142\times10^{17}, 2.49523\times10^{17}\}$

```
In[129]:= 5*Append[Normalize[p], 0]
```

Out[129]= {3., 4., 0}

For low-degree curves, this can be used symbolically to directly produce the signed hemisphere representation of the curve from the equation. It is not, however, an algebraic curve anymore.

Consider the Gauss curve for $z^3 - 3z^2$ in the hemisphere with $r = 5$:

```
In[130]:= f = gaussCurve[z^3-3 z^2, x, y, z]
```

Out[130]= $-3x^2 + x^3 + 3y^2 - 3xy^2$

In[131]:=
```
h = f /. Thread[{x, y} → inverseGHSmap[{u, v}, 5]]
```

Out[131]=
$$-\frac{2700\,u^2\,\text{Tan}\left[\frac{1}{30}\pi\sqrt{\text{Abs}[u]^2+\text{Abs}[v]^2}\right]^2}{\pi^2\left(\text{Abs}[u]^2+\text{Abs}[v]^2\right)}+\frac{2700\,v^2\,\text{Tan}\left[\frac{1}{30}\pi\sqrt{\text{Abs}[u]^2+\text{Abs}[v]^2}\right]^2}{\pi^2\left(\text{Abs}[u]^2+\text{Abs}[v]^2\right)}+\frac{27000\,u^3\,\text{Tan}\left[\frac{1}{30}\pi\sqrt{\text{Abs}[u]^2+\text{Abs}[v]^2}\right]^3}{\pi^3\left(\text{Abs}[u]^2+\text{Abs}[v]^2\right)^{3/2}}-\frac{81000\,u\,v^2\,\text{Tan}\left[\frac{1}{30}\pi\sqrt{\text{Abs}[u]^2+\text{Abs}[v]^2}\right]^3}{\pi^3\left(\text{Abs}[u]^2+\text{Abs}[v]^2\right)^{3/2}}$$

In[132]:=
```
Show[RegionPlot[h > 0 && u^2+v^2 < 225, {u, -15, 15}, {v, -15, 15}],
  ContourPlot[{u^2+v^2 == 225, u^2+v^2 == 25},
   {u, -15, 15}, {v, -15, 15}, ContourStyle → {Red, Pink}]]
```

Out[132]=
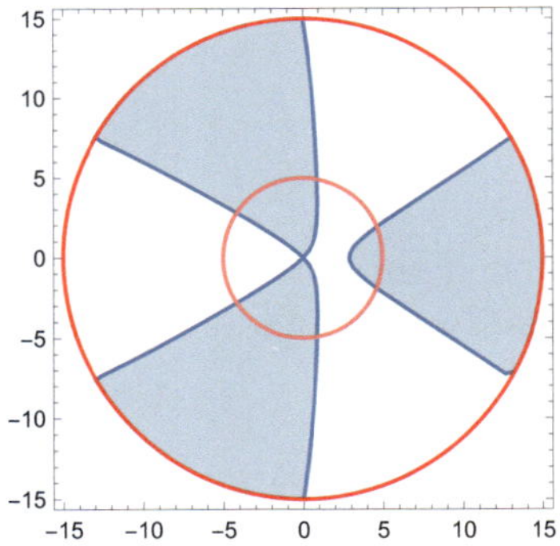

We note that the hemisphere plot is not greatly different for Gauss curves than a straight region plot on a small disk. Gauss probably realized this, but we cannot expect the same for complicated curves like our running example. One still can evaluate our transformed function `h = f /. Thread[{x, y} → inverseGHSmap[{u, v}, r]]` at a noninfinite point of the hemisphere off the curve to determine if that point is positive or negative. This gives us a new Gauss Principle:

Let $f(x, y) = 0$ be a square-free algebraic curve. The point set `ghsMap[f] (x, y) = 0` in the hemisphere is the full boundary of the set `ghsMap[f] (x, y) > 0`. Moreover, each connected component of this noninfinite boundary is the image of a continuously parameterized curve, which either begins and ends at the same point or begins and ends on the infinite circle of the hemisphere.

8.4 Projective Graphs

As mentioned in Section 5.4, each real projective plane curve comes with a (possibly not connected) Eulerian graph, that is, a graph drawn in such a way that each vertex has even degree. This is probably the most efficient way of describing the topology of a projective plane curve. In Chapter 5, it was easy to see how the Gauss components of Gauss curves connected to the infinite points. The previous two sections of this chapter show how to make these connections in general.

For the running example in this chapter, we can take the original critical points for our vertex set together with the infinite points, getting a list of 16 points:

{{3865.95, −14.9506}, {−1.46007, 259.698}, {−0.0584974, −77.3069}, {−28.7503, 2.16897},
{−25.0575, 0.201784}, {0.230938, −25.9504}, {−0.254255, −26.1364},
{−12.8489, −12.7587}, {0.0000290324, −4.90484}, {−1.15024, 0.306826},
{2.22011×10^{-7}, −1.34015}, {−9.97212, 0.058275, 0}, {−9.86118, 0.0217749, 0},
{4.80252, −4.80265, 0}, {0.0305306, −3.23264, 0}, {−0.00440114, −3.22115, 0}}

These points correspond to the numbered points of this hemisphere plot and detail:

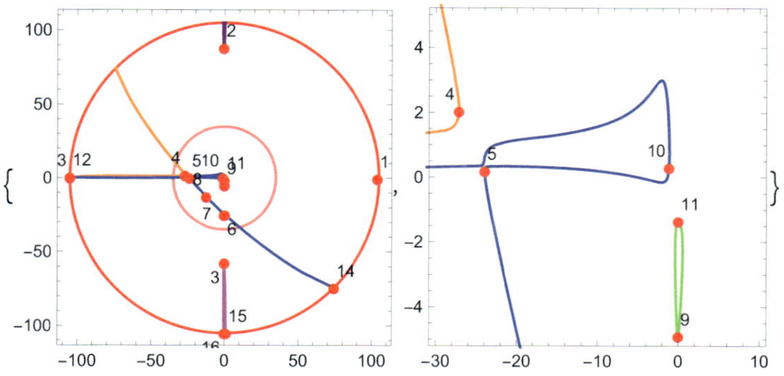

This gives rise to the following directed graph. Note the original orange and blue components meet up at the critical point 1:

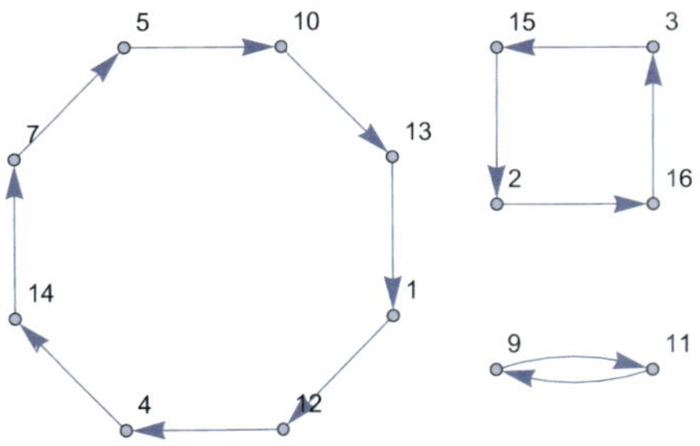

You can see the three projective components of this nonsingular curve.

One of the main things we learned in Section 8.2 is that ultimately, infinite points are just like finite points. One can do Gauss decomposition with any infinite point, or any point near several infinite points, as the origin. This perspective gives a graph where each critical point (not just finite ones) give graphs with even vertices. This is exactly the main content of the Fundamental Theorem of Real Projective Curves:

The projective (directed) graph of any real algebraic curve is an Euler graph, that is, each vertex is even.

The directions are not necessarily natural in the projective case, but we use them so that the Wolfram Language can give the correct picture when there are two curve segments between the same two vertices, for example, vertices 9 and 11 in the previous graph.

We ended Chapter 1 by asking what would be wrong with the following picture from Abhyankar's book (1990) if the image was intended to show a real algebraic curve.

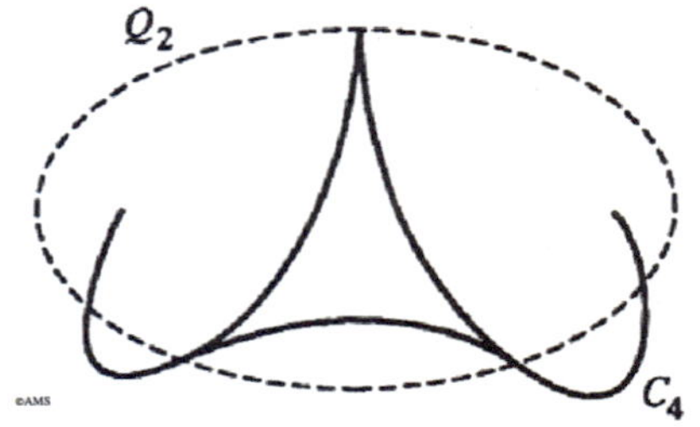

One answer would be that it pictures odd vertices. It also violates the Gauss Principle described at the end of Section 8.3.

8.5 Möbius Band Model of the Real Projective Plane

A useful model of the projective plane is given by the Möbius band, also known as the Möbius strip:

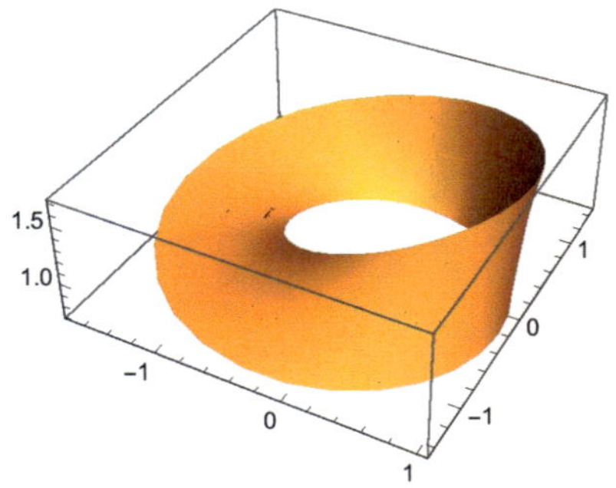

This is a well-known, one-sided surface with boundaries, which can be physically modeled by taking a long skinny strip of paper, making a half twist, and gluing the ends together. The boundary is, topologically speaking, a circle. If we declare that every point on this boundary will represent the affine origin, then we see below that we get an abstract topological space that is topologically equivalent to the projective plane.

The infinite line is the set of points that are midway between the boundary:

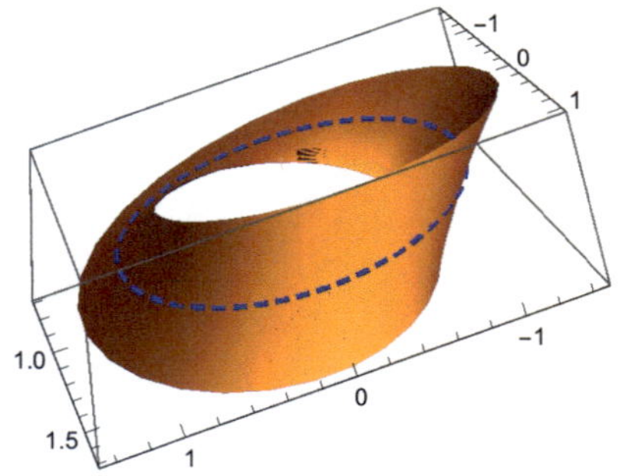

Some geometers prefer to consider the boundary of the Möbius strip a very small circle around the origin. Since it is our intention to plot curves on this surface, this interpretation will also work if we insist that curves in this subsection avoid the origin, that is, the equation of the curve $f = 0$ has a nonzero constant term. A small translation may be necessary to make this happen. The advantage of this interpretation is that it makes the real projective plane minus a tiny disk around the affine origin a subspace of real 3-space.

Plotting a curve on the Möbius band model of the real projective plane can be accomplished by the methods discussed earlier in this chapter. However, there are some tricky issues caused by points with different coordinates having the same physical position or conversely, so we will discuss the practical details of this plotting in Section 8.7. Unless you plan to try some Möbius plots on your own, you can skip this material.

Here are some Möbius plots of lines. The first is a line through the origin—these go straight across the strip from {0, 0} to {0, 0}, which is evenly placed throughout the boundary. The second is a typical line not through the origin.

Both Möbius plots cross the infinite line at one point:

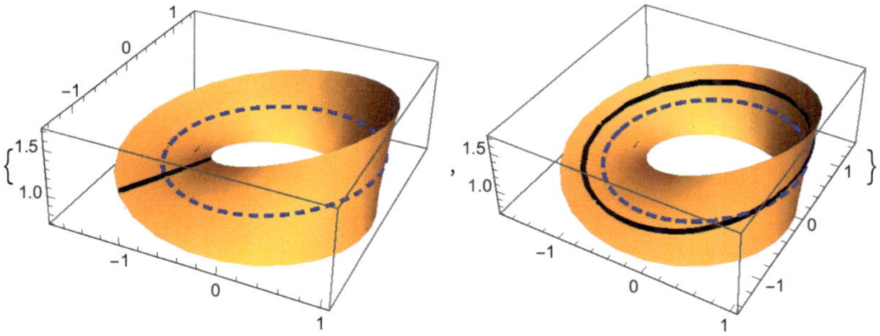

Any two lines through the origin would not meet in the plot, but they meet at the origin, which is the boundary. It is clear that any line that does not go through the origin will meet each line that does go through the origin elsewhere on the band. The Möbius plot below of two typical non-parallel lines not though the origin shows why any two such distinct lines will meet at a singular point. Parallel lines will meet at their common infinite point.

Example A1—Two affinely parallel lines:

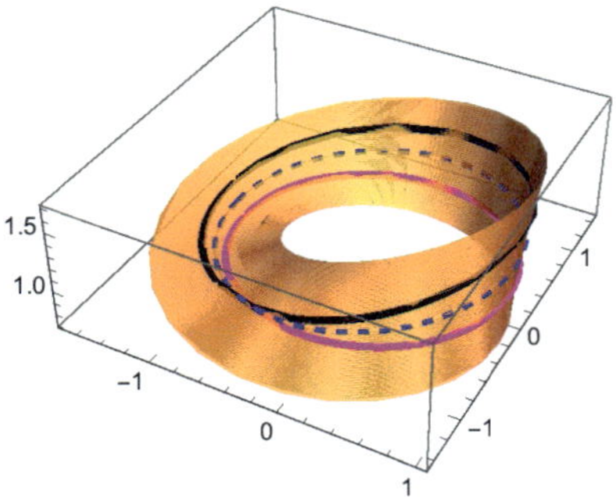

This property (that any two lines meet at exactly one point) is the standard property of the real projective plane. This is one way to see that our Möbius model of the real projective plane is correct, as each point in the standard model can be described as the intersection of two lines. This point will correspond to the intersection of those two lines in our Möbius model. It is not difficult to see that this is actually a topological equivalence.

Next are Möbius plots of a parabola and a hyperbola. We can see here that Kepler was correct when he decided a parabola was a circle missing one point while the hyperbola was missing two infinite points.

Example A2—An affine parabola on the left and the affine hyperbola on the right:

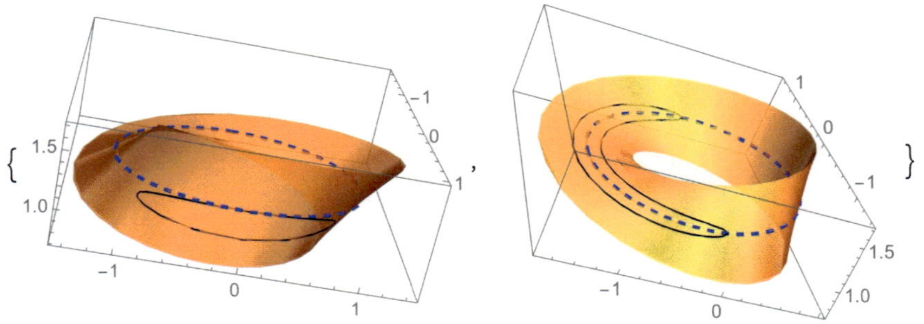

A circle looks different depending on whether it contains the origin inside or outside the circle.

Example A3—Three circles:

```
In[133]:= c1 = {Cos[t], Sin[t] + 3};
          c2 = {Cos[t], Sin[t]};
          c3 = {Cos[t] - 3, Sin[t]};
```

```
In[136]:= ParametricPlot[{c1, c2, c3}, {t, 0, 2 Pi},
            PlotStyle → {Blue, Black, Purple}, PlotLegends → "Expressions"]
```

Out[136]=

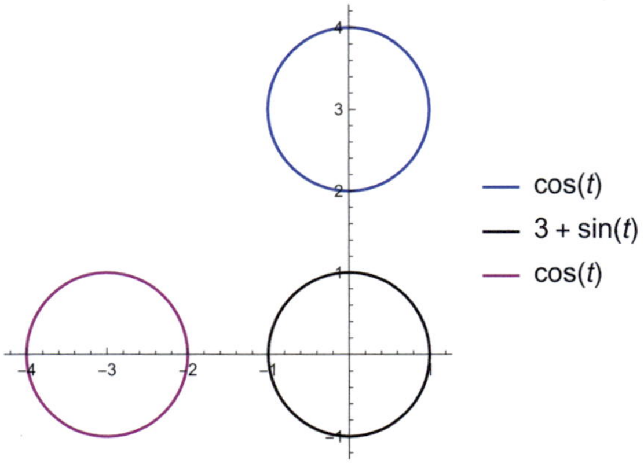

One more example is a simple rational function $y = \frac{4}{x(x^2-4)}$, plotted with Plot on the Möbius band as an algebraic curve.

Note the ordinary singularity at {0, 1, 0} and none at the regular point at {1, 0, 0}.

On the right, Example A3 again—A rational function curve:

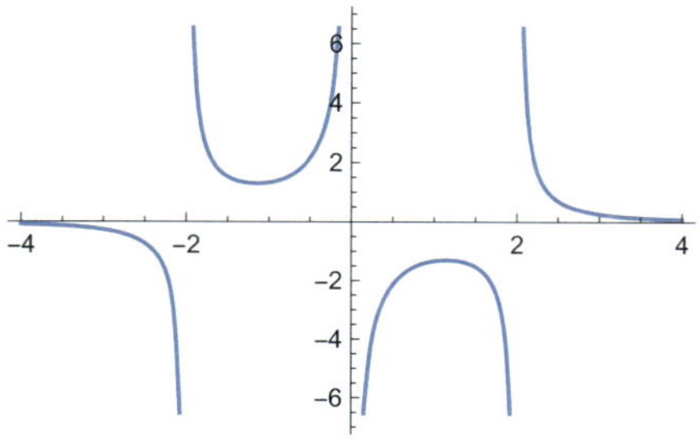

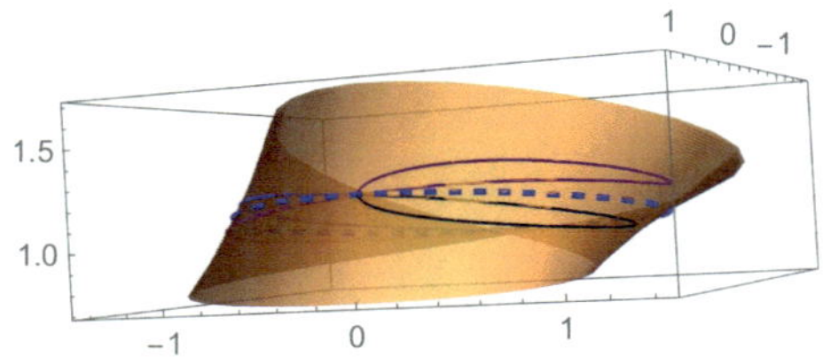

Besides being able to visualize entire projective curves on the Möbius band, there is another important reason to look at curves on the band. Many have seen the demonstration of cutting a Möbius band along the center (infinite) line. Instead of having two disconnected pieces there is one long piece, the resulting topological space is still connected. Also, cutting across the Möbius band along the plot of a line through the origin leaves the resulting space connected. This happens for the plot of any line. On the other hand, cutting along the plot of a circle (with or without the origin) leaves two disconnected pieces. This leads to an important distinction between components of a nonsingular curve.

There are two cases, recalling that each connected component of a nonsingular curve is topologically a simple closed curve, with no self intersections. We will assume the curve does not contain the affine origin, so the Möbius plot does not touch the boundary of the Möbius band. We will use the definitions of components described in Bochnak's book (1998):

1. Cutting along the curve gives two components. We call such a component an oval.
2. Cutting along the curve does not separate the Möbius band into disconnected sets. We call such a component a pseudo-line.

As we noted previously, any line is in the second category, motivating the notation. Different authors may use other notations.

In the first case, the two components are different. One is one-sided like the Möbius band and the other is two-sided like an ordinary piece of paper. The one-sided piece is called the exterior and the two-sided piece is called the interior.

There are two important properties about ovals and pseudo-lines that follow from the Gauss Principle.

Property 1: If Ω *is an oval of curve* $f = 0$ *and* $g = 0$ *is another curve, then the sum of the real intersection multiplicities of* g *with* Ω *is even.*

In other words, if a curve enters the interior of an oval from outside, it must come out again somewhere else. The following example shows that this property does not hold true for pseudo-lines.

The corresponding property for pseudo-lines is:

Property 2: If Λ_1 is a pseudo-line in a real algebraic curve and Λ_2 is a pseudo-line of a potentially different real curve, then they intersect in at least one real point.

Consider the Gauss curve of $z^3 + 3$:

In[137]:=
```
f = gaussCurve[z^3+3, x, y, z]
```

Out[137]= $3 + x^3 - 3 x y^2$

This has infinite points:

In[138]:=
```
infiniteRealPoints[f, x, y]
```

Out[138]= {{0., 18.3651, 0}, {2.45851, 1.41942, 0}, {2.90802, −1.67895, 0}}

Example A4—The ContourPlot is on the left and the Möbius band plot of the Gauss curve is on the right:

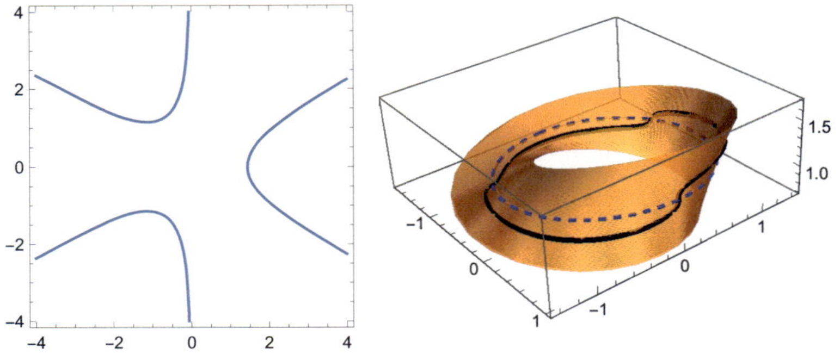

We see $f = 0$, as a projective curve, consists of a single pseudo-line.

Our two properties above show that every nonsingular curve of odd degree must contain a pseudo-line. This is because in order to intersect a line with our curve, there must be (trivially from Bézout's theorem) an odd number of intersection points of the line and curve counted by multiplicity. Only an even number of these can be in ovals, so at least one of these intersection points must be in a pseudo-line of the curve. A similar argument shows that an even-degree, nonsingular curve cannot contain a pseudo-line.

Conversely, a nonsingular curve can have only one pseudo-line because two pseudo-lines would intersect but nonsingular curves have no self-intersection points, which leads to a third property. Thus:

Property 3: A nonsingular curve of even degree has no pseudo-lines, but a nonsingular curve of odd degree has exactly one pseudo-line.

One consequence of these properties is that in order to identify whether a loop is an oval or a pseudo-line, it is necessary to only count infinite points. Ovals have an even number (possibly zero) of infinite points while pseudo-lines have an odd number.

Things are more complicated for singular curves, partly because the decomposition guaranteed by the Fundamental Theorem is not unique. Properties 1 and 2 will still hold, but 3 will not (because two pseudo-lines of the same singular curve can intersect). This will allow decompositions with several pseudo-lines or an even-degree curve with a decomposition containing pseudo-lines. An odd-degree singular curve will still have at least one pseudo-line in any decomposition into simple closed paths.

In identifying potential pseudo-lines, it is useful to note that since Property 1 still holds, (the definition of an oval is topological and not algebraic so singularities do not matter) when a simple closed path inside a curve intersects some line in an odd number of points, that path is a pseudo-line. In particular, if a simple closed path of a curve passes through an odd number of nonsingular infinite points, then it is a pseudo-line.

Consider a reducible curve that is the union of five lines:

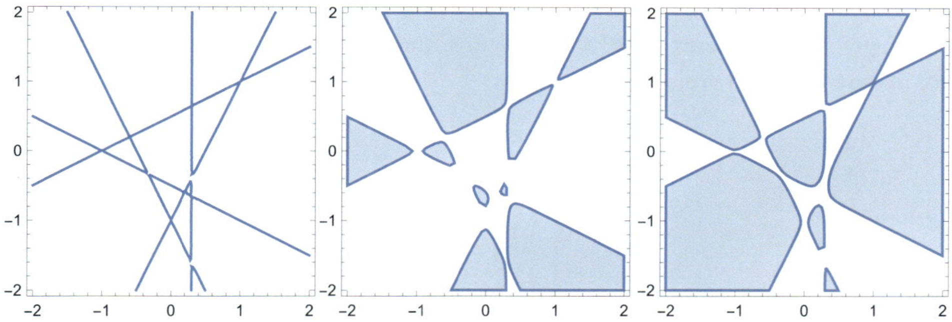

If we consider the algebraic decomposition on the left, we have a curve consisting of five pseudo-lines, but in the Gauss decompositions of f and $-f$ in the middle and right images, we still have exactly one pseudo-line when we continue these affine decompositions into the projective plane. This happens uniquely because there are no singular infinite points. The number of ovals does vary between these two decompositions, as does the pseudo-line (which is the boundary of the shaded infinite regions). Note that the infinite line intersects this latter path in five points, so we know it is a pseudo-line.

One way to decompose a projective curve into simple closed paths is to work directly from the projective graph. Since the graph is an Euler graph, every vertex is even and the graph can be decomposed into edge-disjoint cycles. One way to do this is to use the built-in Mathematica procedure `FindFundamentalCycles` (which may not give a direct answer since different cycles returned may not be edge-disjoint).

The documentation on `FindFundamentalCycles` suggests doing a symmetric difference operation on a pair of cycles with common edges.

There is no such built-in procedure, but it is not hard to write one of your own:

```
In[139]:= graphSD[G_, H_] := With[{GH = GraphIntersection[G, H]}, FindCycle[
              EdgeList[GraphUnion[GraphDifference[G, GH], GraphDifference[H, GH]]]]][[1]];
```

Consider gaussCurve[$z^5 - 5z^3$]:

```
In[140]:= f = 5 x^3 + x^5 - 15 x y^2 - 10 x^3 y^2 + 5 x y^4;
```

In that example, we found and numbered critical points:

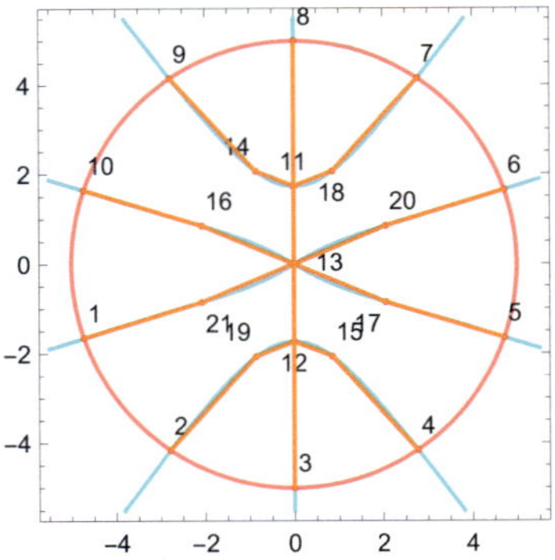

Note that points 1 and 6, 9 and 4, 10 and 5, 8 and 3, and 4 and 2 represent the same infinite points in the undirected projective graph:

PG = 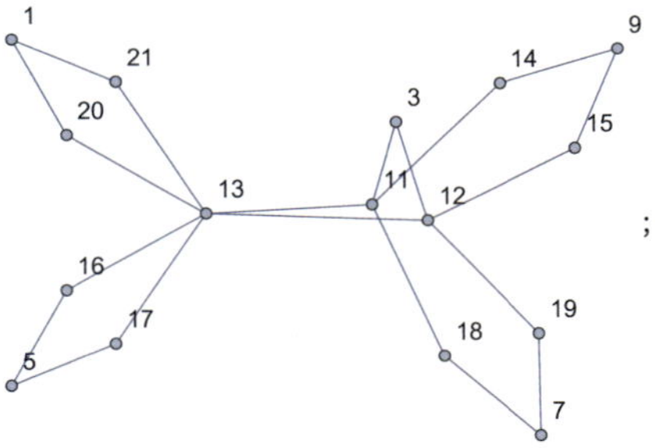 ;

We can attempt to decompose this graph into edge-disjoint cycles by using the following code:

```
In[141]:= FC = FindFundamentalCycles[PG]

Out[141]= {{20 ⟷ 1, 1 ⟷ 21, 21 ⟷ 13, 13 ⟷ 20},
    {14 ⟷ 11, 11 ⟷ 13, 13 ⟷ 12, 12 ⟷ 15, 15 ⟷ 9, 9 ⟷ 14},
    {18 ⟷ 11, 11 ⟷ 13, 13 ⟷ 12, 12 ⟷ 19, 19 ⟷ 7, 7 ⟷ 18},
    {16 ⟷ 13, 13 ⟷ 17, 17 ⟷ 5, 5 ⟷ 16}, {11 ⟷ 13, 13 ⟷ 12, 12 ⟷ 3, 3 ⟷ 11}}
```

The second and third cycles contain common edges 11 ⟷ 13, 13 ⟷ 12, so we do:

```
In[142]:= SD = graphSD[FC[[2]], FC[[3]]]

Out[142]= {7 ⟷ 18, 18 ⟷ 11, 11 ⟷ 14, 14 ⟷ 9, 9 ⟷ 15, 15 ⟷ 12, 12 ⟷ 19, 19 ⟷ 7}
```

We arrive at a simple cycle edge-disjoint decomposition:

```
In[143]:= C1 = {7 ⟷ 18, 18 ⟷ 11, 11 ⟷ 14, 14 ⟷ 9, 9 ⟷ 15, 15 ⟷ 12, 12 ⟷ 19, 19 ⟷ 7};
          C2 = {16 ⟷ 13, 13 ⟷ 17, 17 ⟷ 5, 5 ⟷ 16};
          C3 = {11 ⟷ 13, 13 ⟷ 12, 12 ⟷ 3, 3 ⟷ 11};
          C4 = {20 ⟷ 1, 1 ⟷ 21, 21 ⟷ 13, 13 ⟷ 20};
```

This results in the following plots:

```
In[147]:= Row[{Graph[C1, VertexLabels → "Name"], Graph[C2, VertexLabels → "Name"],
              Graph[C3, VertexLabels → "Name"], Graph[C4, VertexLabels → "Name"]}]
```

Out[147]=

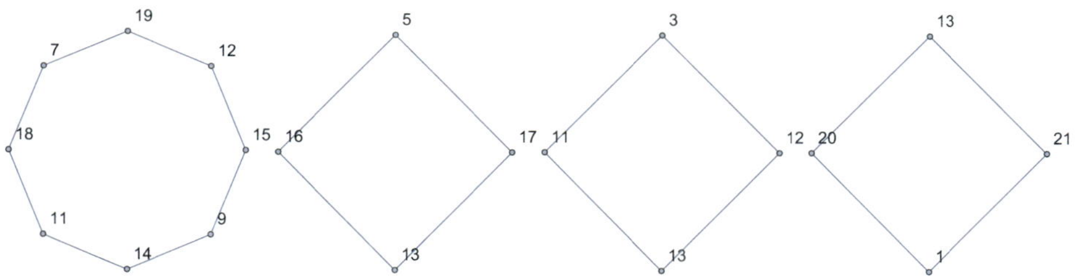

This is the classical ordinary singularity decomposition. Surprisingly, the three 4-vertex graphs all corresponding to pseudo-lines; vertices 1, 3, 5, 7, and 9 are the infinite points (these graphs contain one each), and only the 8-vertex graph corresponds to an oval. It is impossible to decompose this curve into simple closed paths using only one pseudo-line.

8.6 A Paradigm for Analyzing Curves

In social sciences, a paradigm is somewhat like an algorithm but not as precise. Now that we have most of our methods and theory, I will give a paradigm for how to analyze an unfamiliar curve given by a bivariate polynomial. Be prepared to improvise.

1. Find a set of critical points containing at least singular points and infinite points and possibly boundary points on a large circle. You need enough points to accomplish step 3, but discard points that do not give new information.
2. While finding these points, you may discover the curve is not square-free. Find the square-free reduction. You may also discover your curve is reducible. If you start with integer coefficients, you can try to factor. In rare cases, the factors may be numerical or complex. Such factorization may be visible if there are too many singular points. If you have numerical coefficients, the curve is probably not reducible.
3. Visually or using path tracking with an indefinite critical terminal point, find the affine graph. Also inspect around infinite points to connect all branches to finite points. Then produce the projective graph. Break the curve into connected components or, if possible, do a decomposition into simple closed paths.
4. You may wish to plot your curve. Choose an appropriate local or global plot. Sometimes a contour plot is sufficient; in other cases, you may need to break the curve into pieces, do path tracking, and then reassemble your curve (as in the first part of this section).

8.7 Plotting Projective Curves on a Möbius Band

For the reader who is interested in Wolfram Language implementation, we give some more details on how to plot curve components on the Möbius band representation of the projective plane.

Going backwards, we take a Möbius band and consider the transversal line that represents the *x*-axis of the projective plane. We cut, starting at the inside edge (representing the origin) toward the point {– 1, 0, 0} at infinity in the negative direction. We then cut along the midline of the band (that is, the infinite line) until we reach the infinite point {1, 0, 0} again.

What we have left is one long strip half the width of the original Möbius band. Both ends represent the ray along the negative *x*-axis from the origin to the infinite point. The whole bottom of the strip represents 0 while the top of the strip is the infinite line wrapped twice.

Reversing the above, we view the Möbius band as a strip with width, the first coordinate, from -π to π, and height, the second coordinate, from 0 to .5. This is glued together along the sides and top, first half to second half, with a twist to make the Möbius band.

In the Wolfram Language, this is implemented by:

```
In[148]:= moebiusB[t_, s_] := Module[{a = N[Sqrt[2] - 1], mb, MB, tt},
      mb[tt_] :=
        {Cos[2 tt + Pi] / (1 + a Cos[tt]), Sin[2 tt + Pi] / (1 + a Cos[tt]), 1 / (1 + a Cos[tt + Pi / 2])};
      s * mb[t] + (1 - s) * mb[t + Pi]];
```

If they occur as other than the first or last points of M, you may need to split the list before combining with MB:

```
In[149]:= moebiusPLMB :=
    Show[ParametricPlot3D[moebiusB[t, s], {t, -Pi, Pi}, {s, 0, .5}, Mesh → None,
      PlotStyle → Opacity[0.8]], ParametricPlot3D[moebiusB[t, .5], {t, 0, Pi},
      PlotStyle → {Thick, Blue, Dashed}], Boxed → False, Axes → False];
```

In[150]:= `Show[moebiusPLMB, ImageSize → Small]`

Out[150]=

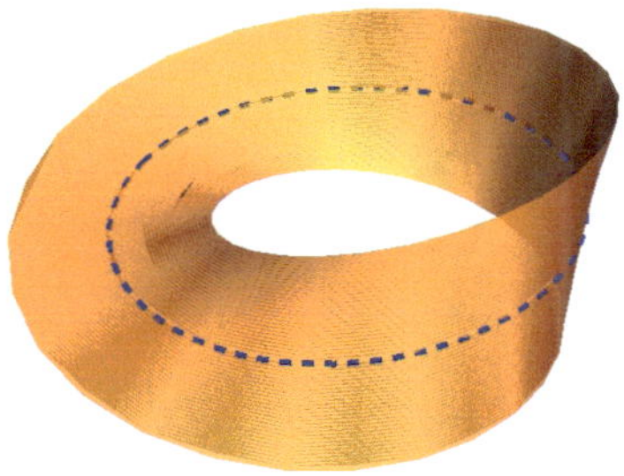

To make the real projective plane, we now squish the entire lower side of our rectangle to one point, which will represent the origin. This must be done mentally since the real projective plane cannot be imbedded in real affine 3-space.

Thus our rectangle model of our projective plane has outlines:

In[151]:=
```
moebiusRMPP := ListLinePlot[{{{-Pi, 0}, {Pi, 0}},
    {{-Pi, .5}, {Pi, .5}}, {{Pi, 0}, {Pi, .5}}, {{-Pi, 0}, {-Pi, .5}}, {{0, 0}, {0, .5}}},
   PlotStyle → {Brown, Orange, Green, Green, Cyan}, AspectRatio → 1/3.25];
```

An annotated version is:

In[152]:=
```
Show[moebiusRMPP, ListPlot[{Labeled[{-1.5, .02}, "zero"], Labeled[{1.5, .02}, "zero"],
    Labeled[{-3, .2}, "negative x-axis"], Labeled[{.2, .2}, "positive x-axis"],
    Labeled[{3, .2}, "negative x-axis"]}, PlotStyle → White],
  ListPlot[{Labeled[{1.8, .43}, "infinite line"], Labeled[{-1.8, .43}, "infinite line"]},
   PlotStyle → White],
  ListPlot[{Labeled[{-2, .25}, "lower half-plane"], Labeled[{2, .25}, "upper half-plane"]},
   PlotStyle → White, LabelStyle → Directive[Red, Medium]]]
```

Out[152]=

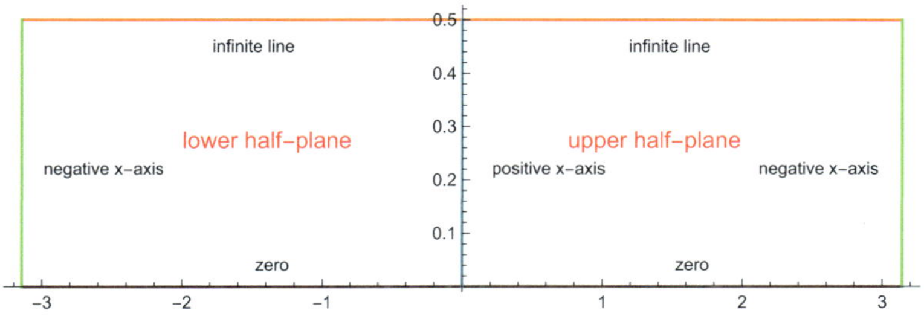

On the other hand, if we choose to not glue together the two copies of the infinite line, we have a hemisphere (which will be more convenient to use). The entire hemisphere plane (except for origin) lies between the orange lines. All points on the lower brown line represent zero, and the upper line is the infinite line (which once was the upper orange line). We can map the affine plane except for points $\{x, y\}$ such that $y = 0$ and $x \leq 0$ to this hemisphere rectangle. Note that points $\{x, 0\}$ with $x < 0$ would map to `{± π, N[ArcTan[Norm[x]]]/π}`, where the user needs to choose the sign of the first coordinate to insure continuity. The case of $\{0, 0\}$ will be considered later.

The function from the projective plane minus the negative x-axis to rectangle hemisphere provides the model for the projective plane. It is assumed projective points will have a third coordinate 0 or 1. If any points are on the negative x-axis, including {1, 0, 0}, then the function will return an error message and {}. As usual, to apply to a list L of points, use M = moebiusPhi /@ L. You can find the position of error points in M by Position[M, {}]. Then fix these points manually.

If they occur as other than the first or last points of M, you may need to split the list before combining with MB:

```
In[153]:= moebiusPhi[p_] :=
    If[Length[p] == 2 && (p[[1]] > 0 || p[[2]] ≠ 0),
      Return[{ArcTan @@ N[p], N[ArcTan[Norm[p]] / Pi]}],
      Echo[p, "ambiguous point "]; Return[{}]];
```

Since in the projective plane, a curve is a union of simple closed topological curves, it is important (for continuity) that the image of the curve in the hemisphere rectangle actually hits the boundary. It is necessary to plot the boundary points accurately. At least one coordinate of these points needs to be exact, so the exact or symbolic value should be given. We add these points manually after applying phi, making sure to get the correct hemisphere signs (since these would be ambiguous to the function phi).

Combining moebiusPhi with moebiusB—as in moebiusB @@ moebiusPhi[p] places the point p on the Möbius band. Typically, to plot a curve on the Möbius band, a representation of the projective plane, we would do the two steps separately as in the examples below. Infinite points and points on the omitted ray will be handled separately.

We now show how to draw some of the Möbius plots in Sections 8.5 and 8.6.

Example A1—Parallel Lines

```
l1 = 2 x - 3 y - 4;
l2 = 2 x - 3 y + 4;
```

The infinite points on the Gauss hemisphere are $\{-3, -2, 0\}$ and $\{3, 2, 0\}$. l1 misses the negative x-axis but l2 does not—it intersects at $\{-2, 0\}$.

We can plot these lines into our rectangular hemisphere by:

```
In[156]:= L1 = Table[{x, (2 x - 4) / 3}, {x, -6, 6, .2}];
    PL1 =
      Append[Prepend[moebiusPhi /@ L1, {N[ArcTan[-3, -2]], .5}], {N[ArcTan[3, 2]], .5}];
    L2n = Table[{x, (2 x + 4) / 3}, {x, -6, -2.1, .1}];
    PL2n = Append[
       Prepend[moebiusPhi /@ L2n, {N[ArcTan[-3, -2]], .5}], N[{-Pi, ArcTan[2] / Pi}]];
    L2p = Table[{x, (2 x + 4) / 3}, {x, -1.9, 6, .1}];
    PL2p =
      Append[Prepend[moebiusPhi /@ L2p, N[{Pi, ArcTan[2] / Pi}]], {N[ArcTan[3, 2]], .5}];
```

Here is the rectangular hemisphere model:

```
In[162]:= Show[moebiusRMPP, ListLinePlot[{PL1, PL2n, PL2p}, PlotStyle → {Cyan, Purple, Purple}],
  Axes → False, Frame → True, AspectRatio → 1/5]
```

Out[162]=

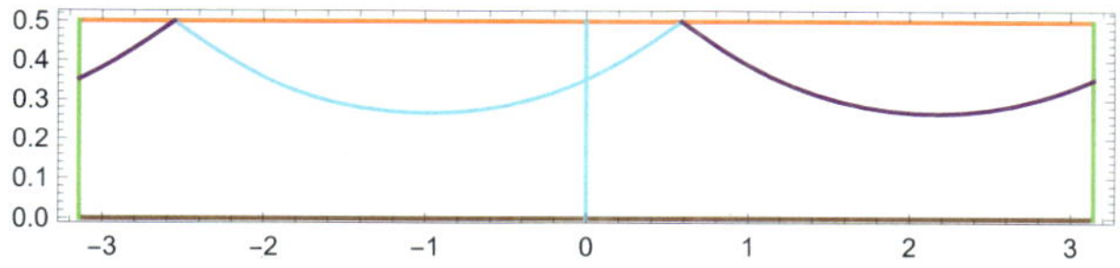

Now we can plot in our Möbius band model of the projective plane.

Example A1—Two affinely parallel lines:

```
In[163]:= Show[moebiusPLMB, Graphics3D[{{Black, Thick, Line[moebiusB@@# &/@PL1]},
     {Magenta, Thick, Line[moebiusB@@# &/@PL2n]},
     {Magenta, Thick, Line[moebiusB@@# &/@PL2p]}}]]
```

Out[163]=

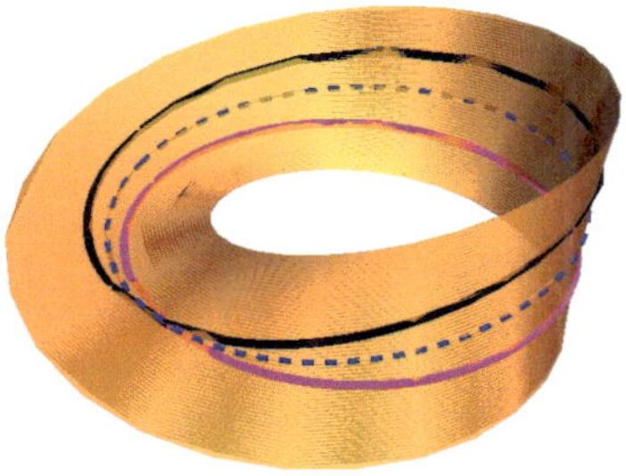

Example A2—Affine Parabola (g) and Hyperbola (h)

```
g = y - x^2 - 1;
h = x^2 - y^2 - 4;
```

The parabola has only one infinite point {0, 1, 0}; the hyperbola has infinite points {1, 1, 0}, {1, − 1, 0}, {− 1, 1, 0}, {− 1, − 1, 0} in the Gauss hemisphere.

Our parabola misses the *x*-axis but not the negative branch of the hyperbola:

```
In[166]:= G = Table[{x, x^2 + 1}, {x, -6, 6, .2}];
PG = Append[Prepend[moebiusPhi/@G, {N[Pi/2], .5}], {N[Pi/2], .5}];
```

```
In[168]:= Show[moebiusRMPP, ListLinePlot[PG, PlotStyle → Blue]]
```

Out[168]=

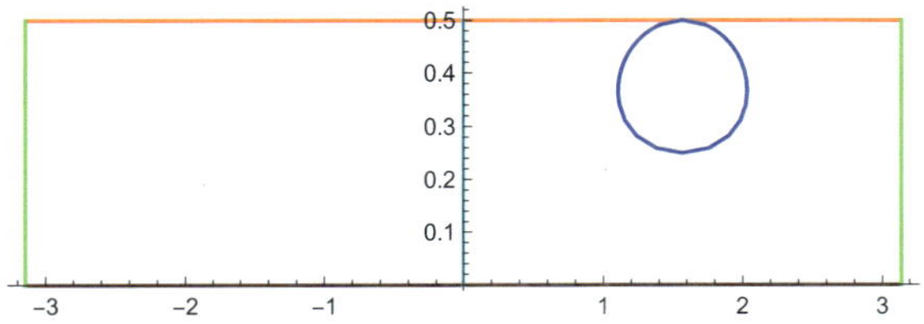

In[169]:=
```
Hp = Table[{Sqrt[y^2+4], y}, {y, -6, 6, .2}];
PHp =
   Append[Prepend[moebiusPhi/@Hp, {N[ArcTan[1, -1]], .5}], {N[ArcTan[1, 1]], .5}];
Hnp = Table[{-Sqrt[y^2+4], y}, {y, .1, 6, .2}];
PHnp =
   Append[Prepend[moebiusPhi/@Hnp, N[{Pi, ArcTan[2]/Pi}]], {N[ArcTan[-1, 1]], .5}];
Hnn = Table[{-Sqrt[y^2+4], y}, {y, -6, -.1, .2}];
PHnn = Append[
     Prepend[moebiusPhi/@Hnn, {N[ArcTan[-1, -1]], .5}], N[{-Pi, ArcTan[2]/Pi}]];
Show[moebiusRMPP, ListLinePlot[{PHnn, PHp, PHnp}, PlotStyle → Black]]
```

Out[175]=

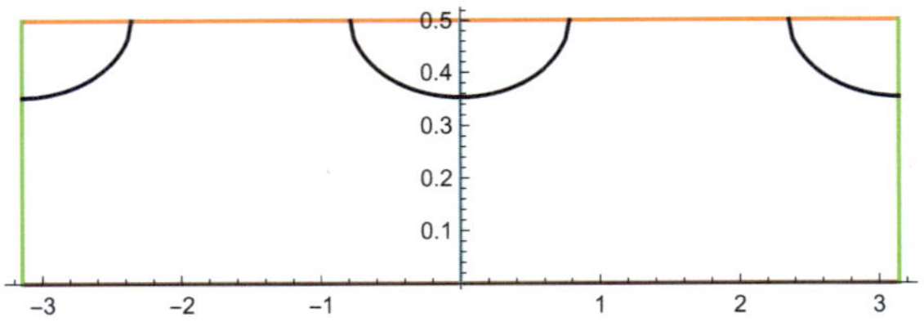

In[176]:=
```
{Labeled[Show[moebiusPLMB,
     Graphics3D[{Black, Thick, Line[moebiusB@@# &/@PG]}], ViewPoint →
       {3.3791867902828234`, -0.13180435786152273`, 0.11715053404397978`}],
   "Example A1: Affine parabola"], Labeled[Show[moebiusPLMB,
     Graphics3D[{{Black, Thick, Line[moebiusB@@# &/@PHp]},
        {Black, Thick, Line[moebiusB@@# &/@PHnp]}, {Black, Thick,
          Line[moebiusB@@# &/@PHnn]}}]], "Example A2: Affine hyperbola"]}
```

Out[176]= {

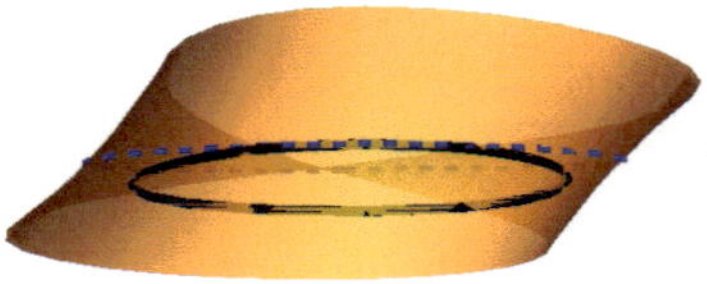

Example A1 : Affine parabola

,

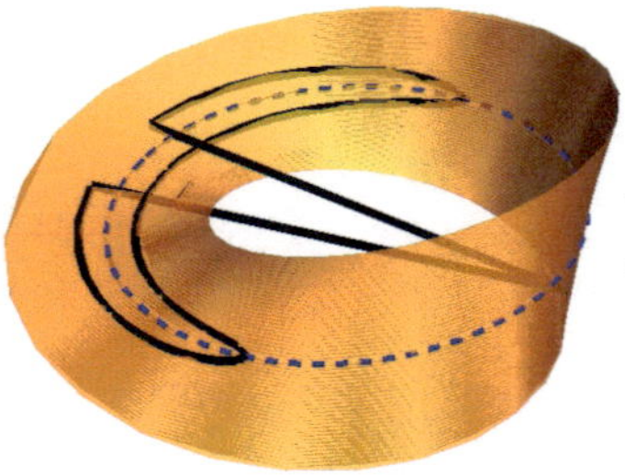

Example A2 : Affine hyperbola

}

Note that the big difference between plots is that the parabola is tangent to the infinite line while the infinite line meets the hyperbola in two distinct points.

Example A3—Circles

We look at three circles: one missing the negative *x*-axis completely, one centered at the origin hitting the negative *x*-axis once, and one hitting the negative *x*-axis twice. We compare their plots in the hemisphere rectangle and Möbius band.

The circles are given parametrically for convenience:

```
In[177]:= c1 = {Cos[t], Sin[t] + 3};
c2 = {Cos[t], Sin[t]};
c3 = {Cos[t] - 3, Sin[t]};
ParametricPlot[{c1, c2, c3}, {t, 0, 2 Pi},
  PlotStyle → {Blue, Black, Purple}, PlotLegends → "Expressions"]
```

Out[180]=

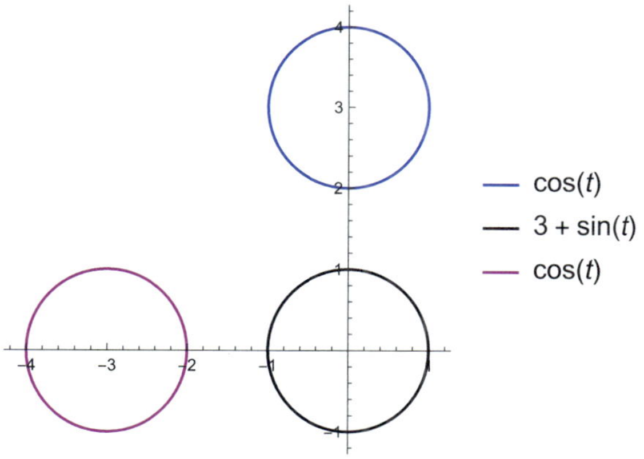


```
In[181]:= C1 = Table[N[c1 /. {t → s}], {s, 0, 2 Pi, Pi/25}];
PC1 = moebiusPhi /@ C1;
C2 = Table[N[c2 /. {t → s}], {s, -Pi + .01, Pi - .01, (Pi - .02)/50}];
PC2 = Append[Prepend[moebiusPhi /@ C2, {-N[Pi], .25}], {N[Pi], .25}];
C3p = Table[N[c3 /. {t → s}], {s, .01, Pi - .01, (Pi - .02)/25}];
PC3p =
   Append[Prepend[moebiusPhi /@ C3p, N[{Pi, ArcTan[2]/Pi}]], N[{Pi, ArcTan[4]/Pi}]];
C3n = Table[N[c3 /. {t → s}], {s, Pi + .01, 2 Pi - .01, (Pi - .02)/25}];
PC3n = Append[
    Prepend[moebiusPhi /@ C3n, N[{-Pi, ArcTan[4]/Pi}]], N[{-Pi, ArcTan[2]/Pi}]];

Show[moebiusRMPP,
 ListLinePlot[{PC1, PC2, PC3p, PC3n}, PlotStyle → {Blue, Black, Magenta, Magenta}]]
```

Out[189]=

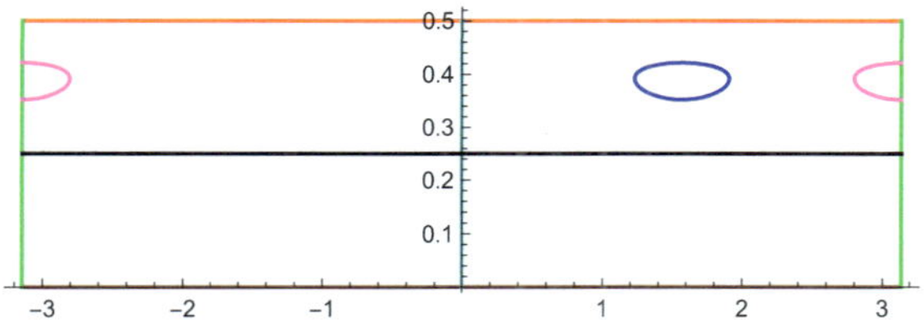

Example A3—Three circles:

In[190]:=
```
Show[moebiusPLMB, Graphics3D[{Blue, Thick, Line[moebiusB@@# &/@PC1]}],
  Graphics3D[{Black, Thick, Line[moebiusB@@# &/@PC2]}],
  Graphics3D[{Magenta, Thick, Line[moebiusB@@# &/@PC3p]}],
  Graphics3D[{Magenta, Thick, Line[moebiusB@@# &/@PC3n]}]]
```

Out[190]=

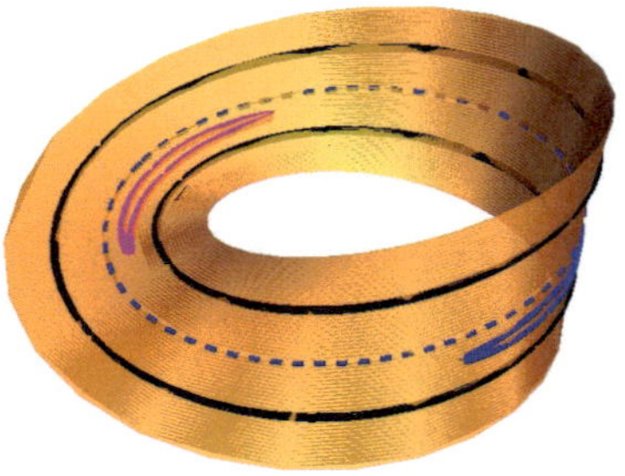

Note that all three circles are ovals; this is evident already in the rectangular hemisphere.

Example A4

The rational function curve $x\,y\,(x^2 - 4) - 4 = 0$ shows infinite singularity.

Here again we write this as a parametric curve $\{x, f\}$ where:

In[191]:=
```
f = 4/(x^3-4x);
Show[Plot[f, {x, -4, -2}, PlotStyle → Blue, PlotRange → {{-4, -2}, {-10, 10}}],
  Plot[f, {x, -2, 0}, PlotStyle → Black], Plot[f, {x, 0, 2}, PlotStyle → Purple],
  Plot[f, {x, 2, 4}, PlotStyle → Magenta, PlotRange → {{-4, 4}, {-10, 10}}], PlotRange → All]
```

Out[192]=

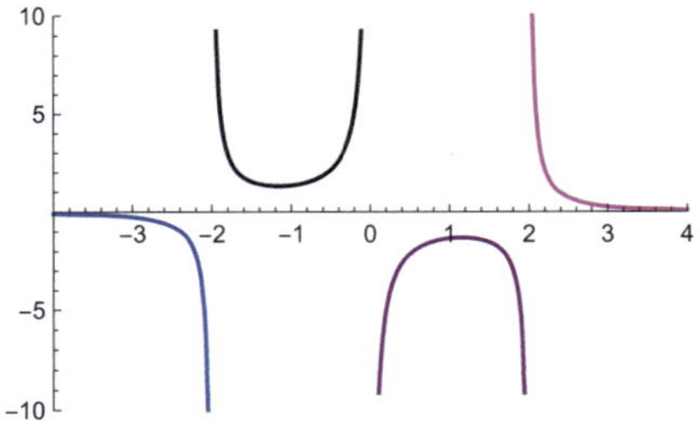

We see infinite points $\{1, 0, 0\}$ and $\{0, 1, 0\}$; the latter will be singular. We handle each affine topological component separately:

In[193]:=
```
F1 = Table[N[{x, f}], {x, -6, -2.05, .025}];
PF1 = Append[Prepend[moebiusPhi/@F1, {-N[Pi], .5}], {-N[Pi/2], .5}];
F2 = Table[N[{x, f}], {x, -1.95, -.05, .05}];
PF2 = Append[Prepend[moebiusPhi/@F2, {N[Pi/2], .5}], {N[Pi/2], .5}];
F3 = Table[N[{x, f}], {x, .05, 1.95, .05}];
PF3 = Append[Prepend[moebiusPhi/@F3, {-N[Pi/2], .5}], {-N[Pi/2], .5}];
F4 = Table[N[{x, f}], {x, 2.05, 6, .025}];
PF4 = Append[Prepend[moebiusPhi/@F4, {N[Pi/2], .5}], {0, .5}];
```

```
Show[moebiusRMPP,
  ListLinePlot[{PF1, PF2, PF3, PF4}, PlotStyle → {Blue, Black, Purple, Magenta}]]
```

Out[201]=

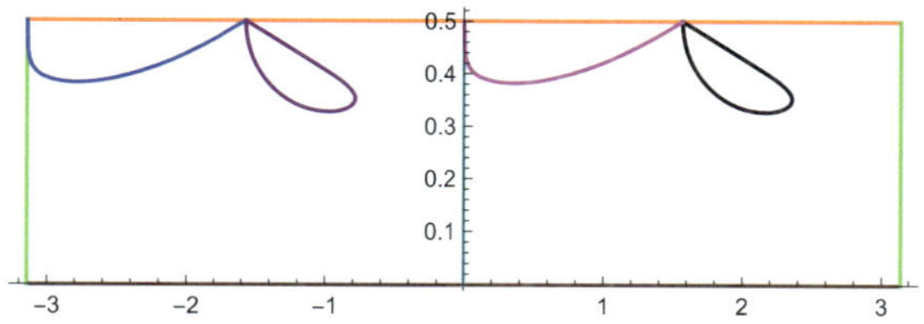

Example A4—Rational function curve:

```
In[202]:= Show[moebiusPLMB, Graphics3D[{Blue, Thick, Line[moebiusB@@# &/@PF1]}],
  Graphics3D[{Black, Thick, Line[moebiusB@@# &/@PF2]}],
  Graphics3D[{Purple, Thick, Line[moebiusB@@# &/@PF3]}],
  Graphics3D[{Purple, Thick, Line[moebiusB@@# &/@PF4]}],
  ViewPoint → {2.8861775939869543`, -0.9323063786376076`, 1.5002612146953958`}]
```

Out[202]=

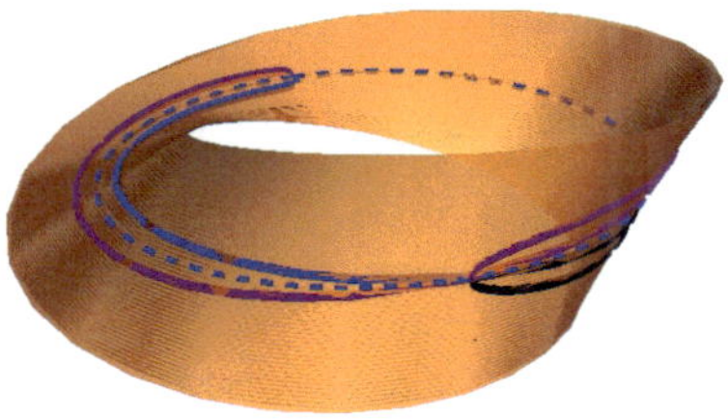

For the previous examples, the parameterization of the given curves was immediate. In general, this is not the case. One needs the techniques from Chapter 4 to find a good piecewise linear approximation of a parameterization.

The following example shows this:

```
In[203]:= g = gaussCurve[z^3+3, x, y, z]
```

Out[203]= $3+x^3-3\,x\,y^2$

```
In[204]:= ContourPlot[g == 0, {x, -4, 4}, {y, -4, 4}]
```

Out[204]=

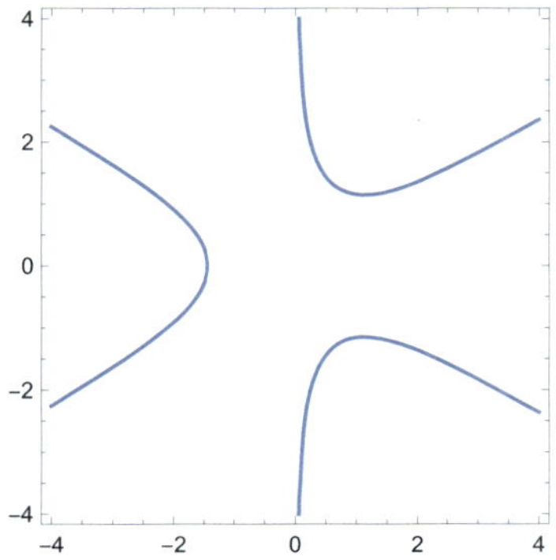

We saw earlier that a Gauss hemisphere plot is about the same as a contour plot in a small region. We can trace and add in the infinite points at the end of each trace.

We start with the gPoints, infinite points, and gaussPlot:

```
In[205]:= gp = gPoints[g, 4, x, y]

Out[205]= {{-3.49494, 1.94561}, {0.0625204, 3.99951}, {3.43242, 2.0539},
  {3.43242, -2.0539}, {0.0625204, -3.99951}, {-3.49494, -1.94561},
  {0.721125, 1.24902}, {-1.44225, 0.}, {0.721125, -1.24902}}

In[206]:= irp = Normalize/@infiniteRealPoints[g, x, y]

Out[206]= {{0.866025, 0.5, 0.}, {0., -1., 0.}, {0.866025, -0.5, 0.}}
```

Because the ordering of these points inside gp and irp could change in a different run, we fixed the order at the time of writing:

```
In[207]:= gpF = {{-3.4949386321816247`, 1.9456114610282333`},
     {0.06252036494620586`, 3.9995113706510463`},
     {3.432418267235486`, 2.0538999096227735`}, {3.432418267235423`,
       -2.0538999096228165`}, {0.06252036494620673`, -3.9995113706510477`},
     {-3.4949386321816074`, -1.9456114610282425`},
     {0.7211247851537057`, 1.2490247664834073`}, {-1.442249570307409`, 0.`},
     {0.7211247851537046`, -1.2490247664834087`}};
  irpF = {{-37.81781954265174`, 21.83412829311467`, 0}, {-1.335064011242463`,
       -0.770799566276224`, 0}, {0.`, -1.5980130051246886`, 0}};
```

We add the negative *x*-axis to our Gauss plot:

```
In[209]:= Show[gaussPlot[g, 4, gp], Graphics[{Magenta, Thick, Dashed, Line[{{-4, 0}, {0, 0}}]}]]
```

Out[209]=

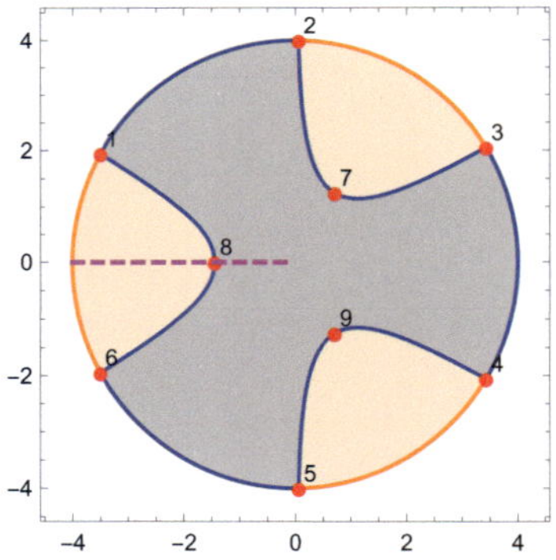

By inspection, we see that points 3 and 6 will connect to a = irpF[[1]], points 1 and 4 to b = irpF[[3]], and 2 and 5 to c = irpF[[2]]. The projective graph will be:

```
In[210]:= Graph[{a → 6, 6 → 8, 8 → 1, 1 → b, b → 4, 4 → 9, 9 → 5, 5 → c, c → 2, 2 → 7, 7 → a},
            VertexLabels → "Name"]
```

Out[210]=

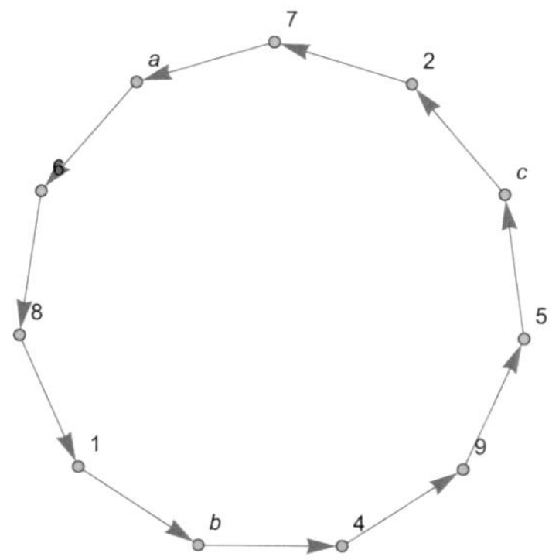

There is one loop that crosses the infinite line three times, so this entire curve is a pseudo-line. Also note that our Gauss plot is symmetric about the horizontal axis, which means that our rectangular hemisphere will have vertical symmetry about the positive x-axis, cutting our work in half.

So computing:

```
In[211]:= L1 = pathFinder[g, gpF[[2]], gpF[[3]], .4, x, y];
          L2 = pathFinder[g, gpF[[8]], gpF[[1]], .2, x, y];
```

These seem to work, so we put these on the hemisphere:

```
In[213]:= K1 = moebiusPhi /@ L1;
          K2 = moebiusPhi /@ L2;
   » ambiguous point {-1.44225, 0.}
```

We have one ambiguous point on K2 to fix:

```
In[215]:= K2[[1]] = N[{Pi, ArcTan[Abs[gpF[[8, 1]]]] / Pi}];
```

As mentioned above, we can go directly from our boundary points to the infinite points, so we add these now:

```
In[216]:= K1 = Append[Prepend[K1, {ArcTan @@ Take[-irpF[[3]], 2], .5}],
                 {ArcTan @@ Take[-irpF[[2]], 2], .5}];
          AppendTo[K2, {ArcTan @@ Take[irpF[[1]], 2], .5}];
```

Using symmetry, we flip over the horizontal axis in hemisphere but also reverse direction:

```
In[218]:= K3 = Reverse[Table[{-p[[1]], p[[2]]}, {p, K1}]];
          K4 = Reverse[Table[{-p[[1]], p[[2]]}, {p, K2}]];
```

Since we have one loop, we join these together:

```
In[220]:= K = Join[K1, K4, K2, K3];
```

We lift to the Möbius band and plot:

```
In[221]:= M = moebiusB@@# &/@K;
Show[moebiusPLMB, Graphics3D[{Black, Thick, Line[M]}],
  ViewPoint → {-1.282969954491907`, -2.4307640696896065`, 1.9736702190024185`},
  ImageSize → Large]
```

Out[221]=

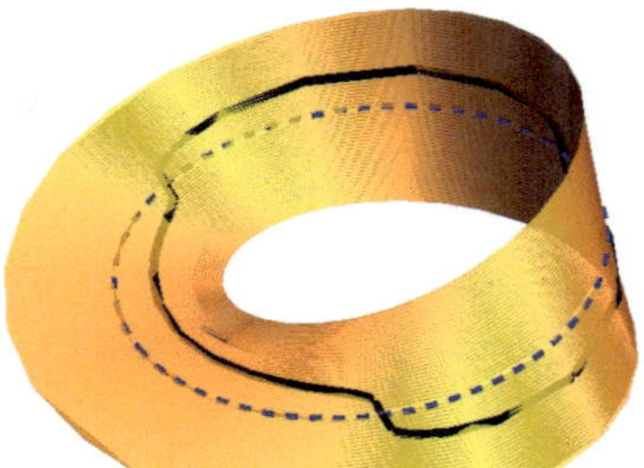

This shows a loop crossing the infinite line in three places, something that can't happen in affine geometry, but it also confirms (although we knew it must happen in degree 3) that this loop is a pseudo-line.

Example A5—Curves Passing Through the Origin

Generally for the Möbius band plots, we avoid curves passing through the origin. It is easy to arrange this since a translation of the curve can easily produce an equivalent curve near the original. However, if we really want to do this, then the ambiguity of $\{0, 0\}$ in this model gives a discontinuity at the origin.

Consider the simple node at the origin $f = -x^3 - x^2 + y^2$. By inspection, we can easily parameterize this with different parameterizations on pieces of the curve between critical points $\{\{0, -1, 0\}, \{-1, 0\}, \{0, 0\}$, and $\{0, 1, 0\}\}$.

Let's set this up so we can plot the discontinuity:

```
In[223]:= f = y^2-x^3-x^2;
N1 = N[Table[{x, Sqrt[x^3+x^2]}, {x, .05, 6, .2}]];
PN1 = Prepend[Append[moebiusPhi/@N1, N[{Pi/2, .5}]], N[{Pi/4, 0}]];
N2 = N[Table[{x, Sqrt[x^3+x^2]}, {x, -.999, -.01, .01}]];
PN2 = Append[Prepend[moebiusPhi/@N2, N[{Pi, .25}]], N[{3 Pi/4, 0}]];
N3 = N[Table[{x, -Sqrt[x^3+x^2]}, {x, .05, 6, .2}]];
PN3 = Append[Prepend[moebiusPhi/@N3, N[{-Pi/4, 0}]], N[{-Pi/2, .5}]];
N4 = N2 = N[Table[{x, -Sqrt[x^3+x^2]}, {x, -.999, -.01, .01}]];
PN4 = Append[Prepend[moebiusPhi/@N4, N[{-Pi, .25}]], N[{-3 Pi/4, 0}]];
```

Example A5—Standard node:

```
In[232]:= Show[moebiusRMPP, ListLinePlot[{PN1, PN2, PN3, PN4}, PlotStyle → Blue]]
```

Out[232]=

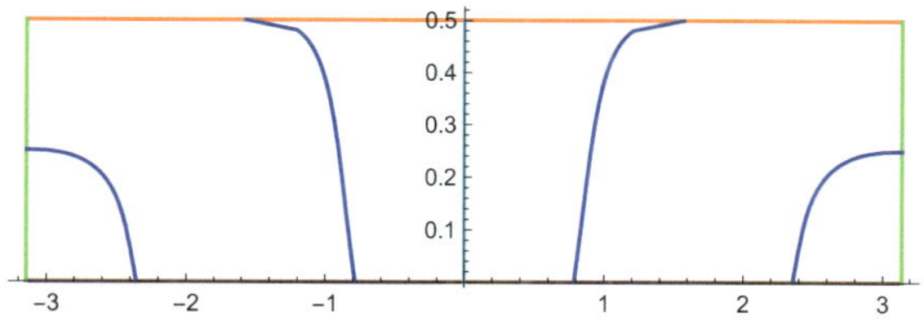


```
In[233]:= {ContourPlot[f == 0, {x, −2, 2}, {y, −2, 2}],
  Labeled[
   Show[moebiusPLMB, Graphics3D[{Black, Thick, Line[moebiusB@@# &/@PN1]}],
    Graphics3D[{Black, Thick, Line[moebiusB@@# &/@PN2]}],
    Graphics3D[{Black, Thick, Line[moebiusB@@# &/@PN3]}],
    Graphics3D[{Black, Thick, Line[moebiusB@@# &/@PN4]}]],
   "Example A5: Standard node"]}
```

Out[233]= {

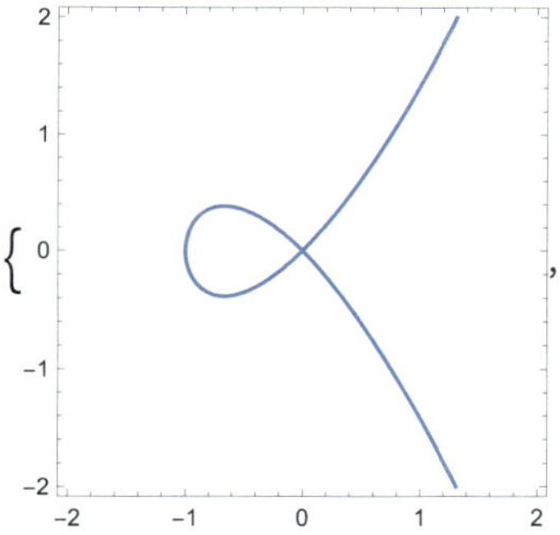

,

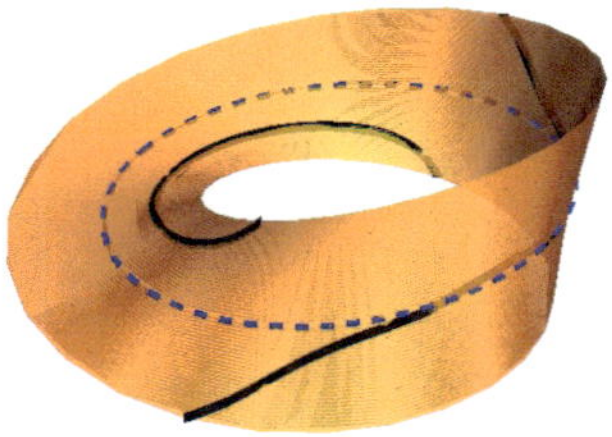

}

Example A5: Standard node

The curve hits the horizontal axis of the rectangular hemisphere at the angle the branches of the singular point make with the x-axis. In algebraic geometry lingo, we have blown up the singularity. This happens, in general, for singularities at the origin. Of course, since all points on the boundary of the Möbius strip are to be squashed together, in the actual projective plane, this looks fine.

More to Explore

For more on the real projective plane, try Wolfram MathWorld (wolfr.am/uOcS1MpF)

For a fun Demonstration on cross-cap surfaces, check out A. Naeemi's entry (wolfr.am/uOcBe9Ec)

9 | Harnack's Theorem and Newton Hyperbolas

In this chapter, we give our version of Harnack's theorem in Section 9.1 and discuss Hilbert's related 16th problem in Section 9.4. In between, we introduce new representations via *diamond diagrams* of nonsingular numerical curves, suggested by the Newton hyperbolas. These give loops (as in the graphs from Chapters 4 and 5) but also indicate geometric position relative to the three axes (x, y, and z, where the z-axis is the infinite line) in addition to the other loops, so we can better describe our curves relative to the 16th problem. In the last section, we give some details on how we constructed the diamond diagrams.

9.1 Harnack's Theorem

For nonsingular curves, each simple closed path is a topological component, and topological components are pseudo-lines and ovals. But for singular curves (including reducible curves), simple closed paths may intersect at singularities to be part of the same topological component. When we perturb a singular curve to create a nearby nonsingular curve, these simple closed paths may split into separate components or not, but a small random perturbation is very unlikely to undo a previous perturbation and glue topological components together. Special consideration is needed for curves with isolated points or singularities on the infinite line; the latter is easily handled by changing to an FLT equivalent curve without infinite singularities.

Consider the three-leaf clover:

```
In[1]:= f = (x^2+y^2)^2+3x^2y-y^3;
        ContourPlot[{f == 0, f-.05 == 0, f+.05 == 0}, {x, -1, 1}, {y, -.8, 1.2}]
```

Out[2]=

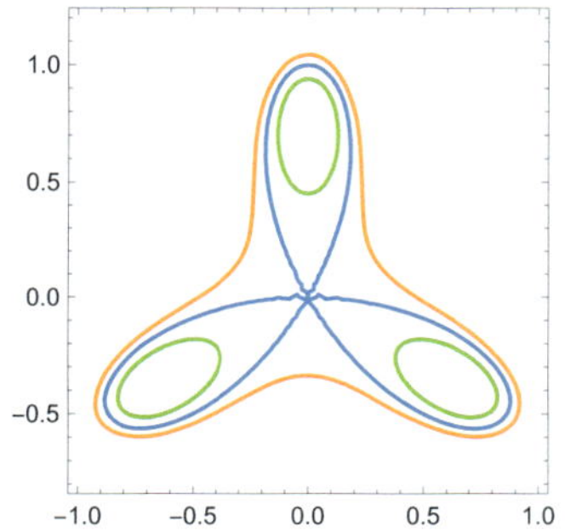

We originally had one topological component: having subtracted a small constant (the number of simple closed paths) but leaving still one topological component. Subtracting a small constant splits the curve into three topological components.

We want to put a bound on the number of topological components possible for a curve of degree d. Since random perturbation does not subtract components, it is enough to look at nonsingular curves. We will use three tools from earlier sections of the book:

1. At the end of Chapter 1, we showed that one can construct a curve of degree d through any affine collection of $\frac{(d+2)(d+1)}{2} - 1$ points.
2. Properties 1, 2, and 3 from Section 8.5.
3. Bézout's theorem from Chapter 5.

The following example will be a warm-up. In Chapter 7, we catalog all possible conics and cubics, but we will give a new argument that a nonsingular cubic can have only two topological components.

Suppose a cubic has three topological components: two ovals and one pseudo-line. We pick one point on each oval, say p and q. There is a unique line ℓ through p and q that, by Property 2 (Section 8.5), must meet the pseudo-line at a point r.

A plot might look like:

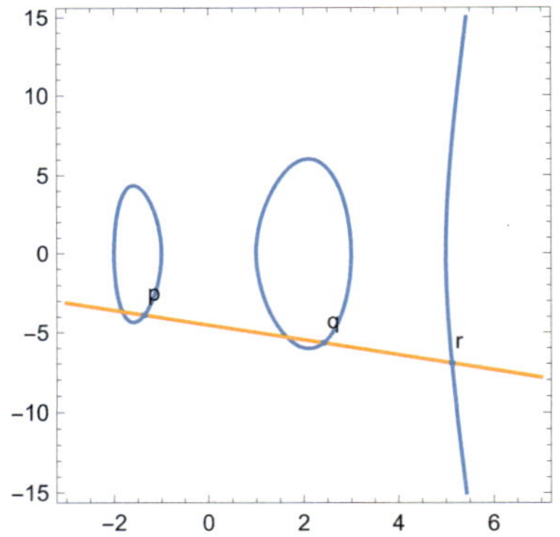

ℓ is the orange line. By a trivial application of Bézout's theorem, a line can only intersect a cubic in three or fewer real points, but this line also intersects the cubic in two other points on the ovals. Therefore, this configuration is impossible for a cubic.

Suppose $f = 0$ is a projective curve of degree $d > 3$, then

$$m = \frac{(d+1)(d+2)}{2} + 1.$$

Suppose the associated projective curve has a simple closed path decomposition with m ovals $\Omega_1, \Omega_2, \ldots, \Omega_m$ and one more simple closed path Γ (which might be an oval if d is even, or pseudo-line if d is odd). In the second case, Γ is certainly infinite. We will assume our curve is not just a collection of isolated points and Γ is an oval (which is not an isolated point and, therefore, infinite in the first case). We pick a random point on each oval Ω_i but not a singular point (therefore not on any other Ω_j).

We now pick $\frac{d(d-1)}{2} - 1 - m$ random points on Γ; this number is positive since $d > 3$. Again, choosing randomly ensures that these points are not also on any of the Ω_i.

So we have $\frac{d(d-1)}{2} - 1$ points, and by our global interpolation method from Chapter 1, this is a curve $g = 0$ of degree $d - 2$ through these points. By our construction, $f = 0$ and $g = 0$ cannot share a common component, but by Property 1 (Section 8.5), the total intersection multiplicity of $g = 0$ with each oval $\Omega_1, \ldots, \Omega_m$ must be even, so $g = 0$ intersects each of these ovals in at least two points.

Altogether, by letting the Wolfram Language do the algebra, we have the following points of intersection of $f = 0$, $g = 0$:

```
In[3]:= Expand[2 m+d (d-1)/2-1-m]
```

$$\text{Out[3]=}\quad -1-\frac{d}{2}+\frac{d^2}{2}+m$$

By Bézout's theorem, we can have the following points:

```
In[4]:= Expand[d (d-2)]
```

$$\text{Out[4]=}\quad -2\,d+d^2$$

We counted at least one more point than we were supposed to have, so our supposition above cannot be true.

This leads to Harnack's theorem: let $f = 0$ be an algebraic curve of degree d, which is not a union of isolated points. The number of topological connected components of this curve in the real projective plane is less than or equal to $\frac{(d-1)(d-2)}{2} + 1$. Moreover, if singular $f = 0$ admits a decomposition into simple closed curves, satisfying Property 3 (Section 8.5), then the number of simple closed curves in this decomposition is also less than or equal to $\frac{(d-1)(d-2)}{2} + 1$.

Our argument follows the argument in Bochnak's book (1998) except—as noted above—we can slightly perturb a singular curve to get our decomposition into ovals, satisfying our properties in Section 8.5, and the second part only requires these properties.

The first term in m in the nonsingular case reminds classical algebraic geometers of the genus. In fact, most classical geometers will prove Harnack's theorem using the genus (see Appendix 6 in Fischer's book [2001]), but Harnack's theorem has little to do with the genus.

Consider the following fourth-degree curves:

```
In[5]:= {ContourPlot[x^4+y^4-1 == 0, {x, -1.5, 1.5}, {y, -1.5, 1.5}],
  ContourPlot[16-20 x^2+4 x^4-20 y^2+17 x^2 y^2+4 y^4 == 0, {x, -2.5, 2.5}, {y, -2.5, 2.5}]}
```

Out[5]= 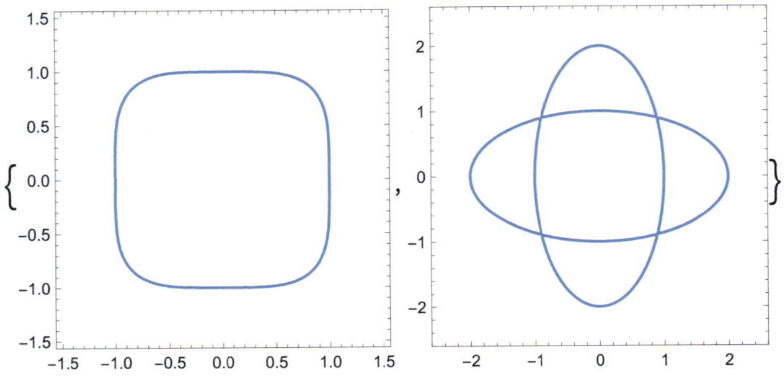

On the left, we have a genus-3 curve with one oval. On the right, a genus-0 curve with decompositions into two to four ovals.

9.2 Descartes's Rule and Viro Diagrams

The most complicated procedure in this book is the exact procedure that draws Viro diagrams. For continuity of our discussion, we save this code for the end of this chapter.

Descartes's Rule

We return to your college algebra class where you learned Descartes's rule. Although it is no longer useful as a tool for solving univariate polynomial equations, it is still a useful theoretical idea. Here, we work exclusively with numerical polynomials which implies, among other things, that no coefficient is exactly zero and there are no multiple zeros. Then Descartes's rule can be formulated in the following combinatorial and visual way:

First, assign to a polynomial $f = a_0 + a_1 x + \cdots + a_d x^d$ a list $s = \{s_1, s_2, \ldots, s_{d+1}\}$, where s_j is zero if a_{j+1} is positive and s_j is 1 if a_{j+1} is negative. Note: the index shift and some of the other adjustments to make this compatible with Mathematica. We now have

$$f = (-1)^{s_1} |a_0| + (-1)^{s_2} |a_1| x + \cdots + (-1)^{s_{d+1}} |a_d| x^d.$$

Then we draw the diagram as follows: on a horizontal axis, if $0 < j \leq d+1$, put a green dot at $\{j-1, 0\}$ if $s_j = 0$ and put a red dot if $s_j = 1$. For $-d \leq j \leq 0$ if j is even, we have the same color as at $-j$ while if j is odd, we put the opposite color.

Some examples:

In[6]:=
```
f = -4.872507552870091` + 3.4368580060422964` x - 0.2018126888217524` x^2 -
    0.22054380664652573` x^3 + 0.030211480362537763` x^4
DD1 = descartesDiagram[coefficientSigns[f, {x}]]
```

Out[6]= $-4.87251 + 3.43686\,x - 0.201813\,x^2 - 0.220544\,x^3 + 0.0302115\,x^4$

Out[7]=

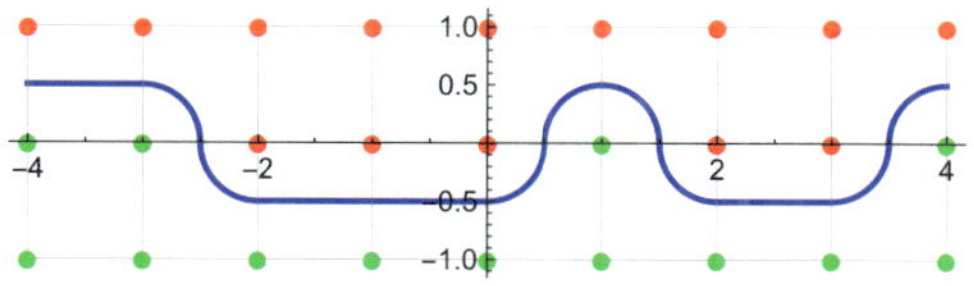

Let *g* be the degree 7 polynomial:

In[8]:=
```
g = 10.070348961821836` - 3.1854981245813794` x + 1.6605231078365708` x^2 -
    0.5956460147354319` x^3 - 0.4997923643670461` x^4 + 0.13414266577361014` x^5 +
    0.004018754186202277` x^6 - 0.0016744809109176153` x^7
DD2 = descartesDiagram[coefficientSigns[g, {x}]]
```

Out[8]= $10.0703 - 3.1855\,x + 1.66052\,x^2 - 0.595646\,x^3 - 0.499792\,x^4 + 0.134143\,x^5 + 0.00401875\,x^6 - 0.00167448\,x^7$

Out[9]=

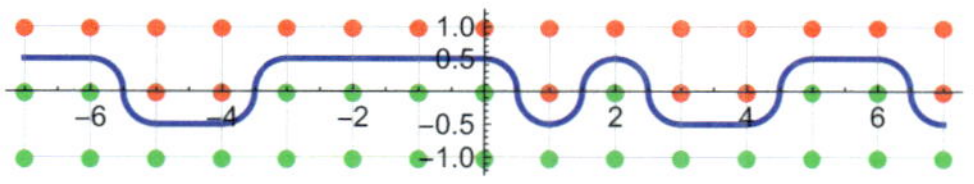

In both examples, the total number of crossings of the *x*-axis (the middle horizontal line in the diagram) is exactly *d*. Descartes's rule can now be stated:

The number of positive zeros of a univariate polynomial is never more than the number of positive crossings in the diagram and differs from this diagram by an even number. Likewise, the number of negative zeros of the polynomial is never more than the number of negative crossings and differs by an even number.

The Descartes's diagram for *f* says we must have one negative crossing and one or three positive ones. For *g*, there can be zero or two negative crossings and one, three, or five positive ones.

Compare the Descartes's diagram below with a clipped plot for a large *x*-range in both examples above:

In[10]:=
```
Column[{DD1, Plot[f, {x, -10, 10}, PlotRange → 1,
    PlotStyle → Blue, ClippingStyle → Blue, AspectRatio → .2]}]
```

Out[10]=

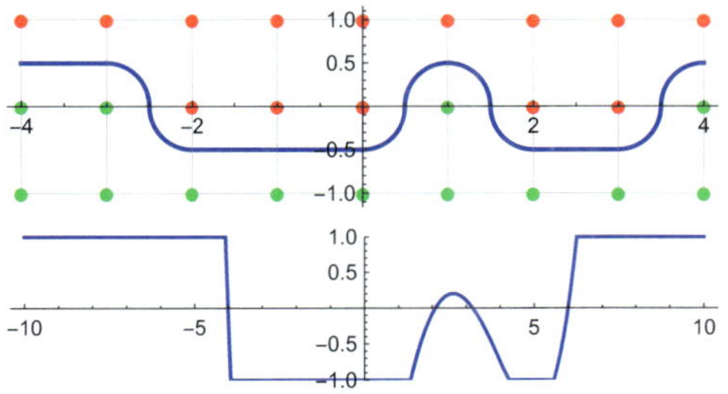

In[11]:=
```
Column[{DD2, Plot[g, {x, −10, 10}, PlotRange → 1,
    PlotStyle → Blue, ClippingStyle → Blue, AspectRatio → .2]}]
```

Out[11]=

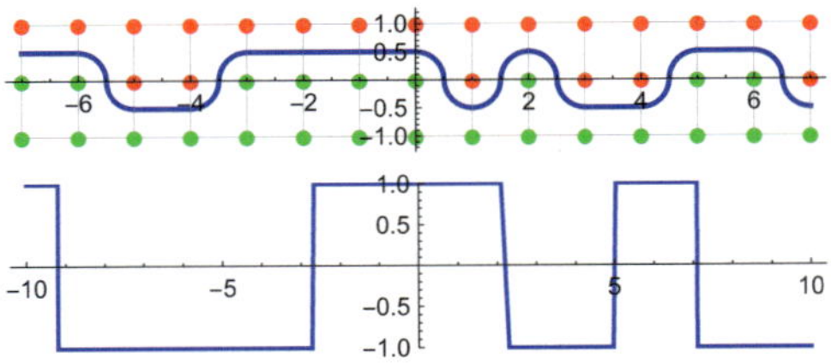

The two match for f, both in number of crossings and asymptotically—that is, the clipped plot is positive at either end if the Desargues's plot is as well. For g, there is again an asymptotic match, but there are two fewer positive crossings.

For later purposes, we will use a sketch rather than a clipped plot (for example, using g above) and techniques from the end of the chapter.

We would kill the first two positive zeros and get:

In[12]:=
```
G = <|{−7} → 0, {−6} → 0, {−5} → 1, {−4} → 1, {−3} → 1, {−2} → 0, {−1} → 0,
    {0} → 0, {1} → 0, {2} → 0, {3} → 1, {4} → 1, {5} → 0, {6} → 1, {7} → 1|>;
Column[{DD2, descartesSketch[G, 7]}]
```

Out[12]=

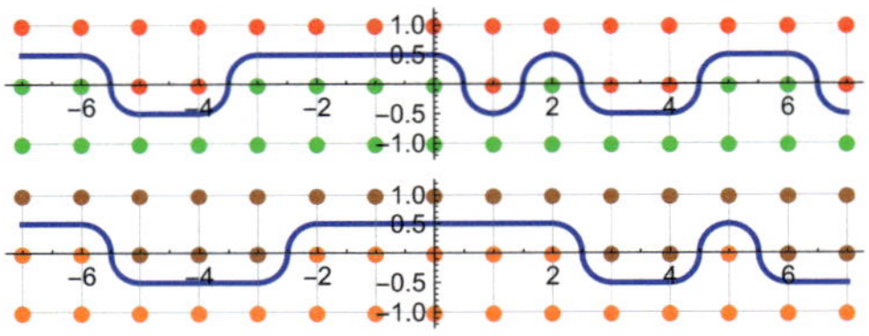

The green-red plot is the prediction from Descartes's rule while the orange-brown plot is a representation of the actual plot. It should be noted that this representation is produced by a human, not a machine, and, therefore, is not unique. We will continue these conventions.

The variable g is called an *association*. This data type will be discussed in Section 9.4. You can also find out more from the Wolfram Language & System Documentation Center or in Stephen Wolfram's online book *An Elementary Introduction to the Wolfram Language* (2017).

Viro Diagrams

A two-dimensional version of this is called a Viro diagram after Oleg Viro, a Russian mathematician who proposed a similar idea in 1978. It is important to note that the diagrams here are different from the diagrams in Viro's famous 1978–1980 papers. In fact, these are abstractions of Gauss's idea: a positive plane region using lattice points and bound by a curve, but (as in the Descartes's diagrams) we can get the first quadrant region from the coefficient signs and try to predict what will happen in the other three quadrants.

We have a bivariate numerical polynomial of degree d,

$$f[x, y] = a_{0,0} + a_{1,0}\, x + a_{0,1}\, y + a_{2,0}\, x^2 + a_{1,1}\, x\, y + a_{0,2}\, y^2 + \cdots + a_{0,d}\, y^d.$$

By "numerical," we mean that all the coefficients $a_{i,j}$ are nonzero machine numbers. Moreover, assume the solution curve $f[x, y] = 0$ is nonsingular (as in Chapter 5) and $\{1, 0, 0\}$ and $\{0, 1, 0\}$ are not infinite points (although there may be infinite points near them). The goal is to count the number of points where the curve crosses the three axes $x = 0$, $y = 0$, and $z = 0$, where the last axis is the infinite line. These axes, in turn, can be divided into positive and negative parts, where, for the infinite line, $\{x, y, 0\}$ is positive if $x\,y > 0$ (that is, points adjacent to quadrants I and III in the high-school plane).

We want to count how many of these crossing points are positive or negative:

In[14]:=
```
f = newtonHyperbola[413, x, y]
viroDiagram[f, x, y]
```

Out[14]= $-1 + x - 0.2\,x^2 + 0.008\,x^3 - y - x\,y - 0.2\,x^2\,y + 0.2\,y^2 - 0.2\,x\,y^2 + 0.008\,y^3$

Out[14]=

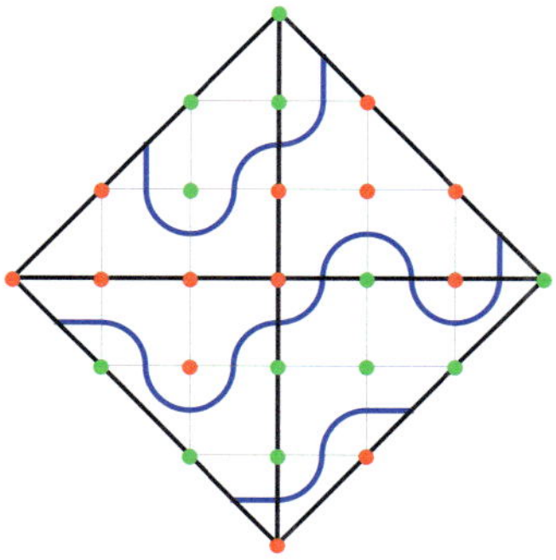

We have a diamond with lattice points $\{i, j\}$ on the boundary and in the interior for $|i| + |j| \leq 3$. Each such point with $0 \leq i, j$ in the first quadrant represents a monomial $x^i\,y^j$ of f. If the coefficient $a_{i,j}$ is positive, the lattice point will be pictured by a green dot. If the coefficient is negative, the coefficient will be given by a red dot. In the diagram above, $\{0, 0\}$ is red since the constant coefficient -1 is negative, but $\{1, 0\}$ is green since the coefficient of x is 1, and so on.

In the first quadrant, the monomial $x^i\,y^j$ will always be positive off the axes, so the sign of the term $a_{i,j}\,x^i\,y^j$ will depend on the sign of the coefficient $a_{i,j}$, but in the other quadrants, $x^i\,y^j$ will not always be positive. This will depend on which quadrant it is and whether i and j are even or odd. The color of a dot will depend on the sign of the term, so when $x^i\,y^j$ is negative, the color will change. In the second quadrant, x is negative and y is positive, so $x^i\,y^j$ is positive if i is even (otherwise $x^i\,y^j$ will be negative). The color changes for odd i but remains the same for even i. In the third quadrant, x and y are both negative, so $x^i\,y^j$ is positive if both i and j are positive or are both negative. In other words, $x^i\,y^j > 0$ if i and j have the same parity. Thus colors are preserved (or off-diagonal lines). The fourth quadrant is like the second but the color changes for odd j.

A pictorial version is obtained from a polynomial with all coefficients positive, e.g. 1 where here Red means change color:

In[16]:=
```
S = Select[Flatten[Table[{i, j}, {i, 0, 6}, {j, 0, 6}], 1], Total[#] ≤ 6 &];
A = Association[Table[p → 0, {p, S}]];
Show[plotAssoc[dvAssoc[A], Green, Red], Axes → True]
```

Out[16]=

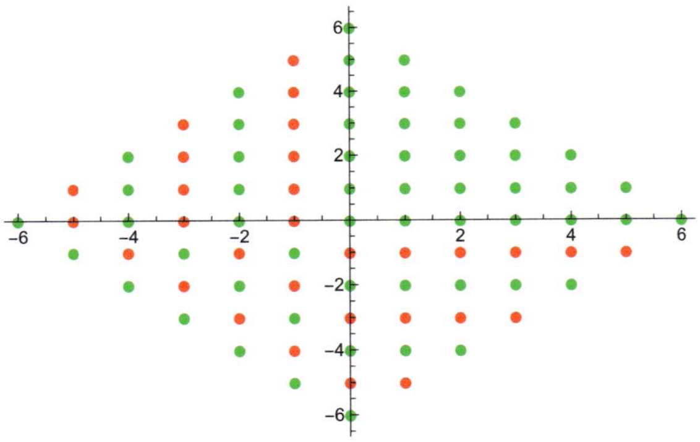

We then draw boundary curves separating the green and red regions. As Gauss pointed out, these curves are continuous and are closed or only have endpoints at the edge of the drawing. We accomplish this in this setting by considering the squares with vertices $\{\{i, j\}, \{i+1, j\}, \{i+1, j+1\}$, and $\{i, j+1\}\}$. If adjacent vertices are a different color, then the curve should go through the midpoint of that edge. Each square must have an even number of such edges: 0, 2, or 4. The cases with zero or two edges are clear. If there are four edges, then opposite vertices are the same color and we have an ambiguous situation.

By default, we draw the two possibilities as dashed lines, one cyan and one magenta:

In[19]:=
```
D1 = viroDiagram[newtonHyperbola[613, x, y], x, y]
```

Out[19]=

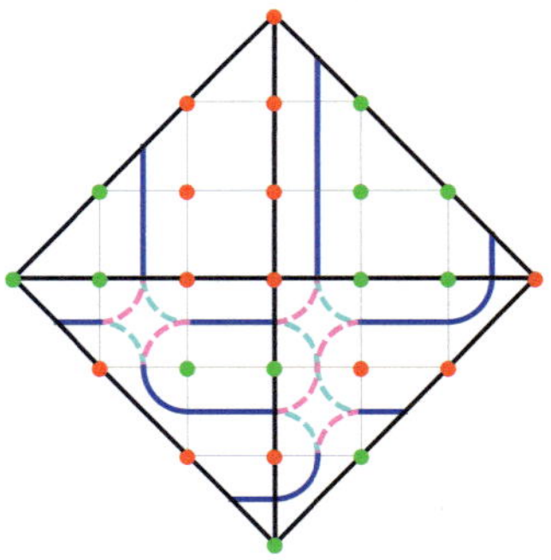

We have some utility functions to allow the user to pick specific choices. In addition, although our standing assumption in this section is that our curves are nonsingular, we can represent a simple crossing as well.

Squares are indexed by their lower-left vertex:

```
In[20]:= D2 = Show[D1, pickMagenta[-2, -1], pickCyan[0, -1], pickSingularity[0, -2]]
```

Out[20]=

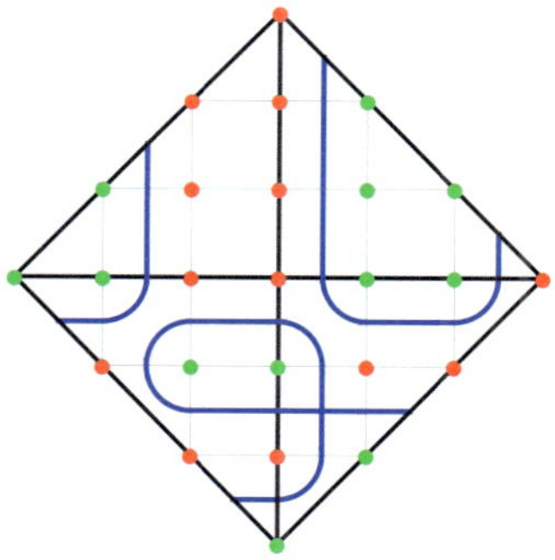

In the diagram above, the number of crossings of the x, y, and infinite (z) axes is exactly three. Here, the infinite axis can be represented by two consecutive outer edges; for example, the edges of the first and second quadrant. From the projective point of view, if a curve hits the outer edge in the first quadrant, it reappears at the antipodal outer edge of the third quadrant (likewise for quadrants two and four).

Optionally, we also have utility functions to label these "infinite" points. There are four of these to give proper placement.

The triangles with diagonal outer edges are indexed by their right-angled vertex:

```
In[21]:= D3 = Show[D2, labelI[0, 2, "A"], labelI[2, 0, "B"], labelII[-1, 1, "C"],
           labelIII[-2, 0, "B"], labelIII[0, -2, "A"], labelIV[1, -1, "C"]]
```

Out[21]=

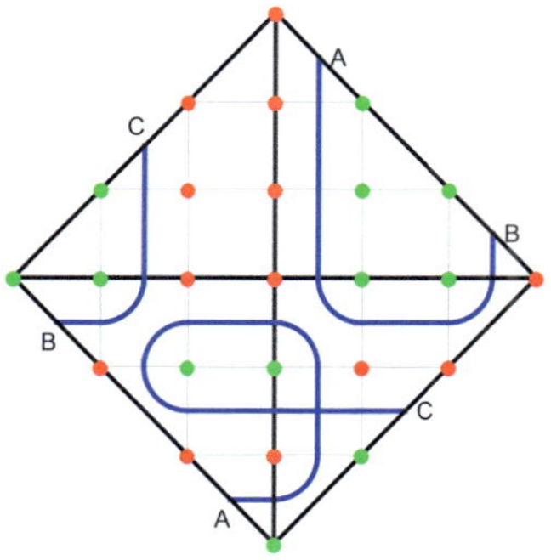

If we label additional points using the Wolfram Language's Labeled function, we can then describe the projective graph (Chapter 5), which only has one topological component:

```
In[22]:= {Show[D3, ListPlot[{Labeled[{.5, -1.5}, "S"],
            Labeled[{1.5, -.5}, "Q"], Labeled[{-1, -.5}, "R"], Labeled[{-1.5, 0}, "P"]}]],
       Graph[{"A" → "Q", "Q" → "B", "B" → "P", "P" → "C", "C" → "S",
          "S" → "R", "R" → "S", "S" → "A"}, VertexLabels → "Name"]}
```

Out[22]=

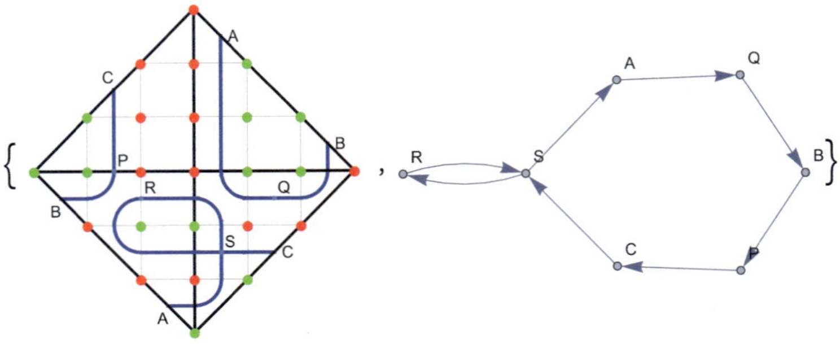

This Viro diagram, especially with our choices at the ambiguous squares, is only a representation of a theoretically possible curve, not the curve on Newton Hyperbola 613 (which is nonsingular, among other things). In the next section, we will give the correct diagram for Newton Hyperbola 613. Like Descartes's rule, our Viro diagrams simply give constraints on the geometry and topology of a curve with given coefficient signs, but any given curve may be missing components, infinite points, or just crossings of the axes.

Newton Hyperbolas

Newton was interested in cataloging real cubic functions sometime between 1676 and 1704 (Struik, 1986), and he called them hyperbolas because of the asymptotes. In Chapter 1, I have generated a numbered list of approximately 2^{28} curves up to degree 6 with asymptotes (which I call as such in honor of Newton's work). These curves incorporate each sign sequence and maximize the number of quadrant changes in their Viro diagrams, but the actual topology of these curves may not be the same as the abstract curve of the Viro diagram.

We start this time with Newton Hyperbola 613:

In[23]:=
```
f = newtonHyperbola[210, x, y]
```

Out[23]= $1 - x + 0.2\,x^2 - 0.008\,x^3 + y - x\,y - 0.2\,x^2\,y + 0.2\,y^2 + 0.2\,x\,y^2 + 0.008\,y^3$

We choose the cyan path. We compare the resulting diagram to a RegionPlot of the curve; the shaded part should correspond somewhat to the green part of the Viro diagram:

In[24]:=
```
{Show[viroDiagram[f, x, y], pickCyan[0, -1]],
  RegionPlot[f > 0, {x, -7, 7}, {y, -7, 7}, Axes → True]}
```

Out[24]=

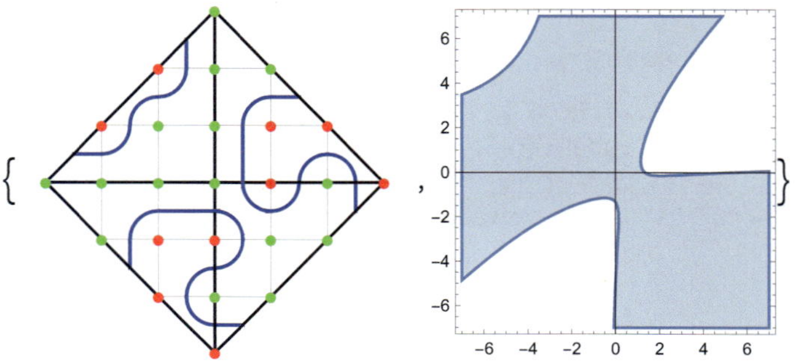

The scale of the region plot makes it hard to see, but f has three zeros on the positive x-axis and negative y-axis.

There is also one asymptote in quadrants I and III and two in quadrants II and IV:

```
In[25]:= {x, y} /. NSolve[{f, y}]
         {x, y} /. NSolve[{f, x}]
         infiniteRealPoints[f, x, y]
```

```
Out[25]= {{1.33975, 0.}, {5., 0.}, {18.6603, 0.}}

Out[26]= {{0., −18.6603}, {0., −5.}, {0., −1.33975}}

Out[27]= {{2.66334, −0.102588, 0}, {1.11393, 1.11393, 0}, {−0.0789843, 2.05055, 0}}
```

In this case, the actual curve matches the topology of the Viro diagram.

However, we can make the positive region smaller by changing the constant term to 0.3 (which does not change the sign of that term and hence gives the same ambiguous Viro diagram).

Here, the match comes by picking the magenta curve in that square:

```
In[28]:= g = f − .7
         {Show[viroDiagram[g, x, y], pickMagenta[0, −1]],
           RegionPlot[g > 0, {x, −7, 7}, {y, −7, 7}, Axes → True]}
```

Out[28]= $0.3 - x + 0.2\,x^2 - 0.008\,x^3 + y - x\,y - 0.2\,x^2\,y + 0.2\,y^2 + 0.2\,x\,y^2 + 0.008\,y^3$

Out[29]= {

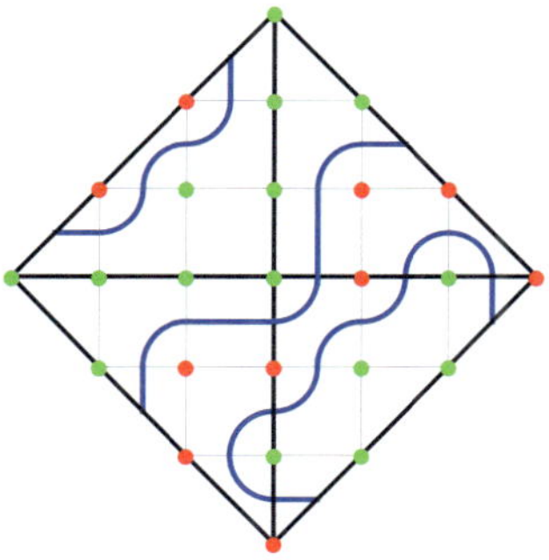

,

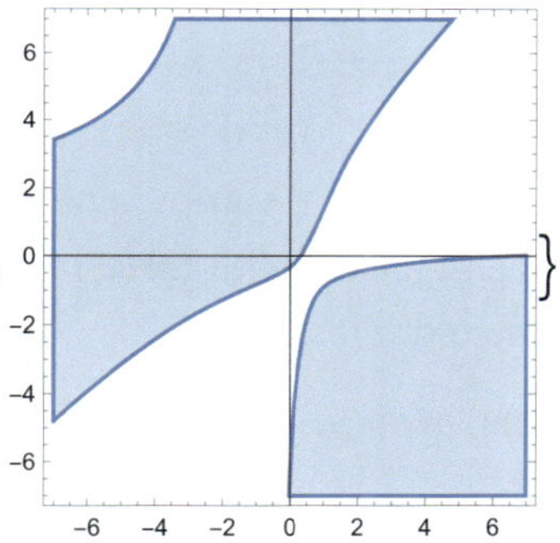

}

```
In[30]:= {x, y} /. NSolve[{g, y}]
         {x, y} /. NSolve[{g, x}]
         infiniteRealPoints[g, x, y]
```

```
Out[30]= {{0.320249, 0.}, {6.40892, 0.}, {18.2708, 0.}}

Out[31]= {{0., −18.2708}, {0., −6.40892}, {0., −0.320249}}

Out[32]= {{−1.44005, 0.0554689, 0}, {−0.697996, −0.697996, 0}, {0.0563905, −1.46398, 0}}
```

Note that this does change the actual zeros on the positive x-axis and negative y-axis but not the number of zeros. The infinite points are unchanged.

Finally, we look at:

```
In[33]:= h = f - .38587
         {Show[viroDiagram[h, x, y], pickSingularity[0, -1]],
           RegionPlot[h > 0, {x, -7, 7}, {y, -7, 7}, Axes → True]}
```

Out[33]= $0.61413 - x + 0.2\,x^2 - 0.008\,x^3 + y - x\,y - 0.2\,x^2\,y + 0.2\,y^2 + 0.2\,x\,y^2 + 0.008\,y^3$

Out[34]= {

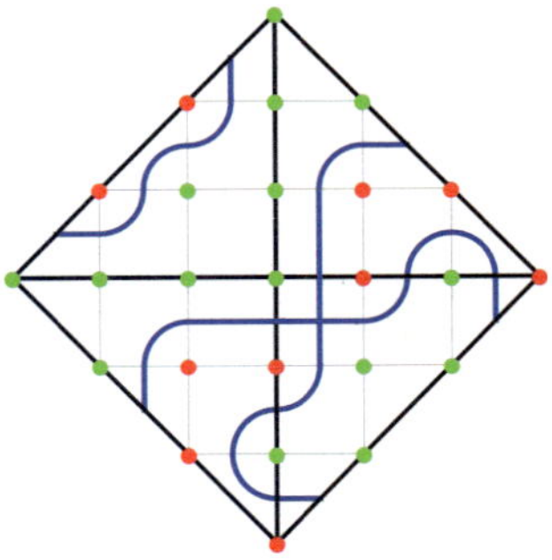

,

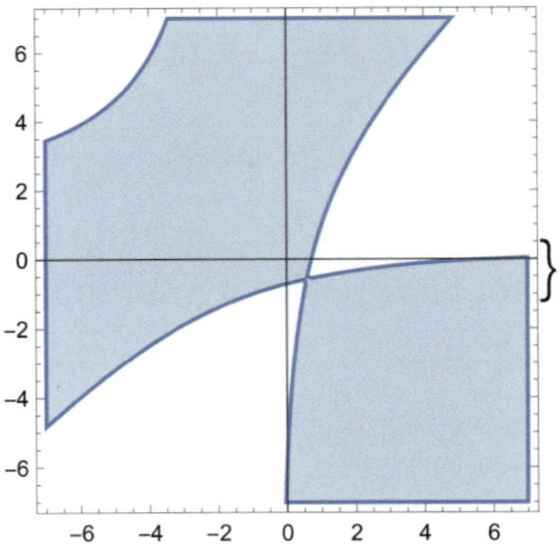

}

This example shows why we must keep the ambiguous squares ambiguous, generally all cases could be possible.

When Harnack stated his curve theorem in 1876, he claimed to have maximal examples (called M-curves) for each degree—that is, a curve of degree d with $\frac{(d-1)(d-2)}{2} + 1$ connected projective components. Viro has a recipe for the diagrams and a proof of existence of curves with these topologies. Here is a Newton-hyperbola-friendly version constructed inductively.

For even degrees, we alternate signs in the highest degree; whereas for odd degrees, all signs are negative:

```
In[35]:= f0 = 1;
         f1 = f0 - x - y; (* NH6 *)
         f2 = f1 + .2 x^2 - x y + .2 y^2; (* NH22 *)
         f3 = f2 - .2^3 x^3 - .2 x^2 y - .2 x y^2 - .2^3 y^3; (* NH982 *)
         f4 = f3 + .2^6 x^4 - .2^3 x^3 y + .2 x^2 y^2 - .2^3 x y^3 + .2^6 y^4; (* NH11222 *)
         f5 = f4 - .2^10 x^5 - .2^6 x^4 y - .2^3 x^3 y^2 - .2^3 x^2 y^3 - .2^6 x y^4 - .2^10 y^5;
         (* NH2075606 *)
         f6 = f5 + .2^15 x^6 - .2^10 x^5 y + .2^6 x^4 y^2 - .2^3 x^3 y^3 +
                 .2^6 x^2 y^4 - .2^10 x y^5 + .2^15 y^6; (* NH90155990 *)
```

```
In[42]:= {{Labeled[viroDiagram[f3, x, y], "d=3"], Labeled[viroDiagram[f4, x, y], "d=4"]},
  {Labeled[viroDiagram[f5, x, y], "d=5"], Labeled[viroDiagram[f6, x, y], "d=6"]}}
```

Out[42]= {{

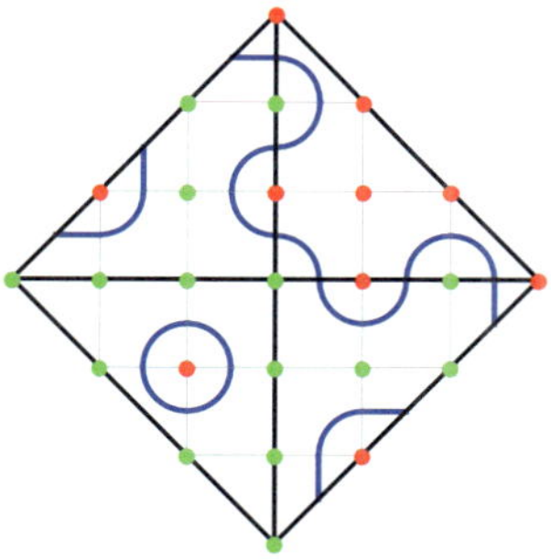

d=3

,

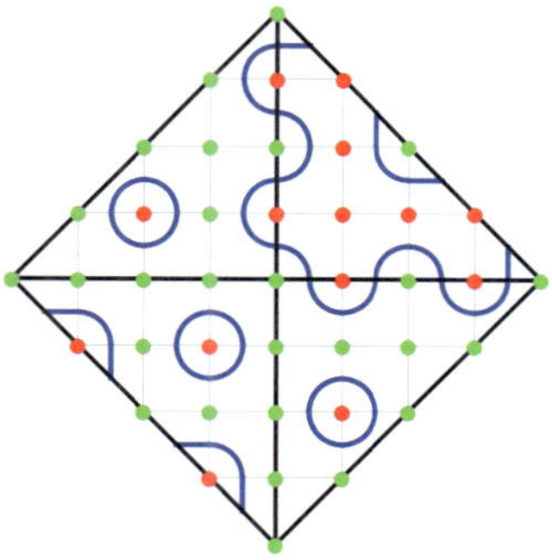

d=4

},

{

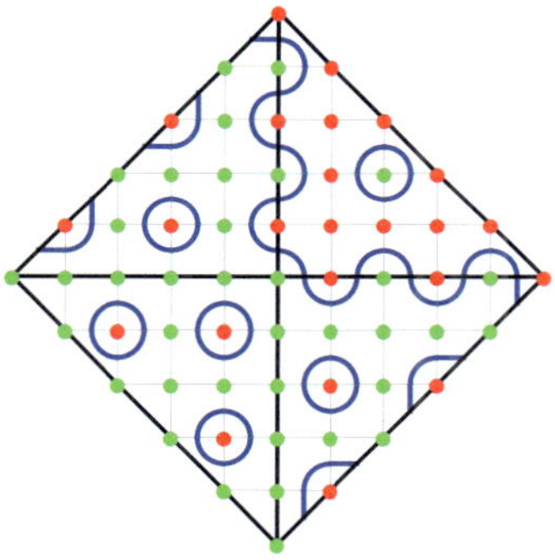

d=5

,

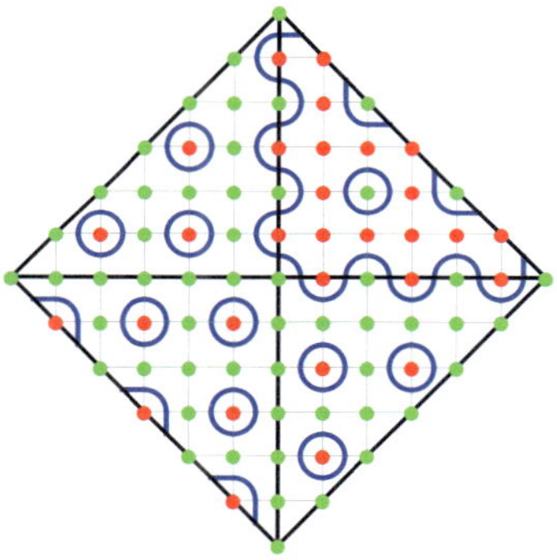

d=6

}}

My Newton hyperbolas do not give curves with the topologies of their Viro diagrams; they only give the large component that intersects the positive x and y axes.

For $d = 3$, we can easily modify $f3$ to get:

```
In[43]:= h3 = f3 - 2 x y
```

Out[43]= $1 - x + 0.2\,x^2 - 0.008\,x^3 - y - 3\,x\,y - 0.2\,x^2\,y + 0.2\,y^2 - 0.2\,x\,y^2 - 0.008\,y^3$

```
In[44]:= {viroDiagram[h3, x, y], RegionPlot[h3 > 0, {x, -40, 40}, {y, -40, 40}, Axes → True]}
```

Out[44]= {

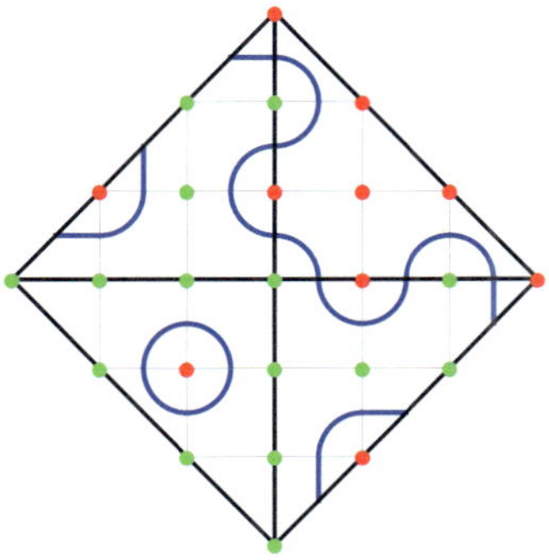

,

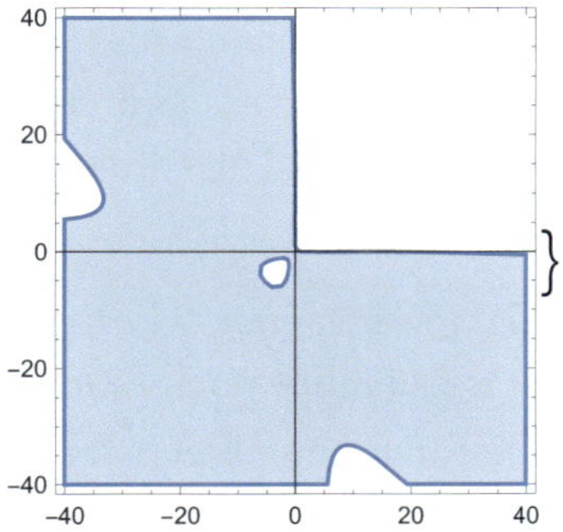

}

For $d = 4$, we take:

```
In[45]:= h4 = f4 - x^2 y - 6 x y - x y^2
         {viroDiagram[h4, x, y], RegionPlot[h4 > 0, {x, -33, 33}, {y, -33, 33}, Axes → True]}
```

Out[45]= $1 - x + 0.2\,x^2 - 0.008\,x^3 + 0.000064\,x^4 - y - 7\,x\,y - 1.2\,x^2\,y - 0.008\,x^3\,y + 0.2\,y^2 - 1.2\,x\,y^2 + 0.2\,x^2\,y^2 - 0.008\,y^3 - 0.008\,x\,y^3 + 0.000064\,y^4$

Out[46]= {

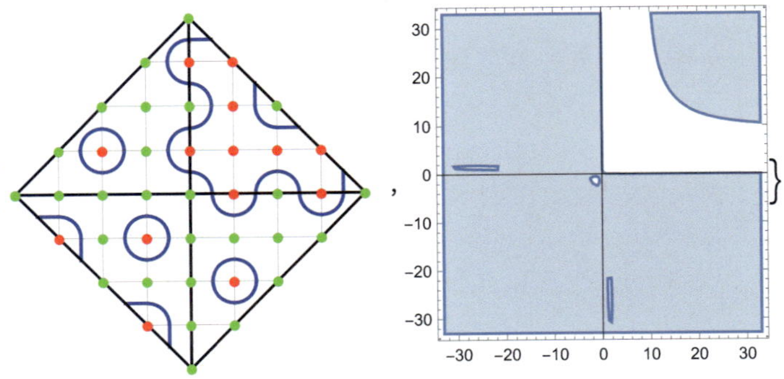

}

While this scale is not large enough to show the asymptotes in the third quadrant or the crossings in the first quadrant, we see that these must exist:

```
In[47]:= {x, y} /. NSolve[{h4, x}]
         {x, y} /. NSolve[{h4, y}]
         infiniteRealPoints[h4, x, y]
```

```
Out[47]= {{0., 1.34015}, {0., 4.90554}, {0., 25.4814}, {0., 93.2729}}

Out[48]= {{1.34015, 0.}, {4.90554, 0.}, {25.4814, 0.}, {93.2729, 0.}}

Out[49]= {{0.10831, 3.73502, 0}, {0.0393735, 3.56234, 0},
           {-1.39901, -0.0405692, 0}, {-1.38894, -0.0153515, 0}}
```

Of course, there are easier, better known M-curves in degree 3 and 4. I will leave it to the reader to come up with explicit examples of curves with the topologies of $f5$ and $g5$.

9.3 Description of Curves by Gauss Diagrams

The stated goal of this book was: given a bivariate polynomial $f(x, y)$, we want to describe the geometry of the real solution set $f(x, y) = 0$ as concisely and accurately as possible, giving the affine and/or projective topology (as well as some positional information). So far, we have considered Euler graphs, Gauss hemisphere plots, as well as plots of the curve (or components on the Möbius band model of the projective plane). I now propose a Gauss diagram as it visually gives the most information, and, in the spirit of Gauss, we view a real algebraic plane curve as the complete boundary of its positive set $f(x, y) = 0$.

A Gauss diagram is a diamond diagram (like the Viro diagrams), but here, the lattice points in the interior represent affine points and the color depends on whether the affine point is in a positive or negative complimentary region of the curve. Points on the x and y axes of the diagram represent points actually on the axes. The points on the outer boundary (z-axis) represent infinite points. The color depends on the parity of nearby affine points. A curve is drawn, separating the positive and negative points. Those cells with diagonal positive and negative points are again ambiguous and the choice of curves inside these cells must be set by the user, as in Viro diagrams. In general, it is not expected that the Viro rules will hold for Gauss diagrams. However, if they hold on to the axes (x, y, z) mod 2 (meaning the number of axis crossings can be an even number less than those in a Viro diagram), we then call the Gauss diagram *plausible.*

I claim that every curve of degree d has a plausible Gauss diagram of size d, but this is not obvious nor do I have a proof. Such a structural result would be at least as deep as Harnack's theorem. I also do not claim that each plausible Gauss diagram actually matches a real plane curve, but it would be a nice theorem if someone could prove this.

In this chapter, we indicate how to attempt a plausible Gauss diagram for a given curve, but first, we introduce two new functions.

The first is axisPoints (points on the *x* or *y* axes):

```
In[50]:= axisPoints[f_, x_, y_] := {x, y} /. NSolve[{f, x y}, {x, y}, Reals];
```

In Chapters 6, 7, and 8, we learned how to examine the projective plane near the infinite points. The following will make this easier (we will give an example later).

The inspectInfPoint function puts together several other functions to allow us to quickly see what is happening at an infinite point and give affine points that directly lead to the infinite points. *ip* is the infinite point and *r* is a positive radius usually less than one. We want a circle that intersects the curve only near the infinite points.

Data shown in inspectInfPoint can be changed by Options. "plot" gives just the plot, "data" gives just the data, and the default All gives both:

```
In[51]:= Options[inspectInfPoint] = {output → All};
inspectInfPoint[f_, ip_, r_, x_, y_, OptionsPattern[]] :=
  Module[{A, g, bp, ap, sol, GP},
   A = ip2z[ip, x, y];
   g = FLT[f, A, x, y];
   sol = NSolve[{g, x^2 + y^2 - r^2}, {x, y}, Reals];
   If[Length[sol] > 0, bp = {x, y} /. sol;
    ap = flti[#, Inverse[A]] &/@ bp,
    bp = {}; ap = {}];
   GP := Show[gaussPlot[g, r, bp], ContourPlot[y == 0, {x, -r, r}, {y, -r, r},
       ContourStyle → Directive[Orange, Dashed]], ImageSize → Small];
   Switch[OptionValue[output], "plot", Return[GP], "data",
    Return[{A, g, bp, ap}], Default, Continue];
   Echo[A, "Transformation"];
   Echo[g, " Transformed equation"];
   Echo[bp, "boundary points coordinates relative to infinite point"];
   Echo[ap, "affine coordinates boundry points"];
   GP]
```

Consider first the cubic f = NH413 from earlier in this chapter (shown again below):

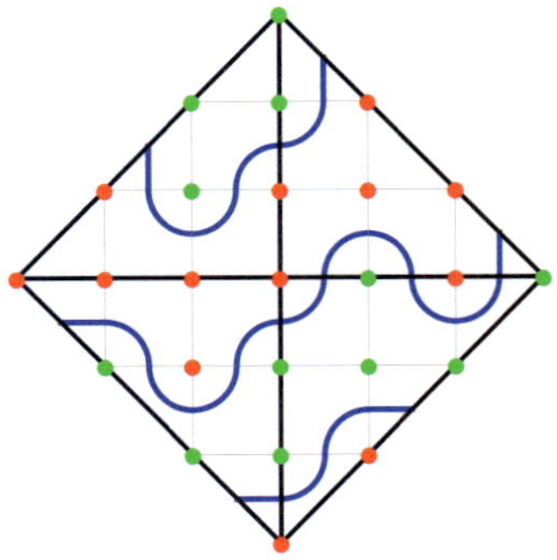

Newton hyperbolas generally have the property that the underlying vertex colors can come, mod 2.

From the Viro diagram:

```
In[53]:= f = newtonHyperbola[413, x, y]
         P = viroDiagram[f, x, y]
```

Out[53]= $-1+x-0.2\,x^2+0.008\,x^3-y-x\,y-0.2\,x^2\,y+0.2\,y^2-0.2\,x\,y^2+0.008\,y^3$

Out[54]=

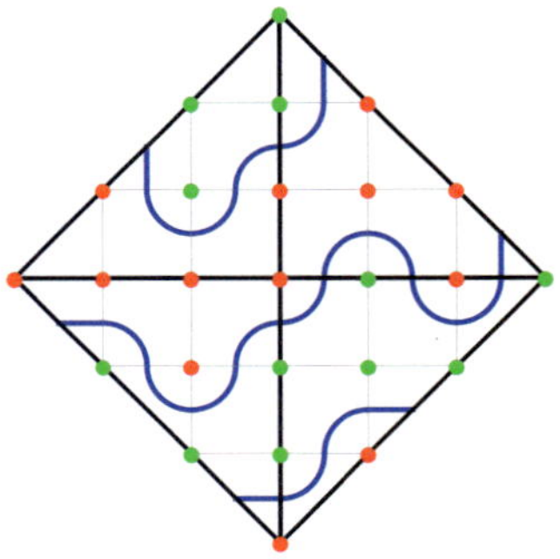

This gives the most general picture possible (based on Descartes's rule), but all features shown may not be present; for example, there could be only one infinite point instead of three. Our next steps should be to compute the infinite points, the default critical points, the zeros on the axes, and the region plot.

If possible, this should be done with a large enough range to contain the critical points and axes zeros; otherwise, several plots may be needed:

```
In[55]:= infiniteRealPoints[f, x, y]
         criticalPoints[f, x, y]
         axisPoints[f, x, y]
```

Out[55]= {{0.647067, 16.7988, 0}, {−2.97058, −0.114423, 0}, {−2.42699, 2.42699, 0}}

Out[56]= {{3.01621, −11.798}, {−1.1593, 2.4495}, {0.383291, −0.440571}}

Out[57]= {{0., −29.1421}, {18.6603, 0.}, {5., 0.}, {0., 5.}, {1.33975, 0.}, {0., −0.857864}}

In[58]:=
```
RegionPlot[f > 0, {x, -35, 35}, {y, -35, 35}]
```

Out[58]=

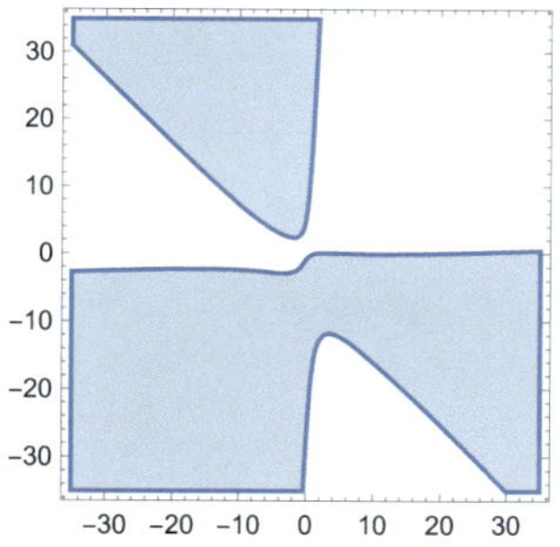

In this case, all the features are accounted for: the three infinite points, the three affine components of the curve, the three positive zeros on the x-axis, and the one positive and two negative zeros on the y-axis.

We can finish our picture by labeling and constructing the graph:

In[59]:=
```
{Show[P, labelI[0, 2, "A={.0385,.999,0}"], labelI[2, 0, "B={.99,.04,0}"],
    labelIV[1, -1, "C={.7071,-.7071,0}"], labelIII[0, -2, "A"],
    labelIII[-2, 0, "B"], labelII[-1, 1, "C"], ImageSize → Medium],
  Graph[{"A" → "C", "C" → "A", "B" → "B"}, VertexLabels → "Name"]}
```

Out[59]=

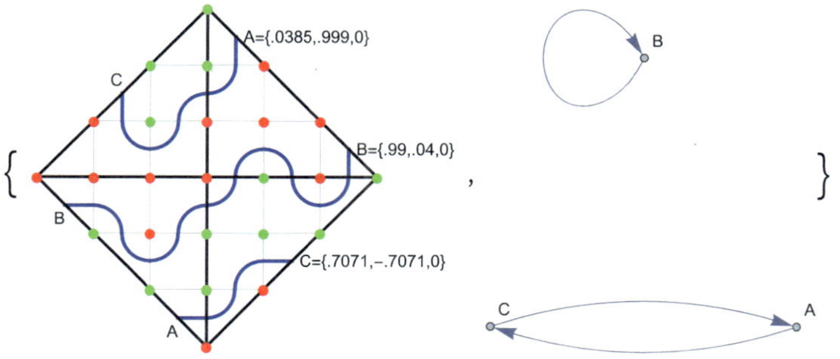

From this final picture, we see this curve has two projective components: the pseudo-line B ↔ B and the oval A ↔ C.

We now reconsider f = Newton Hyperbola 613:

In[60]:=
```
f = newtonHyperbola[613, x, y]
P = viroDiagram[f, x, y]
```

Out[60]= $-1+x+0.2\,x^2-0.008\,x^3-y+x\,y+0.2\,x^2\,y-0.2\,y^2+0.2\,x\,y^2-0.008\,y^3$

Out[61]=

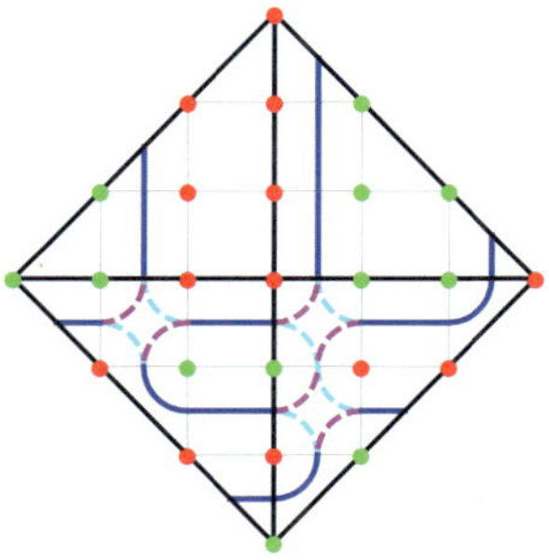

In[62]:=
```
infiniteRealPoints[f, x, y]
criticalPoints[f, x, y]
{x, y} /. NSolve[{f, x y}, {x, y}, Reals]
```

Out[62]= {{2.78305, 0.107199, 0}, {−0.151434, −3.93146, 0}, {1.56676, −1.56676, 0}}

Out[63]= {{−4.80215, −0.472881}, {0.857853, 0.00313321}, {0.0395922, −1.33841}}

Out[64]= {{29.1421, 0.}, {0., −18.6603}, {−5., 0.}, {0., −5.}, {0., −1.33975}, {0.857864, 0.}}

In[65]:=
```
RegionPlot[f > 0, {x, −10, 10}, {y, −10, 10}]
```

Out[65]=

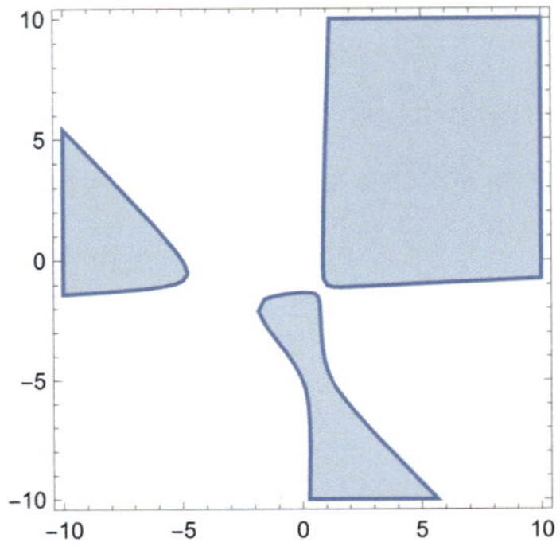

Here we see the negative region f[x, y] < 0, and it should be connected so that the correct choice of ambiguous squares and labels is conveniently the same as before:

In[66]:=
```
P1 := Show[P, pickMagenta[-2, -1], pickCyan[0, -1], pickCyan[0, -2]];
P2 = {Show[P1, labelI[0, 2, "A={.0385,.999,0}"], labelI[2, 0, "B={.99,.04,0}"],
     labelIV[1, -1, "C={.7071,-.7071,0}"], labelIII[0, -2, "A"],
     labelIII[-2, 0, "B"], labelII[-1, 1, "C"], ImageSize → Medium],
   Graph[{"A" → "B", "B" -> "C", "C" → "A"}, VertexLabels → "Name"]}
```

Out[67]= {

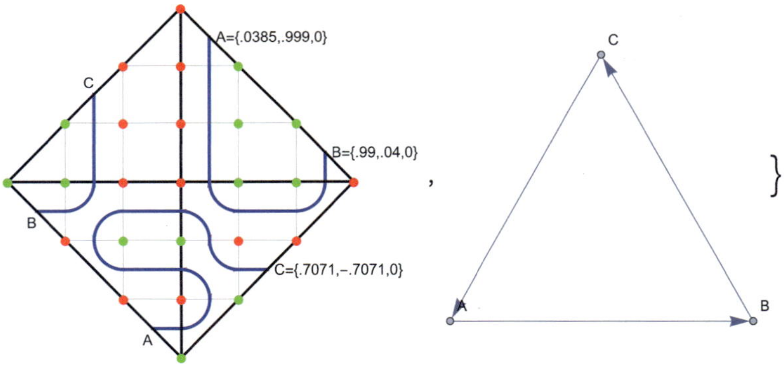

}

Not only do we have a connected green set, but we also have the actual intersections of the curve with the axes. In fact, this corresponds with the correct components. Here, there is only one projective component: the pseudo-line A → B → C → A.

In this example, a perturbation of the function that does not change any signs now has a connected positive region:

In[68]:=
```
RegionPlot[f + .8 > 0, {x, -10, 10}, {y, -10, 10}]
```

Out[68]=

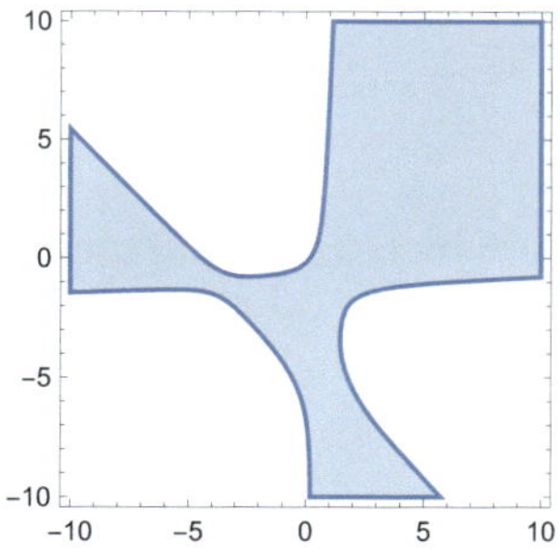

The correct diamond diagram would be:

In[69]:=
```
P1 := Show[P, pickCyan[-2, -1], pickMagenta[0, -1], pickCyan[0, -2]];
P2 = {Show[P1, labelI[0, 2, "A={.0385,.999,0}"], labelI[2, 0, "B={.99,.04,0}"],
      labelIV[1, -1, "C={.7071,-.7071,0}"], labelIII[0, -2, "A"],
      labelIII[-2, 0, "B"], labelII[-1, 1, "C"], ImageSize → Medium],
    Graph[{"A" → "C", "C" → "B", "B" → "A"}, VertexLabels → "Name"]}
```

Out[70]= {

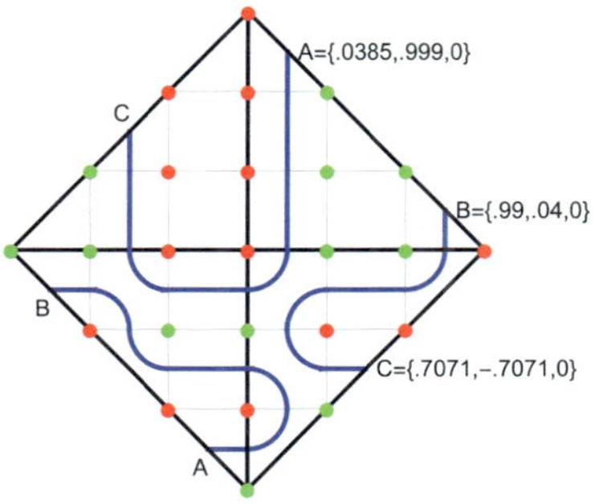

,

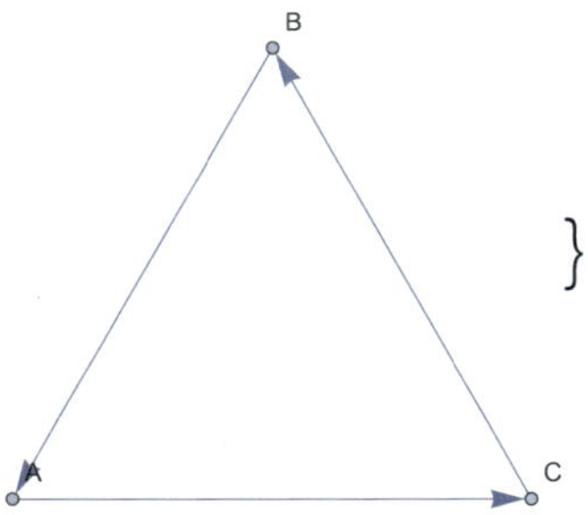

}

We are still using the Viro diagram, so we need not change the coloring:

In[71]:=
```
h = 0.23445264214028833` - 1.5927585520147027` x - 2.580519513261532` x^2 -
    1.1688260976780054` x^3 - 0.2695282534264044` y - 1.302314874550769` x y -
    1.4177824535874757` x^2 y + 0.01151627436977342` y^2 +
    0.05893043878998516` x y^2 + 0.4495559920915271` y^3;
```

The Viro diagram of the polynomial signs has two ambiguous squares:

In[72]:=
```
viroDiagram[h, x, y]
```

Out[72]=

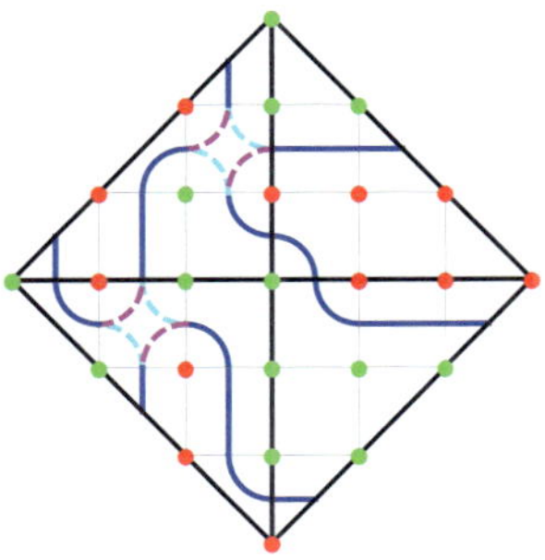

Now we check infinite points and axes zeros:

```
In[73]:= infiniteRealPoints[h, x, y]
         axisPoints[h, x, y]

Out[73]= {{0.878827, 1.79266, 0}}

Out[74]= {{0., -1.05808}, {0.121826, 0.}}
```

We get one asymptote and one axis point on each axis (which are confirmed by using the RegionPlot function).

This is still valid by Descartes's rule because in each case, we are missing an even number of points:

```
In[75]:= RegionPlot[h > 0, {x, -10, 10}, {y, -10, 10}, Axes → True]
```

Out[75]=

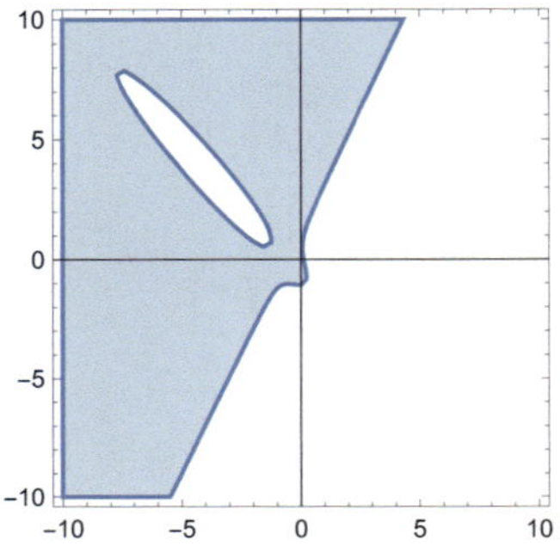

In this case, it is the association that must be changed rather than the ambiguous squares.

We can get by with changing one value on each axis (and quite a few off of the axes) to eliminate the two infinite points:

```
In[76]:= Ah = dvAssoc[coefficientSigns[h, {x, y}]]

Out[76]= <|{0, 0} → 0, {1, 0} → 1, {0, 1} → 1, {-1, 0} → 0, {0, -1} → 0, {2, 0} → 1,
          {1, 1} → 1, {0, 2} → 0, {-1, 1} → 0, {-2, 0} → 1, {0, -2} → 0, {-1, -1} → 1,
          {1, -1} → 0, {3, 0} → 1, {2, 1} → 1, {1, 2} → 0, {0, 3} → 0, {-1, 2} → 1, {-2, 1} → 1,
          {-3, 0} → 0, {0, -3} → 1, {-1, -2} → 1, {-2, -1} → 0, {1, -2} → 0, {2, -1} → 0|>

In[77]:= AssociateTo[Ah,
           {{-2, 0} → 0, {0, 1} → 0, {-1, 1} → 1, {-2, 1} → 0, {-1, 2} → 0, {1, -2} → 1, {2, -1} → 1}];

In[78]:= dDiagram[Ah]
```

Out[78]=

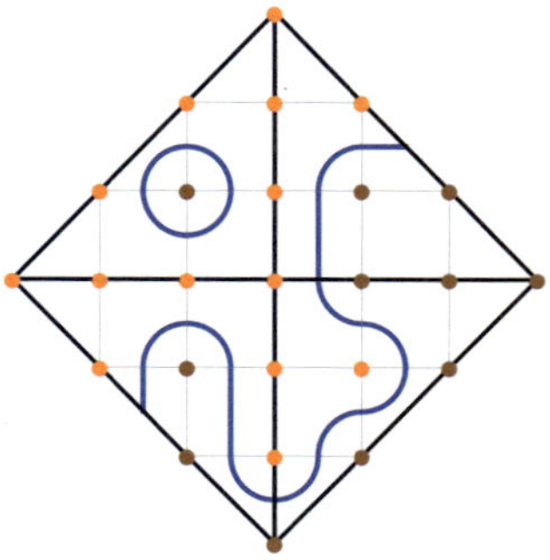

Alternatively to AssociateTo, we have the toggle function, which changes (i.e. toggles) the value from 0 to 1 or 1 to 0, so we only have to list the points to be changed.

Note that if A = toggle[A, K] runs twice, it will return A to the original:

In[79]:=
```
toggle[A_, K_] := Module[{B},
    B = A;
    Do[If[B[p] == 0, B[p] = 1, B[p] = 0], {p, K}];
    B]
```

We can get this last diagram by:

In[80]:=
```
Ah = dvAssoc[coefficientSigns[h, {x, y}]];
Ah = toggle[Ah, {{-2, 0}, {0, 1}, {-1, 1}, {-2, 1}, {-1, 2}, {1, -2}, {2, -1}}];
dDiagram[Ah]
```

Out[82]=

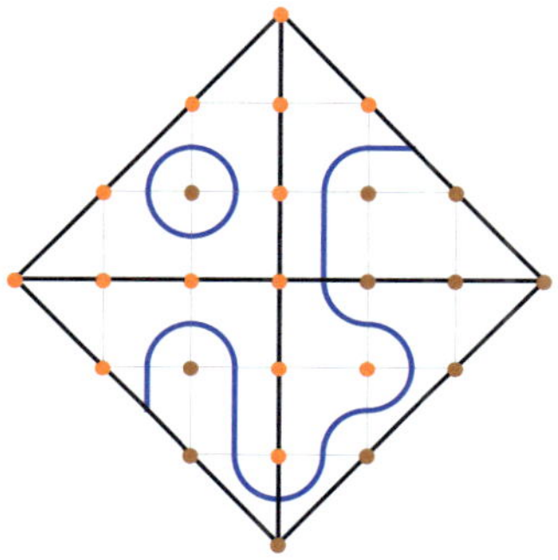

Since this is not a Viro diagram, we can change to the default Orange-Brown grid points, or equivalently, grid points, showing the same data:

In[83]:=
```
Ah = toggle[Ah, {{0, -2}, {1, -1}, {0, -1}}];
```

In[84]:=
```
P1 = dDiagram[Ah];
Show[P1, labelI[1, 1, "A={.44,.90,0}"], labelIII[-1, -1, "A"]]
```

Out[85]=

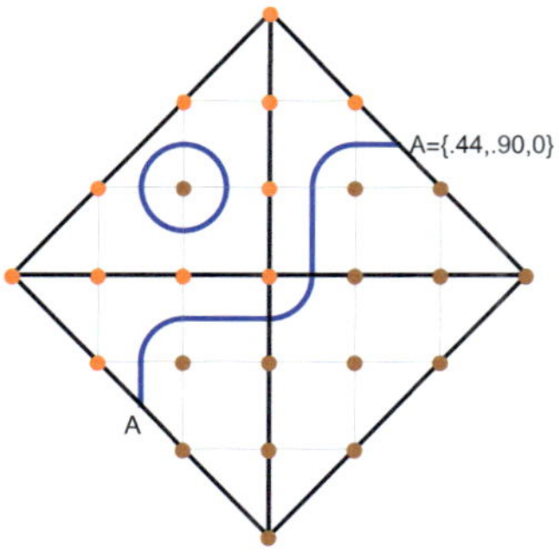

It now looks a little more like the actual curve, but we are far from the original Viro diagram, showing that the Viro diagram may show too much complexity. However, this could be perhaps recovered by a perturbation, leaving the signs unchanged. We now have a finite oval and a pseudo-line projectively.

Now we are going to look at the degree 4 (difficult curve) Newton Hyperbola 18540.

We do the two-step calculation of the Viro diagram, acquiring the association B1 and comparing the region plot:

```
In[86]:= f = newtonHyperbola[18 540, x, y]
         B1 = dvAssoc[coefficientSigns[f, {x, y}]];
         Q1 = dDiagram[B1, pcolr → Green, ncolr → Red];
         {Q1, RegionPlot[f > 0, {x, -30, 30}, {y, -30, 30}, Axes → True]}
```

Out[86]= $1+x-0.2\,x^2-0.008\,x^3+0.000064\,x^4-y+x\,y+0.2\,x^2\,y-0.008\,x^3\,y-0.2\,y^2+0.2\,x\,y^2+0.2\,x^2\,y^2+0.008\,y^3+0.008\,x\,y^3-0.000064\,y^4$

Out[89]= {

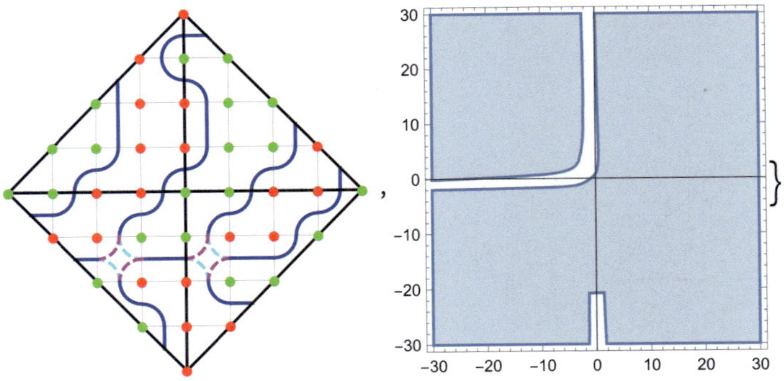

}

The Viro diagram tells us there is one infinite real point in quadrants II and IV and one or three in quadrants I and III. There may be zero or two crossings of each of the positive and negative x-axis, one crossing of the negative y-axis, and one or three crossings of the positive y-axis.

A quick glance at the region plot and the Viro diagram would lead one to guess that we could simply choose the ambiguous cells in the following plot:

```
In[90]:= Show[viroDiagram[f, x, y], pickMagenta[-2, -2], pickMagenta[0, -2],
           label24[-1, 2, "A"], label13[3, 0, "B"], label13[2, 1, "C"], label13[0, 3, "D"]]
```

Out[90]=

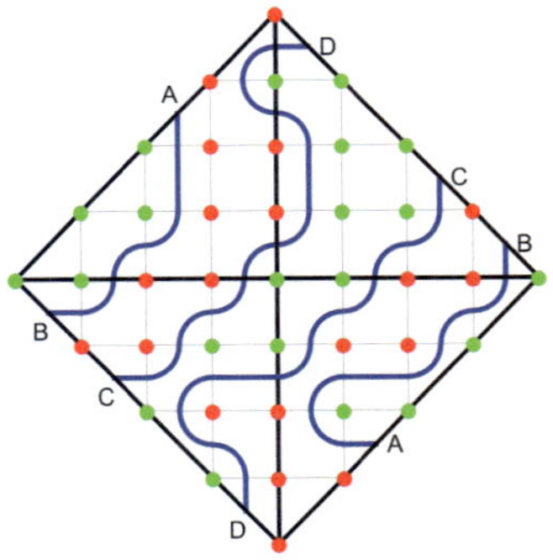

Closer analysis shows this is false:

In[91]:= **infiniteRealPoints[f, x, y]**

Out[91]= {{-1.33001, -0.0147082, 0}, {-1.29971, -0.037549, 0},
{-0.0069171, -1.0123, 0}, {0.0493139, -1.0547, 0}}

We see that there are four infinite points, as is often the case with Newton hyperbolas. Two are near $\{0, 1, 0\}$ and the other two are near $\{1, 0, 0\}$.

We inspect behavior near these points, using first the default $r = 1$:

In[92]:= **in1 = inspectInfPoint[f, {0, 1, 0}, 1, x, y, output -> "plot"]**

Out[92]=

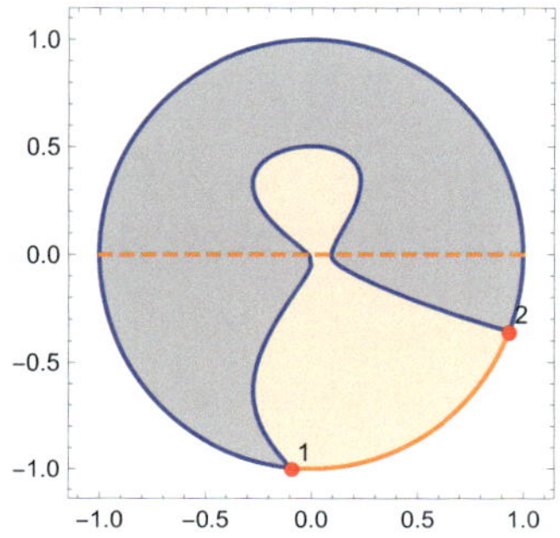

The curve has a clear continuous arc between points 1 and 2 in the upper half plane, one in the first quadrant and the other in the second:

In[93]:= **inspectInfPoint[f, {1, 0, 0}, 1, x, y, output → "plot"]**

Out[93]=

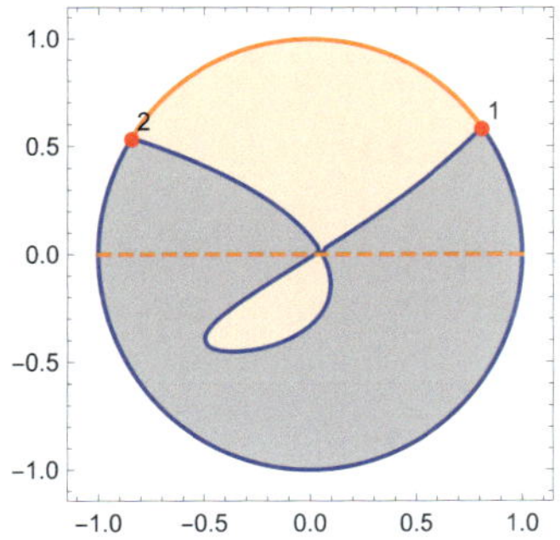

This is not clear. It looks like there is a singularity at $\{1, 0, 0\}$, but that is not likely as infiniteRealPoints recorded two nearby infinite points (and Newton hyperbolas are not known to be singular).

Taking smaller values of r, we find a definite answer (recalling from above that the points given are affine points of the curve):

In[94]:= **inspectInfPoint[f, {1, 0, 0}, .1, x, y]**

» Transformation {{0., -1.41421, 0.}, {0., 0., 1.41421}, {-0.707107, 0., -0.707107}}

» Transformed equation $0.000256 - 0.016\,x + 0.2\,x^2 + 0.004\,x^3 - 0.000016\,x^4 + 0.016512\,y - 0.224\,x\,y + 0.1\,x^2\,y - 0.175616\,y^2 + 0.288\,x\,y^2 - 0.05\,x^2\,y^2 - 0.687872\,y^3 + 0.448\,x\,y^3 - 0.047984\,y^4$

» boundary points coordinates relative to infinite point
{{−0.0784856, −0.0619678}, {0.0904525, 0.042642}, {−0.0134147, 0.0990961}, {0.0824277, −0.0566186}}

» affine coordinates boundry points
{{31.2748, −1.26655}, {−47.9021, −2.12121}, {−21.1824, 0.13537}, {34.3241, 1.45584}}

Out[94]=

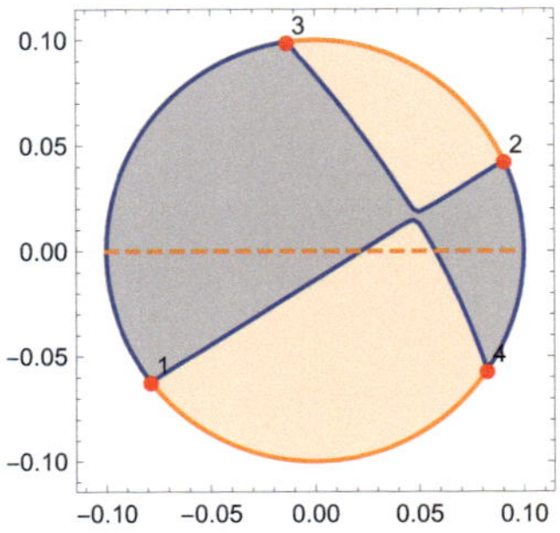


```
In[95]:= in2 = {{31.274814308993278`, -1.2665540243905862`},
    {-47.90208487681591`, -2.1212057690097192`}, {-21.18241973795016`,
     0.13537047756803747`}, {34.324060769362205`, 1.45584105351298`}};
```

The curve has a continuous arc from point 4 in the first quadrant to point 1 in the fourth quadrant, passing through the negative half plane while the arc from point 2 to point 3 stays in the negative half plane.

The critical points are given by:

```
In[96]:= cp = criticalPoints[f, x, y]
```

```
Out[96]= {{-138.817, -3.27735}, {-109.567, -2.68919}, {0.00125133, -4.97536},
  {3.43096, -0.744949}, {-3.4006, 2.53382}, {-0.386779, 0.401243}}
```

We draw two plots. The first is a magnification near the origin. For the second plot, we use a larger *x*-scale and labeled points. The points with positive label are the critical points.

The points with negative labels are the affine points in the set *in*2 calculated above:

```
In[97]:= {ContourPlot[f == 0, {x, -8, 8}, {y, -8, 8}, Axes → True],
  Show[ContourPlot[f == 0, {x, -150, 50}, {y, -10, 10}, Axes → True],
   ListPlot[{Table[Labeled[cp[[i]], i], {i, 6}], Table[Labeled[in2[[i]], -i], {i, 4}]}]]}
```

Out[97]= {

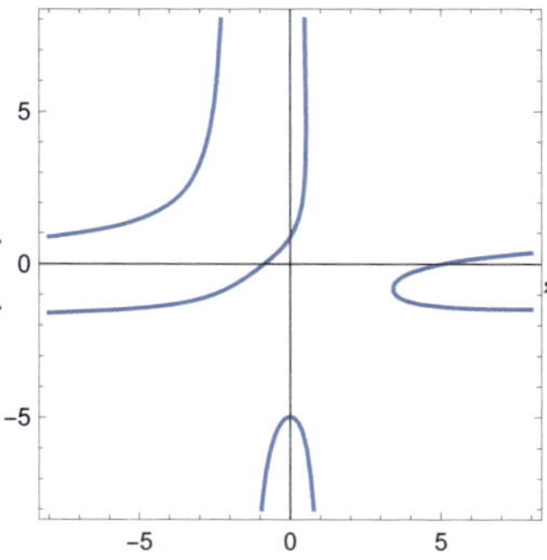

,

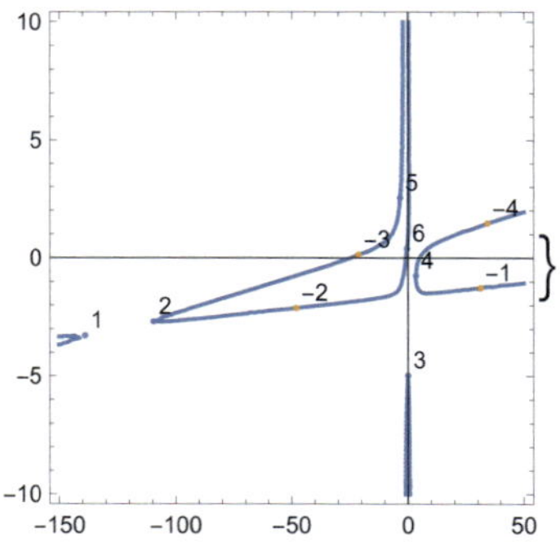

}

Among other things, note that RegionPlot failed to pick up the curve accurately by the x-axis in the first and fourth quadrants.

Contour plots seem to work better:

In[98]:=
```
axisPoints[f, x, y]
```

Out[98]=
```
{{145.706, 0.}, {0., 87.0737}, {0., 42.0439}, {−24.8737, 0.},
  {5.02538, 0.}, {0., −4.97537}, {−0.857891, 0.}, {0., 0.857838}}
```

Finally, the curve crosses twice in the positive x-axis, twice on the negative x-axis, three times in the positive y-axis, and only once on the negative y-axis.

Thus a correct diamond model of the curve is:

In[99]:=
```
B2 = B1;
AssociateTo[B2, {{1, 0} → 1, {−2, −1} → 0, {0, 2} → 0, {0, 3} → 1,
    {−1, −2} → 0, {1, −3} → 0, {1, −1} → 0, {2, −1} → 0, {1, −3} → 1}];
{Show[dDiagram[B2], labelI[3, 0, "A"], pickCyan[−3, −1], pickMagenta[0, 0],
   labelI[2, 1, "B"], labelI[0, 3, "C"], labelII[−1, 2, "D"], labelIII[−3, 0, "A"],
   labelIII[−2, −1, "B"], labelIII[0, −3, "C"], labelIV[1, −2, "D"],
   ListPlot[{Labeled[{−3, −.5}, 1], Labeled[{−1.9, −.4}, 2], Labeled[{.2, −1.5}, 3],
      Labeled[{−.5, .5}, 6], Labeled[{−1.5, .5}, 5], Labeled[{.7, −.9}, 4]}]],
  Graph[{"A" → 4, 4 −> "B", "B" → 1, 1 −> "A", "C" → 6, 6 → 2, 2 → 5,
    5 −> "D", "D" → 3, 3 −> "C"}, VertexLabels −> "Name"]}
```

Out[101]=

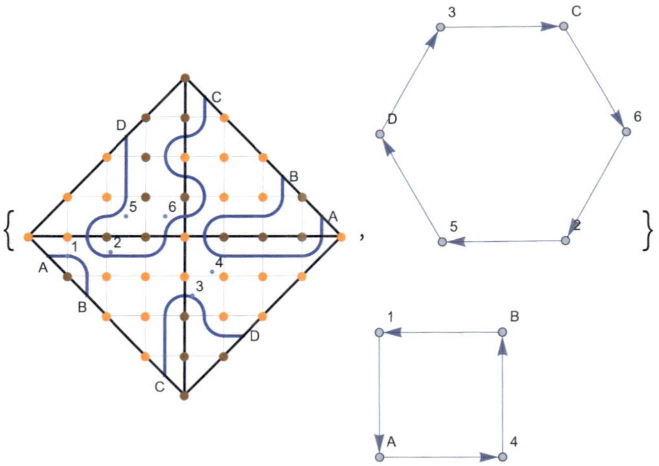

Where A is approximately {1, .01, 0}, B is {1, .02, 0}, C is {.01, 1, 0} and D is {−.05, 1, 0} and the numbers are the indexes of the critical points:

In[102]:=
```
cp
```

Out[102]=
```
{{−138.817, −3.27735}, {−109.567, −2.68919}, {0.00125133, −4.97536},
  {3.43096, −0.744949}, {−3.4006, 2.53382}, {−0.386779, 0.401243}}
```

An example we have seen is Newton Hyperbola 1050752. Its Gauss hemisphere plot has its infinite points labeled, and its Viro diagram plot is to the right:

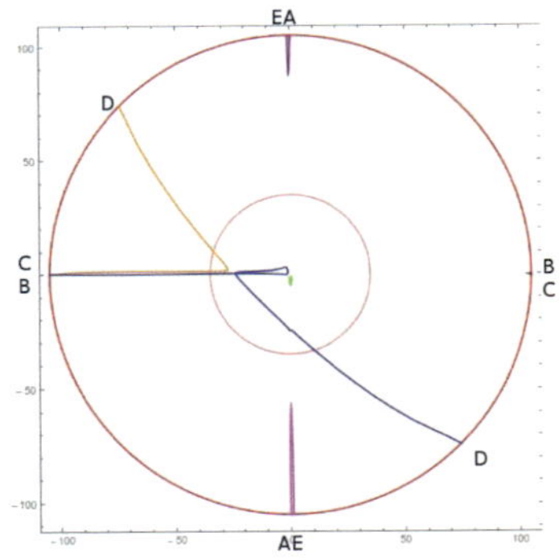

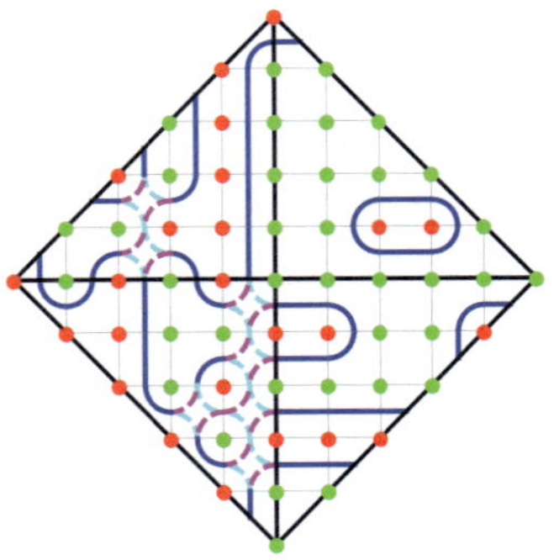

While the Viro diagram looks quite different, it is mostly a matter of choosing ambiguous squares to get the correct diamond diagram. However, the finite oval in quadrant I of the Viro diagram does not exist in this curve. In the first and second quadrant, A should go directly to E without crossing the x-axis.

In order that the reader can see exactly what needs to be done, here is the code to convert the Viro diagram to the correct diamond diagram:

```
In[103]:= f = newtonHyperbola[1 050 752, x, y];
          M1 = dvAssoc[coefficientSigns[f, {x, y}]];
          AssociateTo[M1, {{-1, 2} → 0, {-1, 3} → 0, {2, 1} → 0, {3, 1} → 0}];
          P5 := dDiagram[M1];
          {Show[P5, pickMagenta[-3, 1], pickMagenta[-3, 0], pickMagenta[-1, -1],
             pickCyan[-1, -2], pickMagenta[-1, -3], pickMagenta[-2, -3],
             pickMagenta[-1, -4], labelI[0, 4 , "A"], labelIV[4, 0, "B"], labelIV[3, -1, "C" ],
             labelIV[2, -2, "D"], labelIV[1, -3, "E"], labelIII[0, -4, "A"], labelII[-4, 0, "B"],
             labelII[-3, 1, "C"], labelII[-2, 2, "D"], labelII[-1, 3, "E"]], Graph[
             {"A" → "E", "E" → "A", "B" → "C", "C" → "D", "D" → "B"}, VertexLabels → "Name"]}
```

Out[107]= {

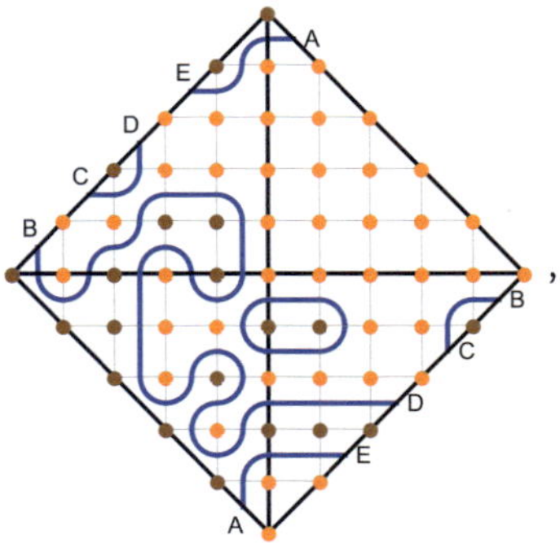

,

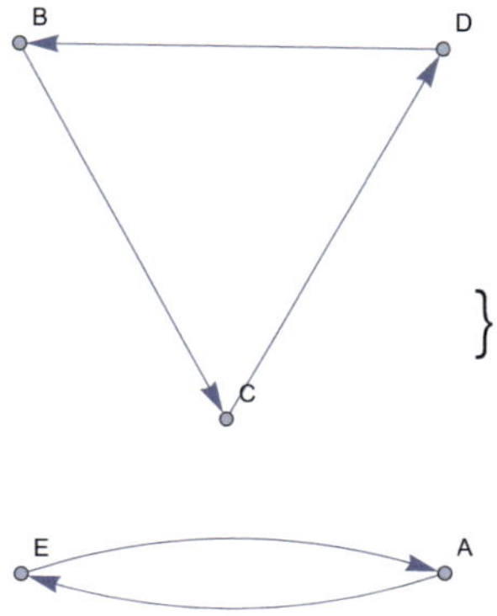

}

Unless the curve is numeric (as explained above) the techniques above will not work directly. Any plane algebraic curve is equivalent to a numeric curve; in some cases, the Gauss diagram of this numeric curve can explain the curve. If the curve has bad singularities, a `dDiagram` is not appropriate.

To get an equivalent numeric curve from an exact curve, one can first pick three small positive random numbers.

Once you've generated them, use the same numbers throughout the problem (maybe even a session):

```
In[108]:= {ar, br, cr} = RandomReal[{.3, .6}, 3]
Out[108]= {0.509921, 0.45308, 0.437721}
```

Then we construct a rotation matrix:

```
In[109]:= Rrot = RotationMatrix[ar, {br, cr, 1}]
Out[109]= {{0.891478, -0.394926, 0.222036},
   {0.431049, 0.890232, -0.147252}, {-0.13951, 0.226981, 0.963855}}
```

Finally, if the curve is given by equation *f*, then a numeric equivalent *fn* is given by:

```
In[110]:= fn = FLT[f, Rrot, x, y]
```

Out[110]= $0.889823 + 1.62036\,x + 0.760045\,x^2 + 0.18383\,x^3 + 0.0646971\,x^4 + 0.0110858\,x^5 + 0.49071\,y + 0.740374\,x\,y + 0.34907\,x^2\,y - 0.0595049\,x^3\,y - 0.030106\,x^4\,y + 0.0366441\,y^2 - 0.192991\,x\,y^2 - 0.0554347\,x^2\,y^2 + 0.010413\,x^3\,y^2 - 0.0786886\,y^3 + 0.100632\,x\,y^3 + 0.0258196\,x^2\,y^3 + 0.0416152\,y^4 - 0.00158432\,x\,y^4 - 0.00343896\,y^5$

We will give two examples, a very easy Gauss curve and a harder one with a surprising answer:

```
In[111]:= f = gaussCurve[z^3-8, x, y, z]
```

Out[111]= $-8 + x^3 - 3\,x\,y^2$

This certainly is not numerical and has infinite point {0, 1, 0}:

```
In[112]:= infiniteRealPoints[f, x, y]
Out[112]= {{-2.49202, -1.43877, 0}, {0., -1.75254, 0}, {3.88213, -2.24135, 0}}
```

In the following examples, we use:

```
In[113]:= {ar, br, cr} = {0.3898838058945685`, 0.48542069774603913`, 0.49793752820382864`};
In[114]:= Rrot = RotationMatrix[ar, {br, cr, 1}]
Out[114]= {{0.936873, -0.299821, 0.179935},
   {0.324275, 0.937495, -0.126286}, {-0.130825, 0.176663, 0.975538}}
In[115]:= fn = FLT[f, Rrot, x, y]
```

Out[115]= $-7.41715 - 4.19095\,x - 0.769501\,x^2 + 0.523062\,x^3 + 3.0007\,y - 0.223002\,x\,y + 2.44458\,x^2\,y - 0.391958\,y^2 - 1.69668\,x\,y^2 - 0.804802\,y^3$

So *fn* is a numerical polynomial with:

```
In[116]:= infiniteRealPoints[fn, x, y]
Out[116]= {{-1.92606, 5.8508, 0}, {-0.868414, -0.973016, 0}, {-1.05031, 0.20056, 0}}
```

Comparing plots:

```
In[117]:= {RegionPlot[f > 0, {x, -10, 10}, {y, -10, 10}, Axes → True],
   RegionPlot[fn > 0, {x, -10, 10}, {y, -10, 10}, Axes → True]}
```

Out[117]=

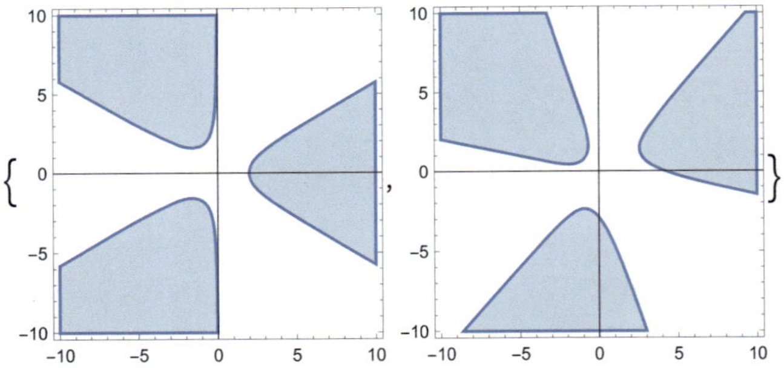

Now we find an initial diamond diagram:

```
In[118]:= AC = coefficientSigns[fn, {x, y}];
DC = dvAssoc[AC];
dDiagram[DC]
```

Out[120]=

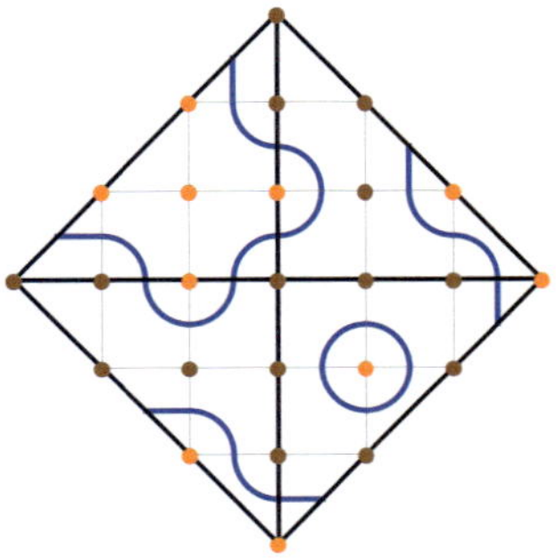

This is not correct on two counts: the component in the second quadrant should not cross either axis and the finite oval shown in the fourth quadrant does not exist.

We modify and label to get the correct Gauss diagram for *fn*, which does not describe axis points (including infinite axis) for *f* but does describe its topology:

```
In[121]:= DC1 = DC;
AppendTo[DC1, {{1, -1} → 1, {-1, 0} → 1, {0, 1} → 1}];
P = dDiagram[DC1];
Show[P, labelI[1, 1, "A={.67,.75,0}"],
  labelIV[2, 0, "B={.98,-.19,0}"], labelIV[0, -2, "C={.31,-.95,0}"],
  labelIII[-1, -1, "A"], labelII[-2, 0, "B"], labelII[0, 2, "C"]]
```

Out[124]=

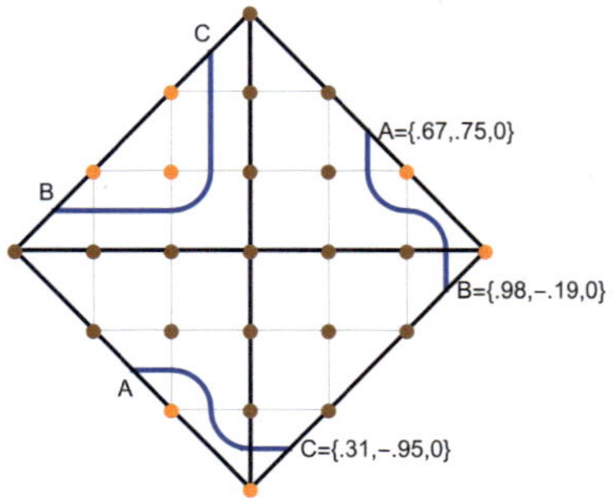

Our second example is the singular Chebyshev quartic, which is also not numeric and has {0, 1, 0} as a point where the curve is tangent to the infinite line.

We use the same Rrot as above but a different *f*:

In[125]:=
```
f = 1 - 8 x^2 + 8 x^4 + 3 y - 4 y^3
fn = FLT[f, Rrot, x, y]
```

Out[125]= $1 - 8x^2 + 8x^4 + 3y - 4y^3$

Out[126]= $1.24825 + 1.96209\,x - 5.67008\,x^2 - 5.83276\,x^3 + 5.95115\,x^4 + 2.26778\,y - 1.84109\,x\,y - 4.03446\,x^2\,y + 8.5237\,x^3\,y - 3.55517\,y^2 + 1.88663\,x\,y^2 + 5.1695\,x^2\,y^2 - 2.78224\,y^3 + 0.015189\,x\,y^3 + 0.485854\,y^4$

Comparing *f* with *fn*, labeling points on *fn* using drawing tools:

In[127]:=
```
{ContourPlot[f == 0, {x, -4, 4}, {y, -3, 7}, ContourStyle → Blue, Axes → True],
 ContourPlot[fn == 0, {x, -4, 4}, {y, -3, 7}, ContourStyle → Blue, Axes → True]}
```

Out[127]=
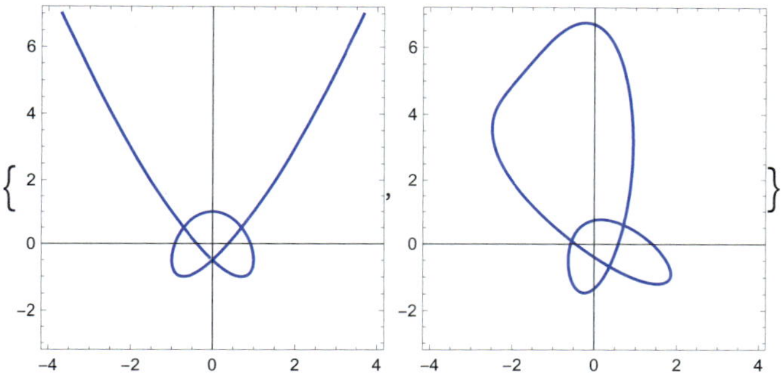

Now we find the Gauss diagram of the right-hand curve *fn*:

In[128]:=
```
AC = coefficientSigns[fn, {x, y}];
DC = dvAssoc[AC];
T = dDiagram[DC]
```

Out[130]=
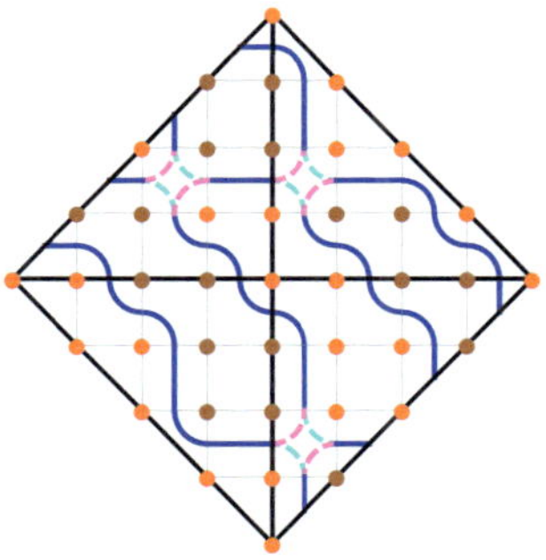

This doesn't look much like the contour plot of *fn*, but the axis points are counted correctly and Descartes is OK with this (since the number of infinite points of *fn*, 0, is an even number less than the diagram).

Since we have a singular curve with normal crossings, we replace the ambiguous squares with singular squares.

Then label infinite points and some of the interior points of both the diagram and contour plot with the drawing tools:

```
In[131]:= Show[T, pickSingularity[-2, 1], pickSingularity[0, 1], pickSingularity[0, -3],
  labelIV[3, 0, "4"], labelIV[2, -1, "5"], labelIV[1, -2, "6"], labelIV[0, -3, "7"],
  labelII[-3, 0, "4"], labelII[-2, 1, "5"], labelII[-1, 2, "6"], labelII[0, 3, "7"]]
```

Out[131]=

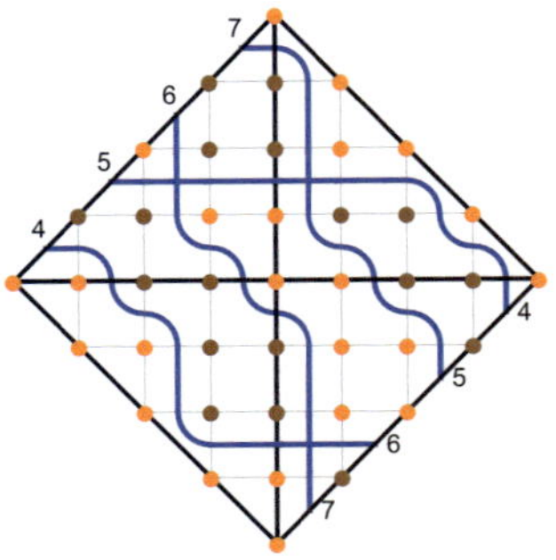

The surprise is that if we construct the projective Euler graph (Chapter 5) from the data from the contour plot or the data from the diamond diagram above, then it is the same graph:

```
In[132]:= Graph[{1 → 5, 5 → 2, 2 → 7, 7 → 3, 3 → 6, 6 → 1, 1 → 8, 8 → 2, 2 → 4, 4 → 3, 3 → 9, 9 → 1},
  VertexLabels → "Name", DirectedEdges → False]
```

Out[132]=

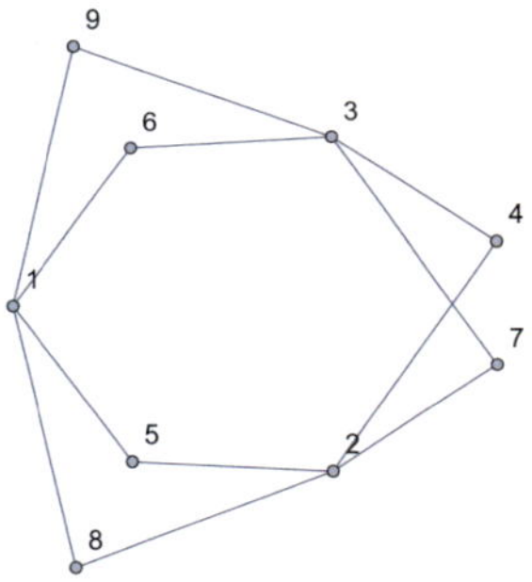

The diamond diagram above has the same projective topology as the curve but not the same affine topology.

9.4 Some More Topology of the Real Projective Plane

In this section, we discuss some more consequences of the topology of the projective plane and some questions that have arisen from these consequences. The general reference for this section is Section 11.6 of Bochnak (1998). This material is also largely available on Wikipedia.

As mentioned in Section 8.5 discussing the Möbius band model of a projective curve loop, an oval is defined to be a loop (or component in the nonsingular case) that separates the projective plane into two topological components. Some deeper topological reasoning shows that one of these components will be topologically a disk (two sided) while the other will be an actual Möbius band (one sided).

As was pointed out in Chapter 8, some geometers would prefer to think of the boundary Möbius band model as a small circle about the origin rather than collapsed to the origin. The full projective plane is a Möbius band with a disk attached, so the previous statement follows by going backwards.

For example, if the curve is a circle about the origin (left-hand picture), then the Möbius band model looks like the right-hand picture:

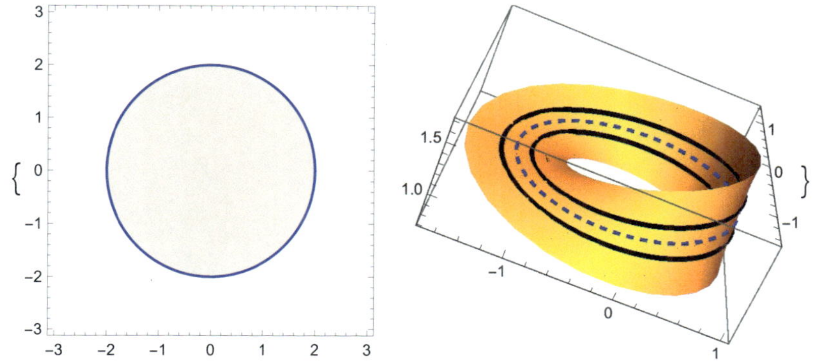

The component of the complement of this curve containing the (dashed) infinite line is an actual Möbius band while the other component contained the original boundary of the band (which is pinched to the origin), so that component is a disk; that is, the light brown interior of the original circle.

Generally, we can call the (topologically speaking) disk component the interior of the oval. Now, restricting to nonsingular curves for the rest of this section, it is possible that a component lies in the interior of another component. It is further possible that another component lies in the interior of the second component, and so on. We get a *nest* of components. For each oval in this nest, its *depth* is the number of ovals containing it, so our outer oval has depth zero.

A simple consequence of the fact that a line (or other curve) must intersect an oval in an even number of points is the following. Suppose we have a nest of ovals. Pick a point p on the outer oval and a second point q on the innermost oval. Then this line must intersect each oval in the nest in a positive even number of points. If the innermost oval has depth k (that is, there are $k+1$ curves in the nest), then this line intersects the curve containing all these ovals in at least $2\,(k+1)$ points. Therefore, this curve has a degree of at least $2\,(k+1)$. Simple algebra shows that the number of ovals in a nest must be less than or equal to $\frac{d}{2}$, where d is the degree of the curve.

A curve of degree 2 or 3 cannot have nested ovals, but a curve of degree 4 could possibly have a nest of two ovals. In fact, we can give a simple family of examples showing that when $d = 2\,n$, it is possible to have a nest of $n = \frac{d}{2}$ ovals. Take the Gauss curve of $p = (z^2 - 1)\,(z^2 - 4) \ldots (z^2 - n^2)$.

The y-axis is disjointed from this Gauss curve, so removing this line with the FLT gives a nest of n curves:

In[133]:=
```
M = {{0, 1, 0}, {0, 0, 1}, {1, 0, 0}}
g4 = gaussCurve[(z^2-1) (z^2-4), x, y, z]
h4 = FLT[g4, M, x, y]
{ContourPlot[g4 == 0, {x, -10, 10}, {y, -10, 10}],
  ContourPlot[h4 == 0, {x, -3, 3}, {y, -2, 2}]}
```

Out[133]= {{0, 1, 0}, {0, 0, 1}, {1, 0, 0}}

Out[134]= $4-5x^2+x^4+5y^2-6x^2y^2+y^4$

Out[135]= $1-6x^2+x^4-5y^2+5x^2y^2+4y^4$

Out[136]=

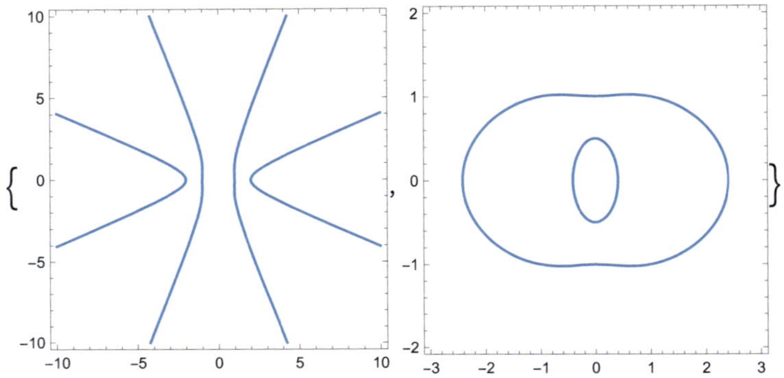

In[137]:=
```
g6 = gaussCurve[(z^2-1) (z^2-4) (z^2-9), x, y, z]
h6 = FLT[g6, M, x, y]
{ContourPlot[g6 == 0, {x, -15, 15}, {y, -15, 15}],
  ContourPlot[h6 == 0, {x, -5, 5}, {y, -3, 3}]}
```

Out[137]= $-36+49x^2-14x^4+x^6-49y^2+84x^2y^2-15x^4y^2-14y^4+15x^2y^4-y^6$

Out[138]= $1-15x^2+15x^4-x^6-14y^2+84x^2y^2-14x^4y^2+49y^4-49x^2y^4-36y^6$

Out[139]=

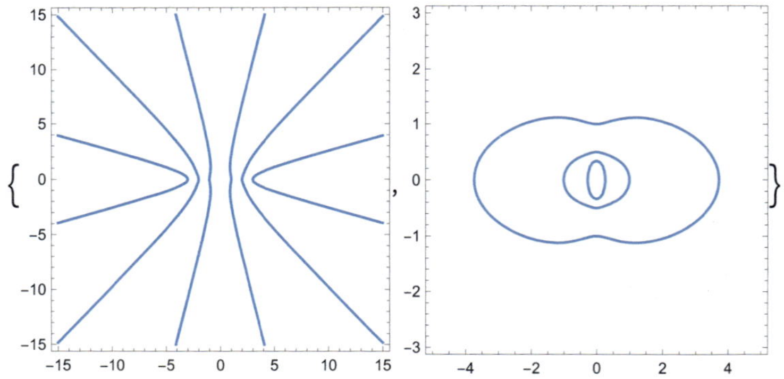

For odd d, we can throw in a $(z + d)$ term and get a nest of $(d - 1)/2$ ovals in a degree d curve:

```
In[140]:= g5 = gaussCurve[(z^2-1) (z^2-4) (z+5), x, y, z]
          h5 = FLT[g5, M, x, y]
          {ContourPlot[g5 == 0, {x, -15, 15}, {y, -15, 15}],
            ContourPlot[h5 == 0, {x, -5, 5}, {y, -3, 3}]}
```

Out[140]= $20+4x-25x^2-5x^3+5x^4+x^5+25y^2+15xy^2-30x^2y^2-10x^3y^2+5y^4+5xy^4$

Out[141]= $1-10x^2+5x^4+5y-30x^2y+5x^4y-5y^2+15x^2y^2-25y^3+25x^2y^3+4y^4+20y^5$

Out[142]= 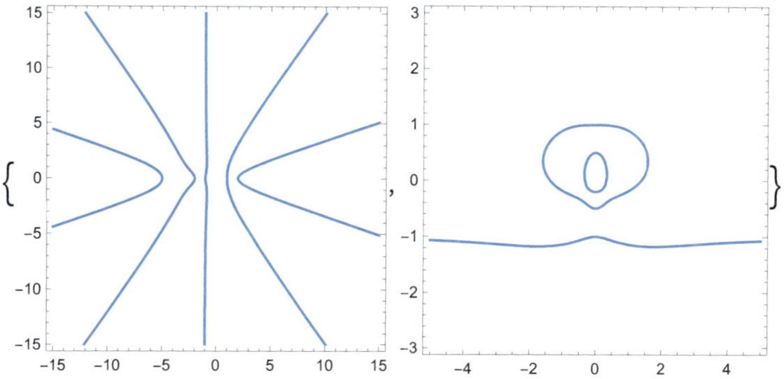

In 1900, the famous mathematician David Hilbert proposed a list of 23 problems he thought were most important for mathematicians to study in the coming century. Problem 16 had two parts. The first had to do with Harnack's theorem (Section 9.1): that a nonsingular, real algebraic plane curve of degree d could have, at most, $m = \frac{(d-1)(d-2)}{2} + 1$ components. Hilbert wanted to know more details about how an M-curve (one with m components) could look.

For example, in Section 9.2 we give Viro diagrams of Harnack's M-curves for $d = 3$, 4, 5, and 6. For degree 3 and 4, there is no nesting. In degree 5 and 6, there is one small oval inside the "big" oval with the other nine ovals in the exterior of the big oval. It is now known that there are two other possibilities: nine ovals of depth 1 in a big oval and one outside or five ovals inside a big oval with five outside.

Some early work on this problem was done by Virginia Ragsdale in the early 1900s. She let p be the number of ovals of even depth and n be the number of odd depth. Harnack's theorem says that for a curve of degree d, $p + n \leq \frac{(d-1)(d-2)}{2} + 1$ with equality for an M-curve. She conjectured that for a curve of $d = 2k$,

$$p \leq \frac{3k(k-1)}{2} + 1 \text{ and } n \leq \frac{3k(k-1)}{2}.$$

Later, Ivan Petrovsky showed that instead

$$p - n \leq \frac{3k(k-1)}{2} + 1 \text{ and } n - p \leq \frac{3k(k-1)}{2}.$$

Ragsdale's conjecture was later disproved by Viro, who was able to use his method to find counterexamples.

A related result for M-curves of degree $d = 2\,k$ is that $(p - n) - k^2$, which could be positive or negative, is a multiple of 8. So if $k = 3$ and $d = 6$, then $n + p = 11$, but $p - n$ must differ from 9 by a multiple of 8. So the only possibilities are $p = 10, n = 1$, $p = 2, n = 9$ or $p = 6, n = 5$ as described above.

Here are some diamond diagrams for the topology of these M-curves. However, since they are not Viro diagrams, they do not imply existence of algebraic curves with these diagrams. That existence is due to David Hilbert in 1891.

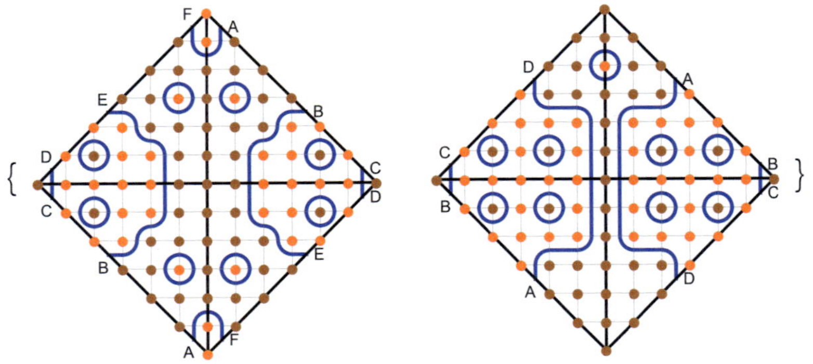

Additional results on the first part of Hilbert's 16th problem have been discovered in this century, but the problem is still not completely solved.

9.5 Another Drawing Method for Diagrams of Symmetric Sparse Curves

We would like to give the maximal nests in the previous section in terms of diamond diagrams, but the method in the Newton hyperbola section above does not do a good job because of the symmetry about the y-axis. Instead, we go back to Gauss, who considered a real algebraic curve as the boundary between positive and negative values of the defining polynomial. Therefore, instead of looking at the coefficients, we look at the values of the polynomial and associate the vertices of our grid with these. The diamond diagram algorithm (with some human help) does a good job of drawing the separating curve.

We will assume our nonsingular polynomial function is positive at the origin—that is, has a positive constant term and has some symmetry there. Let *rr* be a list of d increasing positive numbers, where d is the degree of the curve. Our first function plots the region plot of the function with the diamonds. They should have vertices in the changing positive and negative regions as the distance from the origin increases.

The first function helps to find appropriate radii for these diamonds:

```
In[143]:= plotDiamonds[f_, rr_, x_, y_] := Module[{m},
    m = Max[rr];
    If[Length[rr] ≠ tDeg[f, x, y], Echo[" wrong number of diamonds", "WARNING"]];
    Show[RegionPlot[f > 0, {x, -m, m}, {y, -m, m}], ContourPlot[Abs[x] + Abs[y],
      {x, -m, m}, {y, -m, m}, Contours → rr, ContourShading → None]]];
```

The second function then samples the polynomial along the sides of the diamonds and gives an association assigning these values to the vertices.

These can be used to find a diamond diagram that can be appropriately modified to give a nice diagram of the curve:

```
In[144]:= gaussAssoc[f_, rr_, x_, y] := Module[{A, r, V, d, sgn},
    sgn = Switch[Sign[#], 1, 0, -1, 1, 0, 0] &;
    d = tDeg[f, x, y];
    If[d ≠ Length[rr], Echo[d, "rr size"]; Abort[]];
    r = Prepend[rr, 0];
    V = ddVerts[d];
    <|
     Table[v → sgn[f /. Thread[{x, y} -> r[[Total[Abs[v]] + 1]] / (Total[Abs[v]] + .001) * v]],
      {v, V}]|>
   ];
```

We use the curves h5, h6 as examples:

```
In[145]:= h5 = 1 - 10 x^2 + 5 x^4 + 5 y - 30 x^2 y + 5 x^4 y - 5 y^2 + 15 x^2 y^2 - 25 y^3 + 25 x^2 y^3 + 4 y^4 + 20 y^5;
```

We arrive at the following values for *rr*:

```
In[146]:= rr = {.4, .8, .95, 2.5, 3.5};
          plotDiamonds[h5, rr, x, y]
```

Out[147]=

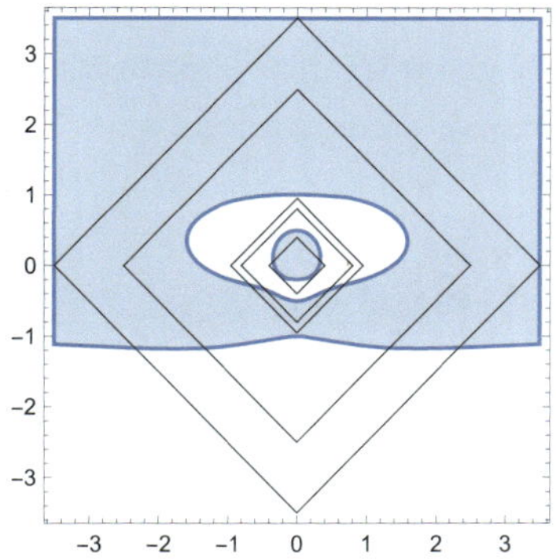

We then get the following diamond diagram:

In[148]:= **Dh5 = dDiagram[gaussAssoc[h5, rr, x, y]]**

Out[148]=

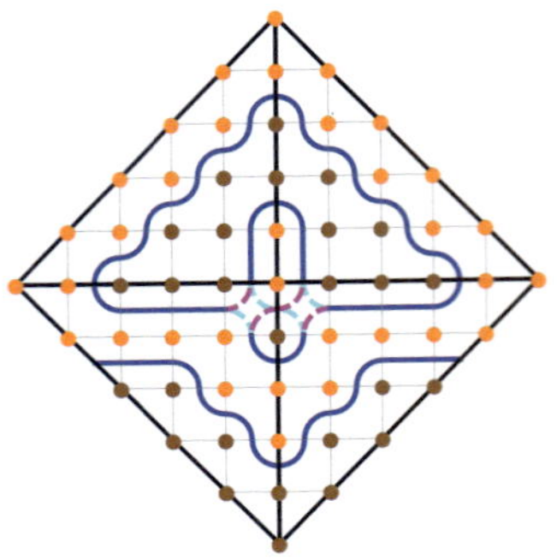

Setting only two ambiguous cells gives the desired diagram:

In[149]:= **Show[Dh5, pickCyan[−1, −1], pickMagenta[0, −1]]**

Out[149]=

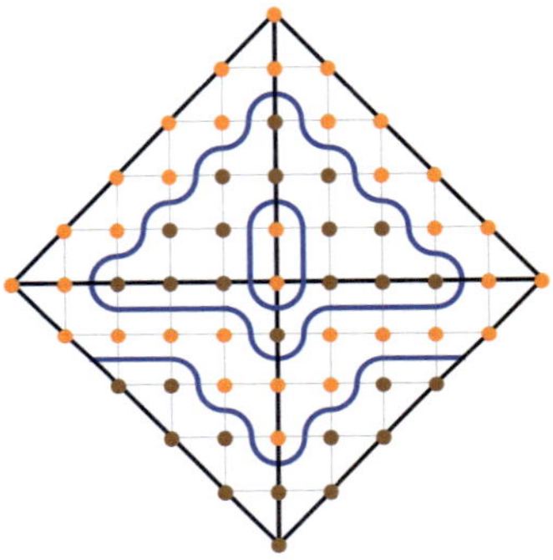

The one infinite point is {1, 0, 0}. This is not properly given by the diagram (in the sense of Viro diagrams) but does give the idea of a horizontal asymptote.

For h6, we use the following:

```
In[150]:= h6 = 1 − 15 x^2 + 15 x^4 − x^6 − 14 y^2 + 84 x^2 y^2 − 14 x^4 y^2 + 49 y^4 − 49 x^2 y^4 − 36 y^6;
          rr6 = {.1, .5, 1, 2, 3, 5};
          plotDiamonds[h6, rr6, x, y]
```

Out[152]=

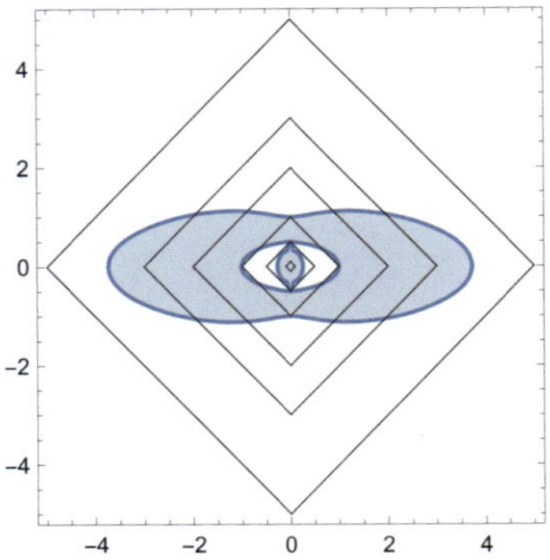

In[153]:= **Dh6 = dDiagram[gaussAssoc[h6, rr6, x, y]]**

Out[153]=

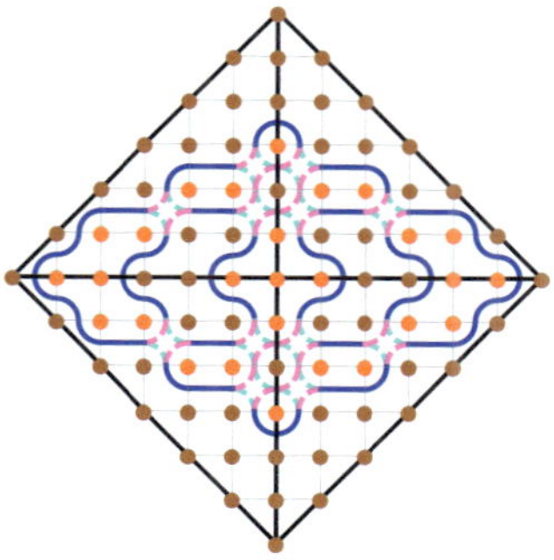

We don't need to toggle any vertices (although sometimes that may be necessary), but we do need to make many decisions on ambiguous cells:

In[154]:=
```
Show[Dh6, pickMagenta[-3, 1], pickMagenta[-1, 2], pickCyan[0, 2], pickCyan[2, 1],
  pickMagenta[2, -2], pickMagenta[0, -3], pickCyan[-1, -3], pickCyan[-3, -2],
  pickMagenta[-1, 1], pickCyan[0, 1], pickMagenta[0, -2], pickCyan[-1, -2]]
```

Out[154]=

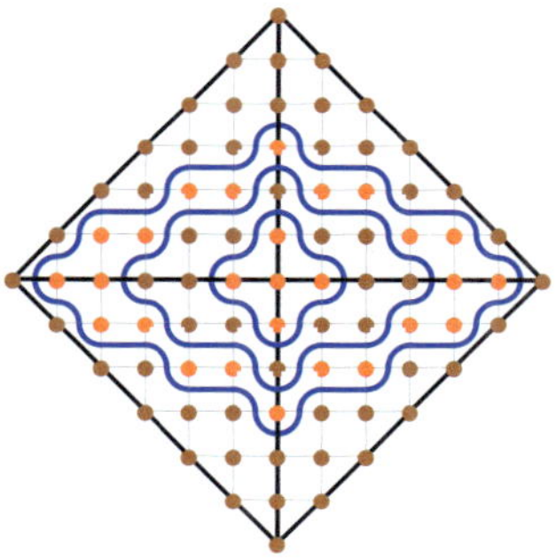

9.6 Appendix: Diamond, Viro, and Gauss Diagrams

A Wolfram Language function we use as a construct is called Association, which has domains and ranges that product lists Keys[A], Values[A], where A is the name of the association. It is not necessary to define these lists in advance.

Any Wolfram Language object can be a key or a value:

In[155]:= **A = Association[{■ → "square", ● → "circle", ▲ → "triangle"}]**

Out[155]= <| ■ → square, ● → circle, ▲ → triangle |>

In[156]:=
```
Keys[A]
Values[A]
```

Out[156]= {■, ●, ▲}

Out[157]= {square, circle, triangle}

...uming each key is unique, then an association does work like a function (which is ...ow our example will be).

To find the value associated with a key, one can write:

```
In[158]:= A[■]
          A[3]
Out[158]= square
Out[159]= Missing[KeyAbsent, 3]
```

The association we will use will have all the lattice points of our diamond as keys.

We can construct this recursively as follows:

```
In[160]:= ddVerts[d_] := Module[{V},
            V[0] = {{0, 0}};
            V[n_] :=
              V[n] = Join[V[n-1], Reverse[Table[{j, n-j}, {j, 0, n}]], Table[{-j, n-j}, {j, 1, n}],
                  Table[{-j, j-n}, {j, 0, n-1}], Table[{j, j-n}, {j, 1, n-1}]];
            V[
              d]];

In[161]:= ddVerts[3]

Out[161]= {{0, 0}, {1, 0}, {0, 1}, {-1, 0}, {0, -1}, {2, 0}, {1, 1}, {0, 2}, {-1, 1}, {-2, 0}, {0, -2}, {-1, -1}, {1, -1},
  {3, 0}, {2, 1}, {1, 2}, {0, 3}, {-1, 2}, {-2, 1}, {-3, 0}, {0, -3}, {-1, -2}, {-2, -1}, {1, -2}, {2, -1}}
```

We now assign keys 0 or 1 to each vertex. One special family of associations we will use is:

```
In[162]:= zeroA[d_] := Association[Table[v → 0, {v, ddVerts[d]}]];

In[163]:= A0 = zeroA[3]

Out[163]= <|{0, 0} → 0, {1, 0} → 0, {0, 1} → 0, {-1, 0} → 0, {0, -1} → 0, {2, 0} → 0,
  {1, 1} → 0, {0, 2} → 0, {-1, 1} → 0, {-2, 0} → 0, {0, -2} → 0, {-1, -1} → 0,
  {1, -1} → 0, {3, 0} → 0, {2, 1} → 0, {1, 2} → 0, {0, 3} → 0, {-1, 2} → 0, {-2, 1} → 0,
  {-3, 0} → 0, {0, -3} → 0, {-1, -2} → 0, {-2, -1} → 0, {1, -2} → 0, {2, -1} → 0|>
```

Starting with this association (which models the empty curve), we can modify to get a nonempty curve by changing some of the keys.

These are two ways to do so:

```
In[164]:= A1 = zeroA[3];
          A1[{1, 1}] = 1;
          AssociateTo[A1, {{-1, 2} -> 1, {0, -3} -> 1, {1, 0} → 1,
              {0, -2} → 1, {-1, -1} → 1, {-2, 0} → 1, {-1, -2} → 1, {-2, -1} → 1}]

Out[166]= <|{0, 0} → 0, {1, 0} → 1, {0, 1} → 0, {-1, 0} → 0, {0, -1} → 0, {2, 0} → 0,
  {1, 1} → 1, {0, 2} → 0, {-1, 1} → 0, {-2, 0} → 1, {0, -2} → 1, {-1, -1} → 1,
  {1, -1} → 0, {3, 0} → 0, {2, 1} → 0, {1, 2} → 0, {0, 3} → 0, {-1, 2} → 1, {-2, 1} → 0,
  {-3, 0} → 0, {0, -3} → 1, {-1, -2} → 1, {-2, -1} → 1, {1, -2} → 0, {2, -1} → 0|>
```

Or we can plot this kind of association with the following procedure:

```
In[167]:= plotAssoc[A_, c1_, c2_] := Module[{K, C1, C2},
            K = Keys[A];
            C1 = Select[K, A[#] == 0 &];
            C2 = Select[K, A[#] == 1 &];
            If[Length[C1] == 0, Return[ListPlot[K, Axes → False, PlotStyle → c2]]];
            If[Length[C2] == 0, Return[ListPlot[K, Axes → False, PlotStyle → c1]]];
            ListPlot[{C1, C2}, PlotMarkers → {●, ●}, PlotStyle → {c1, c2}, Axes → False]];
```

```
In[168]:= plotAssoc[A1, Brown, Orange]
```

Out[168]=

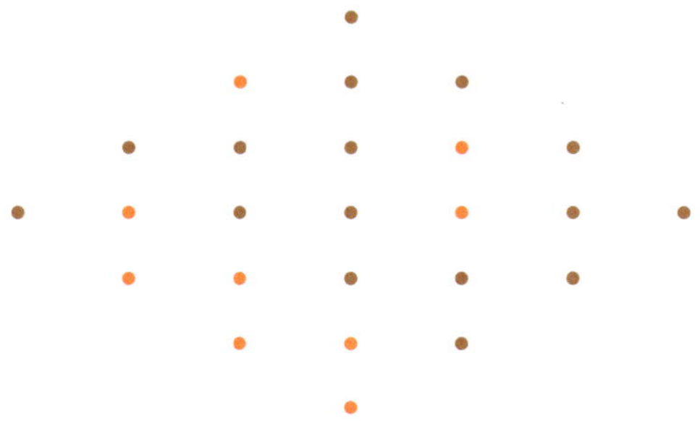

As mentioned earlier in Chapter 4 and Section 9.2, we will view adjacent keys of our association $\{a, b\}$ and $\{a+1, b\}$ or $\{a, b+1\}$ as being connected if they have the same value. The connected component of $\{a, b\}$ is the transitive closure of the relation—that is, all points connected to points connected to $\{a, b\}$, and so on, but if $\{a, b\}$ and $\{a+1, b\}$ have different values, they will be in different components, so the boundary curve of the component must pass between these points. As in Chapter 4, the boundary of each component is a union of continuous curves.

To draw the diamond diagram showing the boundary curves of the association, we divide up our two-dimensional diamond into squares and a right triangle on the outer boundary of the diamond. Each square will be identified by its lower left vertex $\{a, b\}$. This defines the other four vertices $\{a+1, b\}$, $\{a, b+1\}$, $\{a+1, b+1\}$, which, for $\{a, b\}$ to be a valid square, must all be keys of the association. The sequence of values corresponding to these vertices in this order will be called the value sequence of this square.

For the association A1 above, the value sequence for the square identified by {1, 0} is:

```
In[169]:= {A1[{1, 0}], A1[{2, 0}], A1[{1, 1}], A1[{2, 1}]}

Out[169]= {1, 0, 1, 0}
```

We have the function:

```
In[170]:= sqrSeq[p_, A_] := {A[p], A[p+{1, 0}], A[p+{0, 1}], A[p+{1, 1}]};
```

By knowing the value sequence, we usually know what the boundary curve looks like in the square. For the square {1, 0} above, we see the boundary will cut horizontally across the square. This would be the same for the complementary square—all values the opposite, {1, 1, 0, 0}. Thus we have eight possibilities all being defined by the value square except {0, 1, 1, 0} and its complement {1, 0, 0, 1}. As we saw in Section 9.2, in the case of a Viro diagram of an actual curve, a small perturbation can change the choice, so we consider this case to be ambiguous and draw both cases using dashed lines.

We get the following eight graphic primitives:

```
In[171]:= ps0[a_, b_] :=
            Graphics[{Thin, Gray, Line[{{a, b}, {a+1, b}, {a+1, b+1}, {a, b+1}, {a, b}}]}];

In[172]:= ps1[a_, b_] := Graphics[{Thick, Blue, Circle[{a, b}, .5, {0, Pi/2}],
              Thin, Gray, Line[{{a, b}, {a+1, b}, {a+1, b+1}, {a, b+1}, {a, b}}]}];

In[173]:= ps2[a_, b_] := Graphics[{Thick, Blue, Circle[{a+1, b}, .5, {Pi/2, Pi}],
              Thin, Gray, Line[{{a, b}, {a+1, b}, {a+1, b+1}, {a, b+1}, {a, b}}]}];

In[174]:= ps3[a_, b_] := Graphics[{Thick, Blue, Line[{{a, b+.5}, {a+1, b+.5}}],
              Thin, Gray, Line[{{a, b}, {a+1, b}, {a+1, b+1}, {a, b+1}, {a, b}}]}];

In[175]:= ps4[a_, b_] := Graphics[{Thick, Blue, Circle[{a, b+1}, .5, {3 Pi/2, 2 Pi}],
              Thin, Gray, Line[{{a, b}, {a+1, b}, {a+1, b+1}, {a, b+1}, {a, b}}]}];

In[176]:= ps5[a_, b_] := Graphics[{Thick, Blue, Line[{{a+.5, b}, {a+.5, b+1}}],
              Thin, Gray, Line[{{a, b}, {a+1, b}, {a+1, b+1}, {a, b+1}, {a, b}}]}];

In[177]:= ps6[a_, b_] := Graphics[{Thick, Blue, Circle[{a+1, b+1}, .5, {Pi, 3 Pi/2}],
              Thin, Gray, Line[{{a, b}, {a+1, b}, {a+1, b+1}, {a, b+1}, {a, b}}]}];

In[178]:= ps7[a_, b_] :=
            Graphics[{Thin, Gray, Line[{{a, b}, {a+1, b}, {a+1, b+1}, {a, b+1}, {a, b}}], Thick,
              Cyan, Dashed, Circle[{a+1, b+1}, .5, {Pi, 3 Pi/2}], Circle[{a, b}, .5, {0, Pi/2}],
              Magenta, Circle[{a+1, b}, .5, {Pi/2, Pi}], Circle[{a, b+1}, .5, {3 Pi/2, 2 Pi}]}];
```

To associate these primitives with the value sequences, we use a new association:

```
In[179]:= sqAassoc = <| {0, 0, 0, 0} → ps0, {1, 1, 1, 1} → ps0, {1, 0, 0, 0} → ps1, {0, 1, 1, 1} → ps1,
            {0, 1, 0, 0} → ps2, {1, 0, 1, 1} → ps2, {1, 1, 0, 0} → ps3, {0, 0, 1, 1} → ps3,
            {0, 0, 1, 0} → ps4, {1, 1, 0, 1} → ps4, {1, 0, 1, 0} → ps5, {0, 1, 0, 1} → ps5,
            {1, 1, 1, 0} → ps6, {0, 0, 0, 1} → ps6, {0, 1, 1, 0} → ps7, {1, 0, 0, 1} → ps7 |>;
```

For the triangles, we will use the quadrant and the coordinates of the right angle for identification. So pt13[*a*, *b*] will be a triangle in the second quadrant with a right angle at key point {*a*, *b*}.

We need four vertex sequences:

```
In[180]:= triseq1[p_, A_] := {A[p], A[p+{1, 0}], A[p+{0, 1}]};
          triseq2[p_, A_] := {A[p], A[p+{-1, 0}], A[p+{0, 1}]};
          triseq3[p_, A_] := {A[p], A[p+{-1, 0}], A[p+{0, -1}]};
          triseq4[p_, A_] := {A[p], A[p+{1, 0}], A[p+{0, -1}]};

In[184]:= pt10[a_, b_] := Graphics[{Thin, Gray, Line[{{a, b}, {a+1, b}, {a, b+1}, {a, b}}]}];

In[185]:= pt11[a_, b_] := Graphics[{Thin, Gray, Line[{{a, b}, {a+1, b}, {a, b+1}, {a, b}}],
              Thick, Blue, Circle[{a, b}, .5, {0, Pi/2}]}];
```

```
In[186]:= pt12[a_, b_] := Graphics[{Thin, Gray, Line[{{a, b}, {a+1, b}, {a, b+1}, {a, b}}],
              Thick, Blue, Line[{{a, b+.5}, {a+.5, b+.5}}]}];
```

```
In[187]:= pt13[a_, b_] := Graphics[{Thin, Gray, Line[{{a, b}, {a+1, b}, {a, b+1}, {a, b}}],
              Thick, Blue, Line[{{a+.5, b}, {a+.5, b+.5}}]}];
```

```
In[188]:= t1Assoc = <|{0, 0, 0} → pt10, {1, 1, 1} → pt10, {1, 0, 0} → pt11, {0, 1, 1} → pt11,
            {1, 1, 0} → pt12, {0, 0, 1} → pt12, {0, 1, 0} → pt13, {1, 0, 1} → pt13 |>;
```

Likewise, we need four graphic functions and a corresponding association for each of the other three quadrants. These fifteen functions are in Appendix 2, which should be initialized.

We can now write our main drawing routine for diamond diagrams. The dot colors are, by default, orange for positive (0) and brown for negative (1), but these colors can be changed.

Section 7 of Wolfram's book (2017) has basic information about color:

```
In[189]:= Options[dDiagram] = {pcolr → Orange, ncolr → Brown};
          dDiagram[A_, OptionsPattern[]] := Module[{d, sqrs, t1, t2, t3, t4, PS, PT, PB, PA},
              d = Max[Total[#] &/@Keys[A]];
              sqrs = Select[Keys[A], KeyExistsQ[A, # + {1, 0}] &&
                        KeyExistsQ[A, # + {1, 1}] && KeyExistsQ[A, # + {0, 1}] &];
              PS := Show[Table[sqAssoc[sqrSeq[p, A]] @@ p, {p, sqrs}]];
              t1 = Select[Keys[A], #[[1]] + #[[2]] == d - 1 &];
              t2 = Select[Keys[A], -#[[1]] + #[[2]] == d - 1 &];
              t3 = Select[Keys[A], -#[[1]] - #[[2]] == d - 1 &];
              t4 = Select[Keys[A], #[[1]] - #[[2]] == d - 1 &];
              PT := Show[Table[t1Assoc[triseq1[p, A]] @@ p, {p, t1}],
                  Table[t2Assoc[triseq2[p, A]] @@ p, {p, t2}], Table[t3Assoc[triseq3[p, A]] @@ p,
                    {p, t3}], Table[trAssoc[triseq4[p, A]] @@ p, {p, t4}]];
              PB := ListLinePlot[{{{-d, 0}, {d, 0}}, {{0, -d}, {0, d}, {d, 0}, {0, -d}, {-d, 0}, {0, d}}},
                  PlotStyle → Black, Axes → False];
              PA := plotAssoc[A, OptionValue[pcolr], OptionValue[ncolr]];
              Show[PS, PT, PB, PA, PlotRange → d + 1]
          ];
```

```
In[191]:= {dDiagram[A1], dDiagram[A1, pcolr → Purple, ncolr → LightGray]}
```

Out[191]=

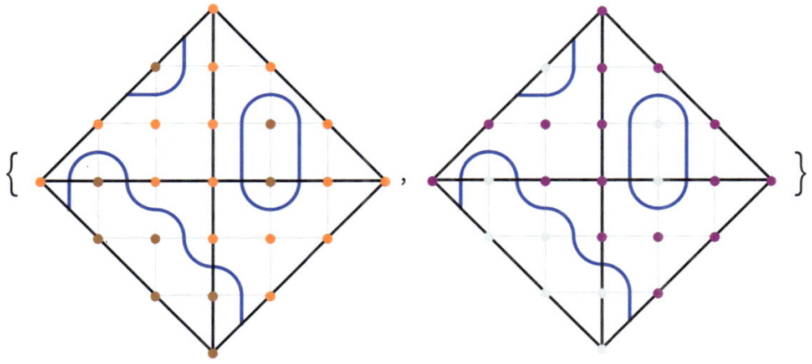

We have several ways to modify this to apply to specific situations. For example, instead of being a diamond, Keys[A] could give lattice points to a rectangle.

Then a simplified version of the above can be used:

```
In[192]:= Options[sDiagram] = {pcolr → Orange, ncolr → Brown};
          sDiagram[A_, OptionsPattern[]] := Module[{d, sqrs, PS, PA},
              sqrs = Select[Keys[A], KeyExistsQ[A, #+{1, 0}]&&
                        KeyExistsQ[A, #+{1, 1}]&&KeyExistsQ[A, #+{0, 1}] &];
              PS := Show[Table[sqAssoc[sqrSeq[p, A]]@@p, {p, sqrs}]];
              PA := plotAssoc[A, OptionValue[pcolr], OptionValue[ncolr]];
              Show[PS, PA, Axes → True]
            ]
```

A nice application of this is the one-dimensional Descartes diagram.

We start with an association, such as:

```
In[194]:= d = 4;
          sa = <|{4} → 0, {3} → 1, {2} → 0, {1} → 0, {0} → 1|>;
```

$\{4\} \to 0$ indicates that the coefficient of x^4 in a polynomial is positive while $\{0\} \to 1$ indicates that the constant coefficient is negative, etc. We do demand an assignment for all degrees up to some degree d.

We now extend this for negatives using Descartes's rule that we use the opposite for $\{-j\}$ when j is odd and $\{j\}$ when j is even:

```
In[196]:= ssa = sa;
          AssociateTo[ssa, Table[{-j} → If[Mod[j, 2] == 0, sa[{j}], 1 - sa[{j}]], {j, d}]]

Out[196]= <|{4} → 0, {3} → 1, {2} → 0, {1} → 0, {0} → 1, {-1} → 1, {-2} → 0, {-3} → 0, {-4} → 0|>
```

We now embed this in a rectangle:

```
In[197]:= K = Flatten[Table[{i, j}, {j, -1, 1}, {i, -d, d}], 1];
          As = Association[Table[p → Switch[p[[2]], -1, 0, 0, ssa[{p[[1]]}], 1, 1], {p, K}]]

Out[198]= <|{-4, -1} → 0, {-3, -1} → 0, {-2, -1} → 0, {-1, -1} → 0, {0, -1} → 0, {1, -1} → 0,
            {2, -1} → 0, {3, -1} → 0, {4, -1} → 0, {-4, 0} → 0, {-3, 0} → 0, {-2, 0} → 0, {-1, 0} → 1,
            {0, 0} → 1, {1, 0} → 0, {2, 0} → 0, {3, 0} → 1, {4, 0} → 0, {-4, 1} → 1, {-3, 1} → 1,
            {-2, 1} → 1, {-1, 1} → 1, {0, 1} → 1, {1, 1} → 1, {2, 1} → 1, {3, 1} → 1, {4, 1} → 1|>
```

```
In[199]:= sDiagram[As, pcolr → Green, ncolr → Red]
```

Out[199]=

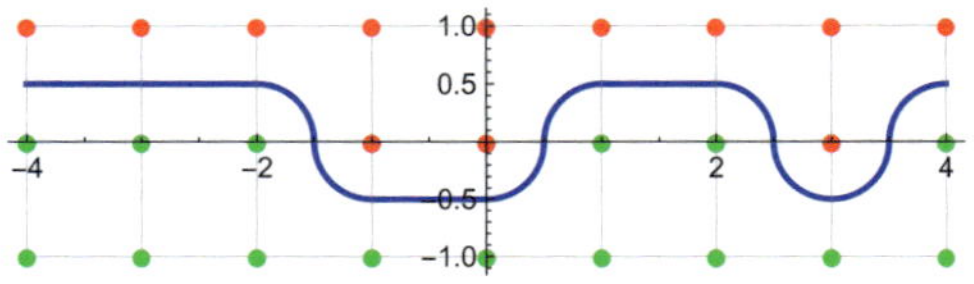

We can also generalize this to a descartesSketch as described herein:

```
In[200]:= Options[descartesSketch] = {pcolr → Orange, ncolr → Brown};
descartesSketch[A_, d_, OptionsPattern[]] := Module[{SA, T, B, AA},
    T = Association[Table[{x, 1} → 1, {x, -d, d}]];
    B = Association[Table[{x, -1} → 0, {x, -d, d}]];
    AA = Association[Table[{x, 0} → A[{x}], {x, -d, d}]];
    SA = Union[B, AA, T];
    sDiagram[SA, pcolr -> OptionValue[pcolr], ncolr -> OptionValue[ncolr]]
  ]
```

```
In[202]:= descartesSketch[ssa, 4]
```

Out[202]=

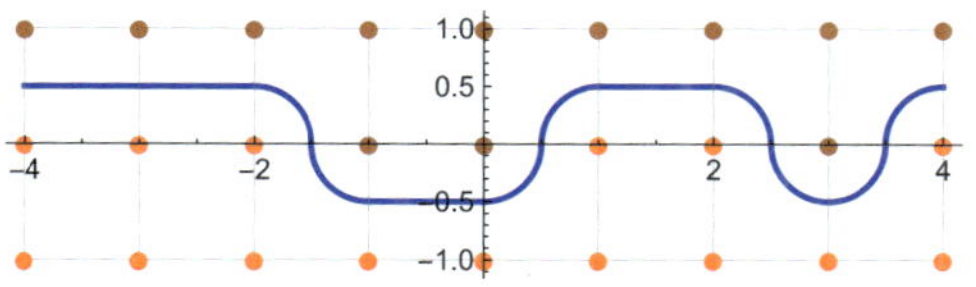

For earlier purposes, we put descartesDiagram in one procedure:

```
In[203]:= descartesDiagram[sa_] := Module[{d, sqrs, PS, PA, ssa, K, A},
    d = Max[Flatten[Keys[sa]]];
    ssa = sa;
    AssociateTo[ssa, Table[{-j} → If[Mod[j, 2] == 0, sa[{j}], 1 - sa[{j}]], {j, d}]];
    descartesSketch[ssa, d, pcolr → Green, ncolr → Red]
  ]
```

Further, we have the following useful procedure to construct the association sa from a univariate polynomial, which will work for bivariate polynomials as well.

In both cases, no monomial with total degrees less than the degree of the polynomial should have a zero coefficient:

```
In[204]:= coefficientSigns[f_, X_] := Module[{A, As},
    A = Association[CoefficientRules[f, X]];
    As = Association[Table[p → -(Sign[A[p]] - 1)/2, {p, Keys[A]}]]
  ]
```

```
In[205]:= f = -3 + 2 x + 4 x^2 - 3 x^3 + x^4;
descartesDiagram[coefficientSigns[f, x]]
```

Out[206]=

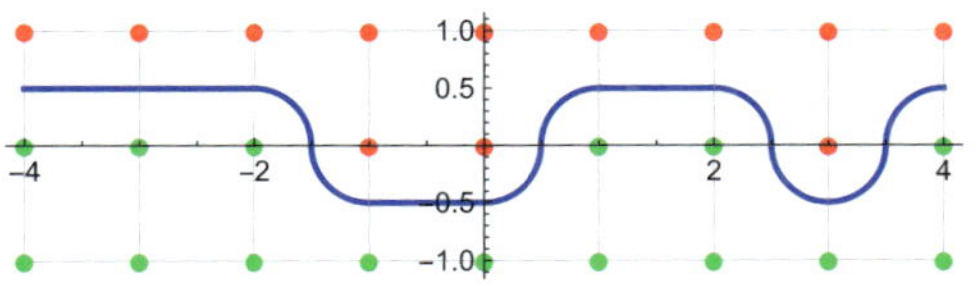

In the bivariate case, we would start with an association with keys $\{i, j\}$ for non-negative i, j so that $i + j \leqq d$ and values 0 (for a positive coefficient) and 1 (for a negative coefficient). We can then use the Descartes–Viro rules to extend this for $\{i, j\}$ with Abs[*i*] + Abs[*j*] ≤ *d*. Since (earlier in the text) we have started with a sign sequence of s, we first give an auxiliary function for d less than 10.

This converts a sign sequence to an association with non-negative keys:

```
In[207]:= s2A[s_] := Module[{n, d, A, K},
    n = Length[s];
    d = Which[n == 3, 1, n == 6, 2, n == 10 , 3, n == 15, 4,
       n == 21, 5, n == 28, 6, n == 36, 7, n == 45, 8, n == 55, 9, True, Fail];
    K = Select[verts[d], (#[[1]] ≥ 0 && #[[2]] ≥ 0) &];
    A = Association[Table[K[[i]] → s[[i]], {i, n}]]];
```

Now we convert non-negative keys to a full diamond association:

```
In[208]:= dvAssoc[A_] := Module[{d, K, V, v, pver},
   K = Keys[A];
   d = Max[Table[Total[p], {p, K}]];
   v = ddVerts[d];
   V = <| |>;
   Do[Which[p[[1]] ≥ 0 && p[[2]] ≥ 0, AppendTo[V, p → A[p]],
      p[[1]] < 0&&p[[2]] >= 0&&Mod[-p[[1]], 2] == 0, AppendTo[V, p → A[Abs[p]]],
      p[[1]] < 0&&p[[2]] >= 0&&Mod[-p[[1]], 2] == 1,
      AppendTo[V, p → (1-A[Abs[p]])],
      p[[2]] < 0&&p[[1]] >= 0&& Mod[-p[[2]], 2] == 0, AppendTo[V, p → A[Abs[p]]],
      p[[2]] < 0&&p[[1]] >= 0&& Mod[-p[[2]], 2] == 1,
      AppendTo[V, p → (1-A[Abs[p]])],
      Sign[(-1)^p[[1]]]*Sign[(-1)^p[[2]]] == 1, AppendTo[V, p → A[Abs[p]]],
      Sign[(-1)^p[[1]]]*Sign[(-1)^p[[2]]] == -1,
      AppendTo[V, p → (1-A[Abs[p]])]], {p, v}];
   V]
```

Putting these together, we recover the procedure viroDiagram[f, x, y] of the previous section:

```
In[209]:= viroAssociation[f_, x_, y_] := dvAssoc[coefficientSigns[f, {x, y}]];
```

```
In[210]:= viroDiagram[f_, x_, y_] := dDiagram[viroAssociation[f, x, y], pcolr → Green, ncolr → Red]
```

If we start with a bivariate function, we can calculate the Viro diagram associated to that function by dDiagram[dvAssoc[coefficientSigns[f, {x, y}]]]:

```
In[211]:= f = newtonHyperbola[413, x, y];
  { dDiagram[dvAssoc[coefficientSigns[f, {x, y}]]],
    RegionPlot[{f > 0, f < 0}, {x, -20, 20},
     {y, -20, 20}, PlotStyle → {LightBrown, LightOrange}]}
```

Out[212]= {

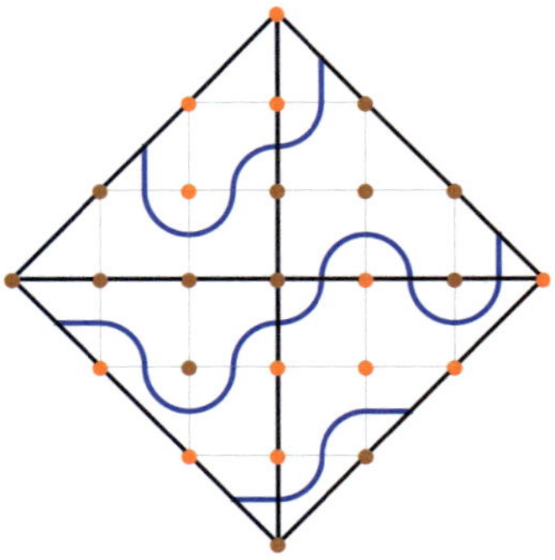

,

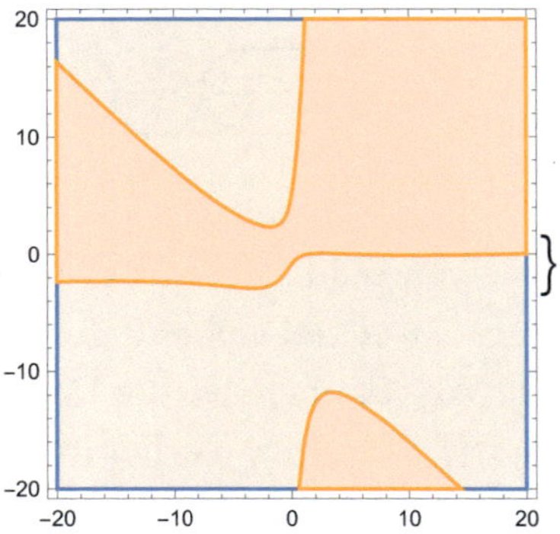

}

As we will see, this will not always work out so that the Viro diagram gives the correct topology and Descartes's values of the function.

We now describe some additional utility functions to enhance the diagrams. First, replace ambiguous squares with the user's choice: the cyan curve, the magenta curve, or a normal crossing. In general, diamond diagrams do not work well in the singular case, but the normal crossing situation (where the curve crosses itself in independent directions) can be handled with care.

The index {*a*, *b*} is the lower-left corner of the square:

```
In[213]:= pickcyan[a_, b_] := Graphics[{White, Rectangle[{a+.1, b+.1}, {a+.9, b+.9}], Thick,
              Blue, Circle[{a+1, b+1}, .5, {Pi, 3 Pi/2}], Circle[{a, b}, .5, {0, Pi/2}]}];

In[214]:= pickmagenta[a_, b_] := Graphics[{White, Rectangle[{a+.1, b+.1}, {a+.9, b+.9}], Thick,
              Blue, Circle[{a+1, b}, .5, {Pi/2, Pi}], Circle[{a, b+1}, .5, {3 Pi/2, 2 Pi}]}];

In[215]:= picksingularity[a_, b_] :=
            Graphics[{White, Rectangle[{a+.1, b+.1}, {a+.9, b+.9}], Thick, Blue,
              Line[{{a, b+.5}, {a+1, b+.5}}], Line[{{a+.5, b}, {a+.5, b+1}}]}]
```

Try this example: consider Viro Diagram 613 but give the plot a variable name:

```
In[216]:= P1 = viroDiagram[newtonHyperbola[613, x, y], x, y]
```

Out[216]=

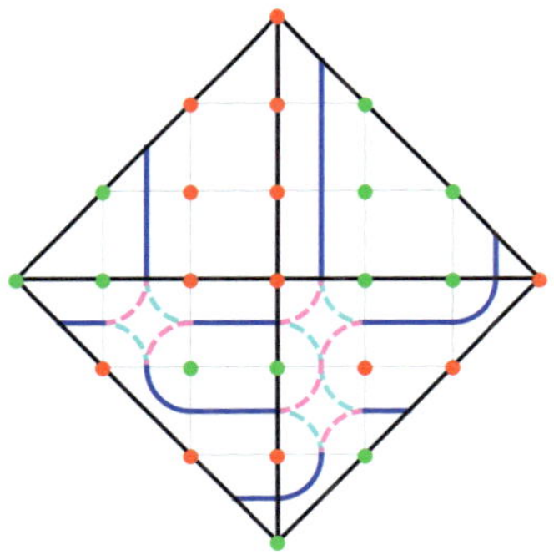

```
In[217]:= P2 = Show[P1, pickcyan[-2, -1], pickmagenta[0, -1], picksingularity[0, -2]]
```

Out[217]=

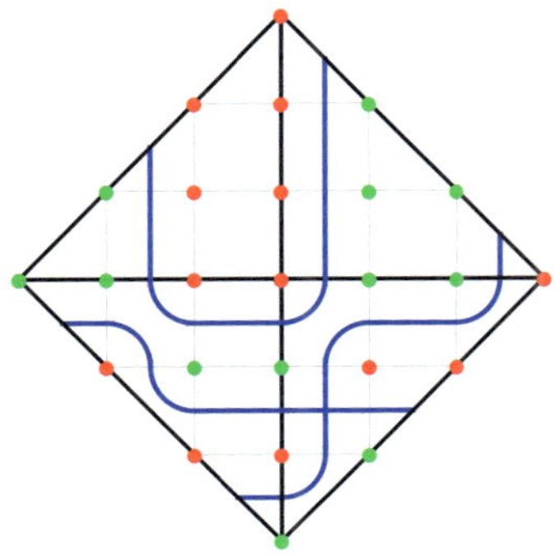

While one could use drawing tools to label the boundary (infinite) points, utility procedures place labels nicely either by letters or coordinates.

We do give separate ones for each quadrant:

```
In[218]:= labelI[a_, b_, txt_] := ListPlot[{{a+.5, b+.5}}, PlotStyle → Directive[Transparent, Thin],
            PlotLabels → Placed[txt, After], Axes → None]
```

```
In[219]:= labelII[a_, b_, txt_] := ListPlot[{{a-.5, b+.5}}, PlotStyle → Directive[Transparent, Thin],
            PlotLabels → Placed[txt, Above], Axes → None]
```

```
In[220]:= labelIII[a_, b_, txt_] := ListPlot[{{a-.5, b-.5}}, PlotStyle → Directive[Transparent, Thin],
            PlotLabels → Placed[txt, Below], Axes → None]
```

```
In[221]:= labelIV[a_, b_, txt_] := ListPlot[{{a+.5, b-.5}}, PlotStyle → Directive[Transparent, Thin],
            PlotLabels → Placed[txt, After], Axes → None]
```

```
In[222]:= label13[a_, b_, txt_] := If[a ≥ 0&&b ≥ 0,
            ListPlot[{{{a+.5, b+.5}}, {{-a-.5, -b-.5}}}, PlotStyle → Directive[Transparent, Thin],
              PlotLabels → {Placed[txt, After], Placed[txt, Below]}, Axes → None],
            Echo["label13 error:must have a,b positive"]]
```

```
In[223]:= label24[a_, b_, txt_] := If[a ≤ 0&&b ≥ 0,
            ListPlot[{{{a-.5, b+.5}}, {{-a+.5, -b-.5}}}, PlotStyle → Directive[Transparent, Thin],
              PlotLabels → {Placed[txt, Above], Placed[txt, After]}, Axes → None],
            Echo["label24error: must have a≤0,b≥0"]]
```

Thus we can fully label the plot P2 above. Note that I use different variables because it allows me to change my mind after I see the results. The right angle vertex and quadrant identify the triangle containing the point.

Thinking projectively, we have:

```
In[224]:= P = Show[P2, labelI[0, 2, "A"], labelI[2, 0, "B"], labelIV[1, -1, "C"],
            labelIII[0, -2, "A"], labelIII[-2, 0, "B"], labelII[-1, 1, "C"], ImageSize → Small]
```

Out[224]=

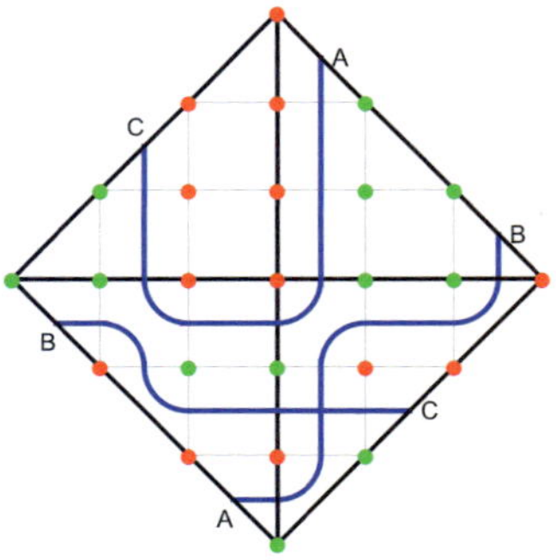

We can traverse this projective curve A → C → B → A so we have only one projective component. In fact, a cubic curve with this diagram would be a node.

More to Explore

Try looking up David Hilbert's problems, in particular part 1 of the 16th problem, on Wolfram MathWorld (wolfr.am/vt4KVIU1)

For more information about David Hilbert, see Wolfram|Alpha (wolfr.am/vt4p8ZGr)

A. Kozlowski's Demonstration helps illustrate the branching of a type of algebraic curve using its Newton polygon (wolfr.am/vt4lf3Ig)

10 | Final Examples

We look at the curve $f = (x^2 + y^2)^2 - 4\,x^3\,y^2$ from the introduction to answer the second "Looking Ahead" exercise from Chapter 1, showing possible answers as well as describing Newton Hyperbola 336941. As usual, if you are reading this as a notebook, initialize the global functions notebook (Appendix 2).

10.1 Curve from the Introduction

We first do a contour plot:

```
In[1]:= f = (x^2+y^2)^2-4x^3y^2;
        ContourPlot[f == 0, {x, -5, 5}, {y, -5, 5}]
```

Out[2]=

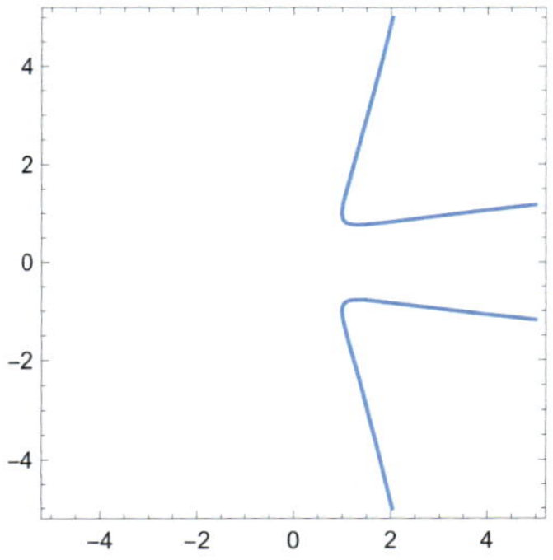

This doesn't look quite right for a 5th degree curve; something must be missing.

We check for critical points:

```
In[3]:= cp = criticalPoints[f, x, y]
```

Out[3]= $\{\{1.04167, 0.850517\}, \{1.04167, -0.850517\}, \{-1.19147 \times 10^{-40}, 1.23696 \times 10^{-40}\}\}$

The last one is the origin $\{0, 0\}$, but there is nothing on the plot near $\{0, 0\}$, even though $\{0, 0\}$ is an obvious solution of $f = 0$. It might be an isolated singular point. We can see that the minimum form at the point $\{0, 0\}$ (see Section 3.4) is $(x^2 + y^2)^2$. This clearly has its only real zero at $\{0, 0\}$, confirming the isolated singular point.

We next look for infinite points:

In[4]:=
```
infiniteRealPoints[f, x, y]
```

Out[4]= {{0., 11.2714, 0}, {−3.60821, 0., 0}}

We find the usual suspects, { 0, 1, 0}, {1, 0, 0}, but a curve of degree 5 should have an odd number of infinite points counted by multiplicity.

We inspect both points:

In[5]:=
```
P1 = inspectInfPoint[f, {0, −1, 0}, .5, x, y, output → "plot"];
P2 = inspectInfPoint[f, {1, 0, 0}, .5, x, y, output → "plot"];
Row[{P1, P2}]
```

Out[7]=

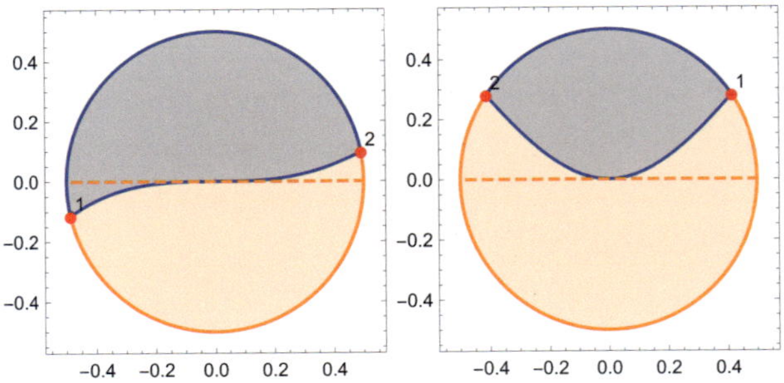

Both appear to intersect the infinite line in multiple points.

Let's check it, noting that transformation ip2z puts the infinite line at $y = 0$:

In[8]:=
```
g1 = FLT[f, ip2z[{0, −1, 0}, x, y], x, y]
```

Out[8]= $-2.82843\,x^3 + 2.82843\,y + 1.41421\,x^2\,y - 2.82843\,x^3\,y + 0.176777\,x^4\,y + 5.65685\,y^2 + 1.41421\,x^2\,y^2 - 0.707107\,x^3\,y^2 + 4.24264\,y^3 + 0.353553\,x^2\,y^3 + 1.41421\,y^4 + 0.176777\,y^5$

In[9]:=
```
g2 = FLT[f, ip2z[{1, 0, 0}, x, y], x, y]
```

Out[9]= $-5.65685\,x^2 + 2.82843\,y - 7.07107\,x^2\,y + 0.176777\,x^4\,y + 5.65685\,y^2 - 2.82843\,x^2\,y^2 + 4.24264\,y^3 - 0.353553\,x^2\,y^3 + 1.41421\,y^4 + 0.176777\,y^5$

In[10]:=
```
intersectionMultiplicity[g1, y, {0, 0}, 1.*^−10]
```

Out[10]= 3

In[11]:=
```
intersectionMultiplicity[g2, y, {0, 0}, 1.*^−10]
```

Out[11]= 2

We have accounted for all five expected infinite points by multiplicity:

In[12]:=
```
ap1 = inspectInfPoint[f, {0, 1, 0}, .5, x, y, output → "data"][[4]];
ap2 = inspectInfPoint[f, {1, 0, 0}, .5, x, y, output → "data"][[4]];
cp = criticalPoints[f, x, y];
pp = Join[ap1, ap2, cp]
```

Out[15]= {{4.26868, 16.537}, {5.07968, −21.7087}, {8.11553, 1.4712}, {8.11553, −1.4712}, {1.04167, 0.850517}, {1.04167, −0.850517}, $\{-1.19147\times10^{-40}, 1.23696\times10^{-40}\}$}

A larger scale contour plot:

```
In[16]:= ContourPlot[f == 0, {x, −25, 25}, {y, −25, 25},
           Epilog → {PointSize[Medium], Red, Point[pp]}]
```

Out[16]=

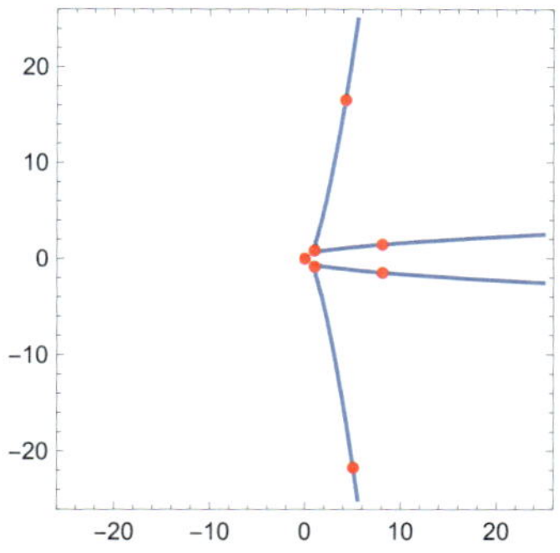

The plot above shows that we have now accounted for all real components and infinite behaviour. We see that, topologically, this real curve consists of an isolated zero and one pseudo-line intersecting the infinite line in five (counting multiplicity) points. The whole pseudo-line is best seen in hemisphere and the Möbius band model of the projective plane. As in Chapter 8, the dashed line is the infinite line for the Möbius band model. I leave the work for these plots to the reader using the methods of Chapter 8.

Note that pathfinding is best done with the naive pathFinder:

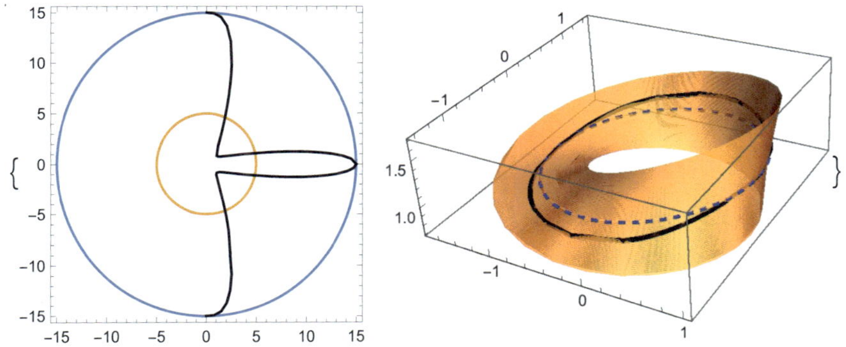

10.2 Curve from Chapter 1: Looking Ahead

The question was to describe the possibilities of completing the picture below in the real projective plane in general and describe this curve in particular.

The second question, posed at the end of Chapter 1, concerned this curve:

```
In[17]:= f = newtonHyperbola[336 941, x, y]
```

Out[17]=

$$-1+x-0.2\,x^2+0.008\,x^3-0.000064\,x^4+1.024\times10^{-7}\,x^5-y+x\,y+0.2\,x^2\,y+0.008\,x^3\,y-0.000064\,x^4\,y-0.2\,y^2+0.2\,x\,y^2+0.2\,x^2\,y^2+0.008\,x^3\,y^2+0.008\,y^3-0.008\,x\,y^3-0.008\,x^2\,y^3+0.000064\,y^4+0.000064\,x\,y^4+1.024\times10^{-7}\,y^5$$

Recall the plot from Chapter 1:

```
In[18]:= ContourPlot[f == 0, {x, -20, 20}, {y, -20, 20}, PlotLabel → "Plot 1"]
```

Out[18]=

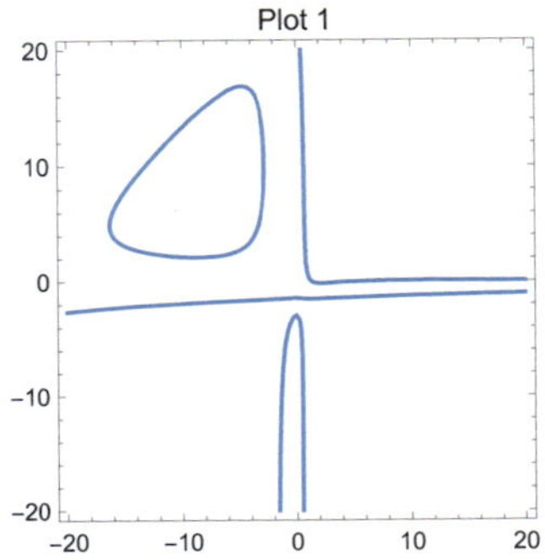

We calculate:

```
In[19]:= infiniteRealPoints[f, x, y];
         axisPoints[f, x, y];
```

To make sure indices stay the same, we set indices for later purposes:

```
In[21]:= ip = {{-0.0013663242136289378`, 0.9999990665786359`, 0.`},
             {0.00944405923828806`, 0.9999554038781449`, 0.`},
             {0.7070975075941625`, 0.7071160546573124`, 0.`},
             {0.9999829255065721`, 0.005843688502762102`, 0.`},
             {0.9999975620703512`, 0.002208133454830325`, 0.`}};
         ap = {{466.3649461309457`, 0.`}, {127.38865636036184`, 0.`},
             {24.999999999999883`, 0.`}, {4.906245327150322`, 0.`},
             {1.340152181644999`, 0.`}, {0.`, -435.36783116589487`},
             {0.`, -210.22357587125362`}, {0.`, -2.9730252537553894`},
             {0.`, -1.4355677090938972`}, {0.`, 24.999999999992397`}};
```

```
In[23]:= K = {"A", "B", "C", "D", "E"};
         ipAss = Association[Table[K[[i]] → ip[[i]], {i, 5}]];
         apAss = Association[Table[i → ap[[i]], {i, 10}]];
```

```
In[26]:= {Labeled[Show[RegionPlot[f > 0, {x, -20, 20}, {y, -20, 20}, Axes → True],
             ListPlot[Table[Labeled[apAss[i], i], {i, 10}]]], "Plot 1"],
           Labeled[Show[viroDiagram[f, x, y], label13[1, 3, "B"], label13[2, 2, "C"],
             label13[3, 1, "D"], label13[4, 0, "E"], label24[0, 4, "A"], ListPlot[
               {Labeled[{4.5, 0}, 1], Labeled[{3.5, 0}, 2], Labeled[{2.5, 0}, 3], Labeled[{1.5, 0}, 4],
                 Labeled[{.5, 0}, 5], Labeled[{0, -4.5}, 6], Labeled[{0, -3.5}, 7], Labeled[
                   {0, -1.5}, 8], Labeled[{0, -.5}, 9], Labeled[{0, 2.5}, 10]}]], "Diagram 1"]}
         ipAss
         apAss
```

Out[26]= {

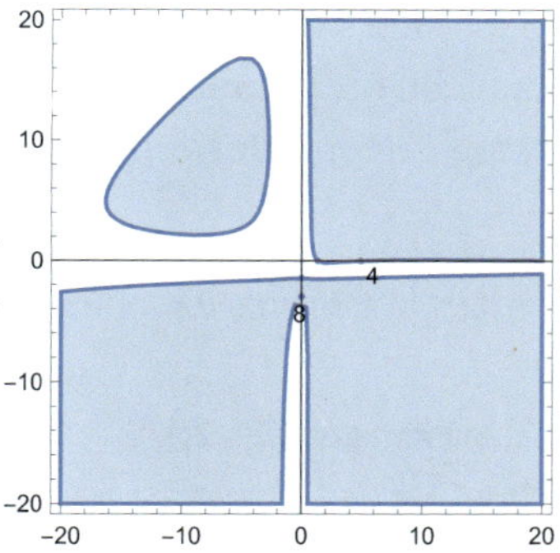

Plot 1

,

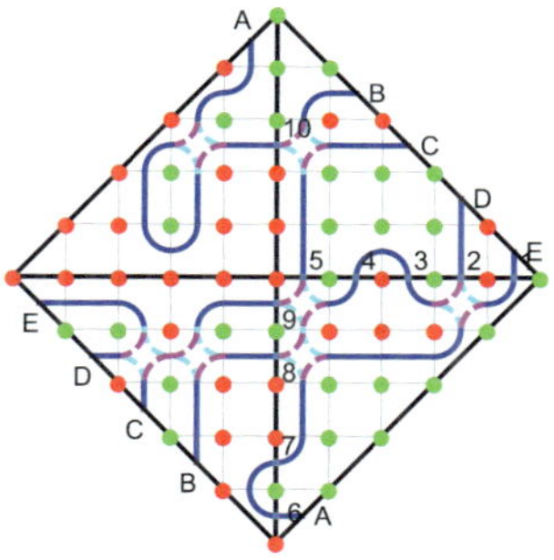

Diagram 1

}

```
Out[27]= <|A → {-0.00136632, 0.999999, 0.},
  B → {0.00944406, 0.999955, 0.}, C → {0.707098, 0.707116, 0.},
  D → {0.999983, 0.00584369, 0.}, E → {0.999998, 0.00220813, 0.}|>
```

```
Out[28]= <|1 → {466.365, 0.}, 2 → {127.389, 0.}, 3 → {25., 0.},
  4 → {4.90625, 0.}, 5 → {1.34015, 0.}, 6 → {0., -435.368},
  7 → {0., -210.224}, 8 → {0., -2.97303}, 9 → {0., -1.43557}, 10 → {0., 25.}|>
```

From the plot, Viro diagram and ip, ap, we can conclude that all infinite and axis points of the Viro diagram exist and the two smallest axis points on each *x*, *y* axis are in the plot. The finite component of the plot requires a finite component of the Viro diagram. In addition, the behavior between *x*-axis zeros 5 and 4 and *y*-axis zeros 8 and 9 is clear from the previous contour plot.

We can infer the following ambiguous cells:

```
In[29]:= VD0 = viroDiagram[f, x, y];
  VD1 = Show[VD0, pickCyan[0, -2], pickCyan[0, -1], pickCyan[-2, 2], label13[1, 3, "B"],
    label24[0, 4, "A"], label13[2, 2, "C"], label13[3, 1, "D"], label13[4, 0, "E"]]
```

Out[30]=

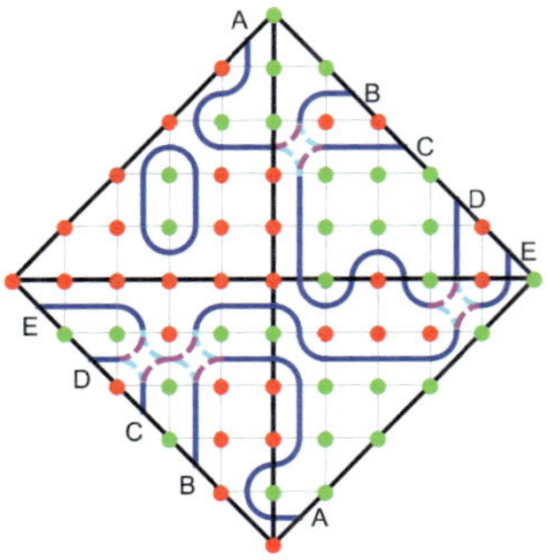

This leaves four ambiguities, hence the following sixteen possible choices:

```
In[31]:= bn = {{0}, {1}};
  Do[bn = Join[Table[Prepend[b, 0], {b, bn}], Table[Prepend[b, 1], {b, bn}]], {3}];
  bn
```

```
Out[32]= {{0, 0, 0, 0}, {0, 0, 0, 1}, {0, 0, 1, 0}, {0, 0, 1, 1}, {0, 1, 0, 0}, {0, 1, 0, 1}, {0, 1, 1, 0}, {0, 1, 1, 1},
  {1, 0, 0, 0}, {1, 0, 0, 1}, {1, 0, 1, 0}, {1, 0, 1, 1}, {1, 1, 0, 0}, {1, 1, 0, 1}, {1, 1, 1, 0}, {1, 1, 1, 1}}
```

The output from the next set of code has been turned off. The output from a previous run is shown. Attempting to properly format the output may cause your computer to crash. You can turn it back on by clicking the pull-down menu **Cell** and selecting **Cell Properties ▸ Evaluatable**.

It is best to run this code in a different notebook after properly initializing VD1 and bn above:

In[33]:=
```
Grid[Partition[Table[
    Labeled[Show[VD1, If[b[[1]] == 0, pickCyan[0, 2], pickMagenta[0, 2]],
      If[b[[2]] == 0, pickCyan[3, -1], pickMagenta[3, -1]],
      If[b[[3]] == 0, pickCyan[-3, -2], pickMagenta[-3, -2]],
      If[b[[4]] == 0, pickCyan[-2, -2], pickMagenta[-2, -2]]],
     b[[1]]*8+b[[2]]*4+b[[3]]*2+b[[4]]+1], {b, bn}], 4]]
```

Out[33]=

These diagrams are all plausible Gauss diagrams for curves matching matching Plot 1 above, but only one of them is actually a Newton hyperbola. Thus, some would also give additional matching curves by eliminating some components or complementary regions. The odd-numbered diagrams do not look like they would give a plot that matches Plot 1 above.

But look at the following example *h* with the same original Viro diagram:

```
In[34]:= h = -0.6592249684891619` + 0.41984463252599646` x -
    0.05847170393474383` x^2 + 0.0014434473234573883` x^3 -
    2.035408876021089`*^-6 x^4 + 2.082939155942196`*^-6 x^5 -
    0.5360674151589624` y + 0.2942113284492145` x y + 0.04358844913164157` x^2 y +
    0.0014576493787909805` x^3 y - 7.3709388570770925`*^-6 x^4 y -
    0.09483961076768323` y^2 + 0.029932148016941676` x y^2 +
    0.03308059628336358` x^2 y^2 + 0.0011093870567532476` x^3 y^2 +
    0.006657032213909227` y^3 - 0.003967318460561958` x y^3 -
    0.0007508723272270821` x^2 y^3 + 0.002106975723096353` y^4 +
    0.0005191436199134571` x y^4 + 0.000029691691033816 63` y^5;
```

```
In[35]:= {Show[RegionPlot[-20 < x < 20 && -20 < y < 20, {x, -40, 40},
    {y, -40, 40}, PlotStyle → LightYellow], ContourPlot[h == 0,
    {x, -40, 40}, {y, -40, 40}, ContourStyle → Black]], viroDiagram[h, x, y]}
```

Out[35]=

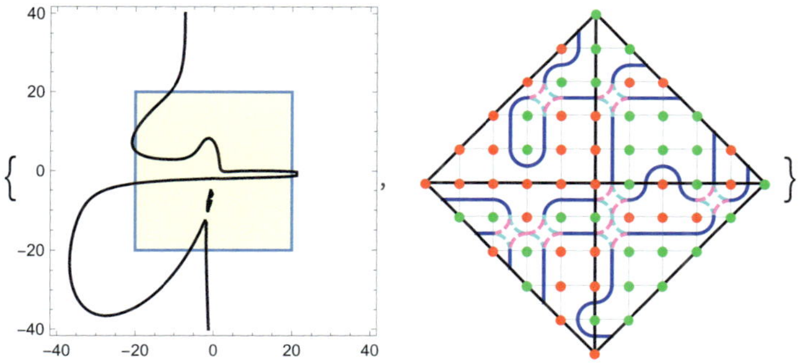

The radius 20 region (in yellow) would not look exactly the same as Plot 1 because the finite component merges with the the component in the first quadrant. In other words, the ambiguous square at $\{-2, -2\}$ would be magenta rather than cyan. Otherwise, this curve would be similar to the curve in Plot 1 inside radius 20. An analysis (such as in Chapter 9) would give a Gauss diagram. Unlike a Newton hyperbola, there is only one infinite point and only three axis points on each axis.

Gauss diagram for h:

In[36]:=
```
hass = viroAssociation[h, x, y];
AssociateTo[hass, {{1, 3} → 0, {2, 3} → 0, {−2, −2} → 0, {−2, −3} → 1, {0, −1} → 1,
    {5, 0} → 0, {4, 0} → 0, {−1, −1} → 0, {0, −1} → 0, {0, −2} → 0, {−1, −2} → 1,
    {0, −3} → 0, {−1, −3} → 0, {4, 1} → 0, {−3, −1} → 1, {−4, −1} → 1}];
DDH = dDiagram[hass];
GDH = Show[DDH, pickMagenta[−2, 2],
  pickCyan[0, −1], pickMagenta[−2, −2], pickCyan[−2, −3]]
```

Out[39]=

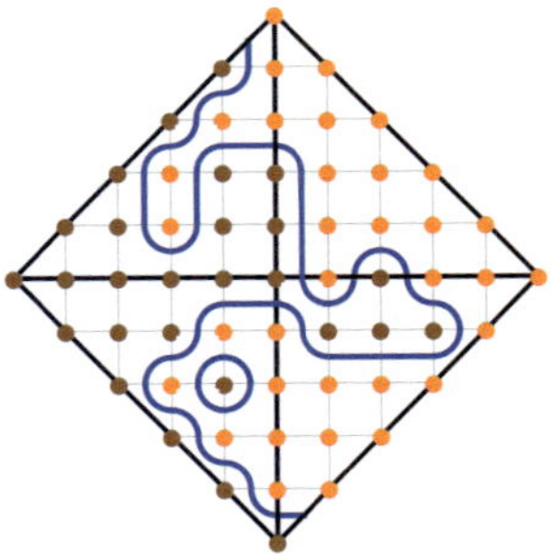

To finish the diagram, we need to specify the last four ambiguous cells. The easiest way to do this is by inspecting the infinite points.

As is often the case with Newton hyperbolas, the infinite points are near {0, 1, 0}, {1, 0, 0} and {1, 1, 0}, so we can inspect these points even though they are not on the curve.

Trying various values of r, we get:

In[40]:=
```
inspectInfPoint[f, {0, 1, 0}, .3, x, y, output → "plot"]
```

Out[40]=

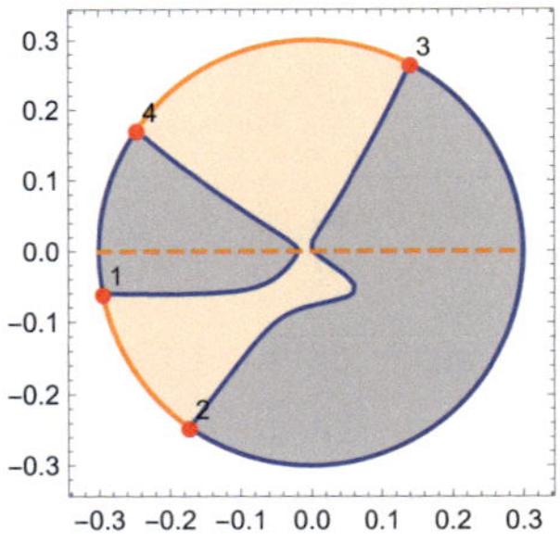

For the second, we try:

In[41]:=
```
inspectInfPoint[f, {1, 0, 0}, .5, x, y, output → "plot"]
```

Out[41]=

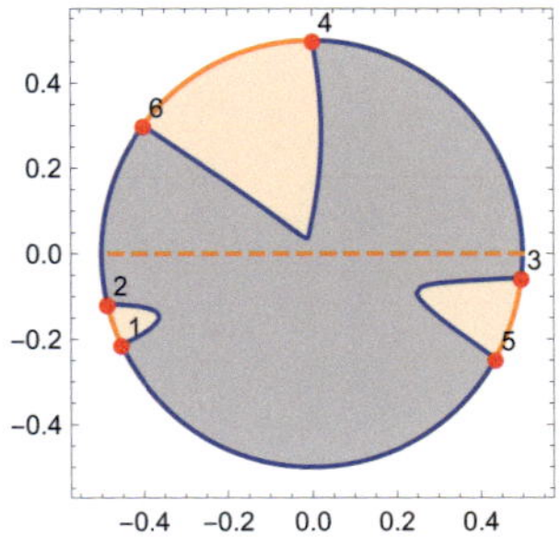

We don't see an infinite point, but our infiniteRealPoints function did give us two points near there. The problem is that RegionPlot is not good at finding small regions.

So we use much smaller values of r, finally choosing r = .015:

In[42]:= **inspectInfPoint[f, {1, 0, 0}, .015, x, y, output → "plot"]**

Out[42]=
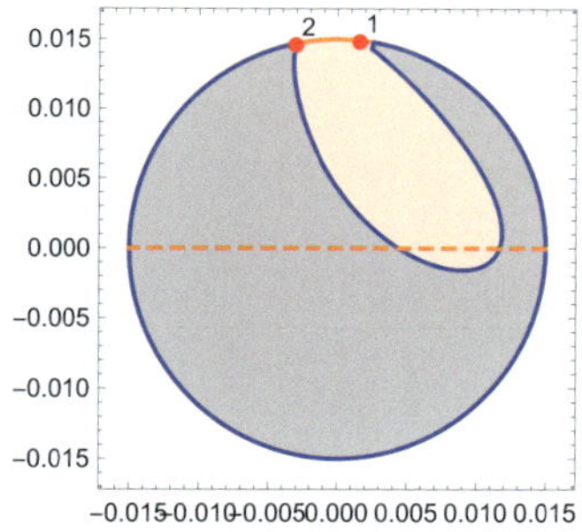

We now see these infinite points, but the component is far away.

The affine components of the numbered points above in the component are:

In[43]:= **inspectInfPoint[f, {1, 0, 0}, .015, x, y, output → "data"][[4]]**

Out[43]= {{135.092, 0.106817}, {137.184, −0.207895}}

Near the third point {0, 1, 0}:

In[44]:= **inspectInfPoint[f, {1, 1, 0}, 1, x, y, output → "plot"]**

Out[44]=
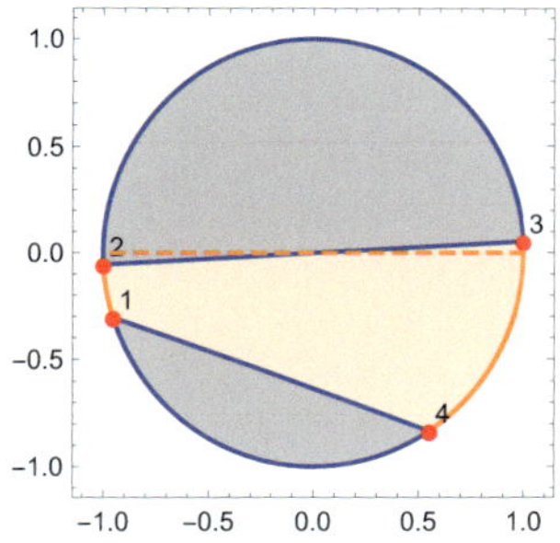

Now collecting affine points that appeared in these plots:

In[45]:= **ins1 = SortBy[inspectInfPoint[f, {0, 1, 0}, .5, x, y, output → "data"][[4]], Norm]**

Out[45]= {{0.748013, 3.99524}, {0.476205, −5.43039}, {−1.17259, −7.16436},
{−2.80493, 10.9114}, {−4.34381, 16.8297}, {9.08957, 35.5777}}

In[46]:= **ins2 = SortBy[inspectInfPoint[f, {1, 0, 0}, .5, x, y, output → "data"][[4]], Norm]**

Out[46]= {{5.00003, 0.00397448}, {−7.12872, −1.76911}, {7.64997, −1.32811},
{−8.42188, 2.13266}, {−16.1403, 4.16676}, {−34.3834, −8.78915}}

In[47]:= **ins3 = SortBy[inspectInfPoint[f, {1, 1, 0}, 1, x, y, output → "data"][[4]], Norm]**

Out[47]= {{−0.520918, −1.45919}, {−6.15608, −1.72751}, {−37.9017, −12.1884}, {13.8117, 39.9443}}

We can collect our points:

```
In[48]:= J = Join[ins1, ins2, ins3];
         allpts = <| Table[i → J[[i]], {i, Length[J]}] |>

Out[49]= <| 1 → {0.748013, 3.99524}, 2 → {0.476205, -5.43039},
   3 → {-1.17259, -7.16436}, 4 → {-2.80493, 10.9114}, 5 → {-4.34381, 16.8297},
   6 → {9.08957, 35.5777}, 7 → {5.00003, 0.00397448}, 8 → {-7.12872, -1.76911},
   9 → {7.64997, -1.32811}, 10 → {-8.42188, 2.13266}, 11 → {-16.1403, 4.16676},
   12 → {-34.3834, -8.78915}, 13 → {-0.520918, -1.45919},
   14 → {-6.15608, -1.72751}, 15 → {-37.9017, -12.1884}, 16 → {13.8117, 39.9443} |>
```

Where points 1–6 are in the region near {0, 1, 0}, 7–12 are near {1, 0, 0}, and {13–15} are near {1, 1, 0}.

Thus we have three components: the finite one in the third quadrant with points 10, 11, 5, and 4 above; the large component coming from {1, 1, 0} at point 15 then 6 that goes to {0, 1, 0} and down to point 1; and then the jog 7 out and back to 8 down to 16 that heads straight to {1, 1, 0}.

Let's look at the plot and label the points:

```
In[50]:= Show[ContourPlot[f == 0, {x, -45, 45}, {y, -40, 40}],
           ListPlot[Table[Labeled[allpts[i], i], {i, Length[J]}]], Graphics[
             {Text["B", {2.3, 39}], Text["A", {-2, 39}], Text["C", {18, 39.9}], Text["A", {2.3, -39}],
               Text["B", {-3, -39}], Text["C", {-44, -20}]}], ImageSize → Large]
```

Out[50]=

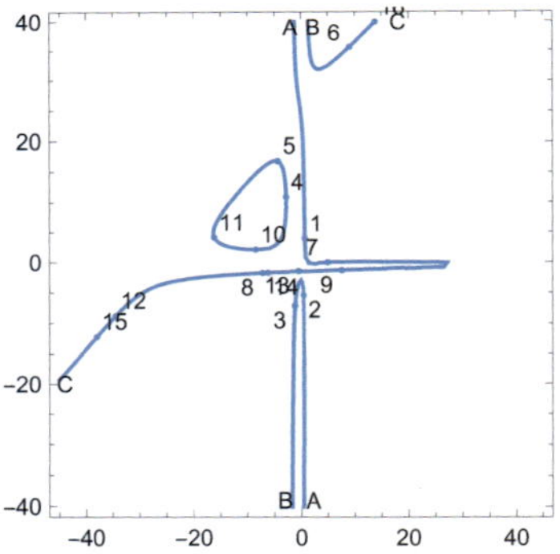

We can now pick out our correct Gauss diagram from the 16 plausible ones previously mentioned. It is number 4:

```
In[51]:= Row[{Show[dDiagram[dvAssoc[coefficientSigns[f, {x, y}]]], pickCyan[0, -2],
             pickCyan[0, -1], pickCyan[-2, 2], pickCyan[0, 2], pickCyan[3, -1],
             pickMagenta[-3, -2], pickMagenta[-2, -2], label13[1, 3, "B"],
             label24[0, 4, "A"], label13[2, 2, "C"], label13[3, 1, "D"], label13[4, 0, "E"],
             Graphics[{Text[6, {.9, 3.7}], Text[16, {2.5, 2.2}], Text[1, {0.7, 1.7}], Text[7, {2, .7}],
                 Text[9, {2, -1.7}], Text[15, {-2.2, -2.2}], Text[3, {-1.3, -2.2}], Text[2, {.7, -2.2}],
                 Text[4, {-1.2, 1.5}], Text[10, {-2, .1}], Text[11, {-2.8, 1.5}], Text[0, {4, -.7}]}],
             ImageSize → Medium], Graph[{"A" → 1, 1 → 7, 7 → 9, 9 → 15, 15 → "C",
               "C" → 16, 16 → 6, 6 → "B", "B" → "A", 4 → 10, 10 → 11, 11 → 4,
               0 → "D", "D" → "E", "E" → 0}, VertexLabels → "Name"]}]
```

Out[51]=

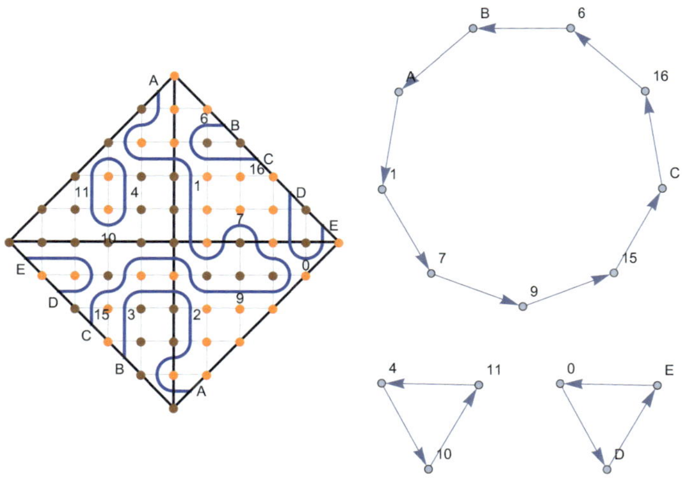

The numbered points are the relative positions of a few of the numbered points in the previous plot. Point 0 refers to the point at approximately $\{137, -.2\}$ on the distant component through infinite points D, E.

Neither a hemisphere nor a Möbius band plot does justice to this curve. We summarize instead by an augmented contour plot, that is, we plot in a big square and add the faraway component using dashed lines:

In[52]:=
```
Show[RegionPlot[-20 < x < 20 && -20 < y < 20, {x, -60, 60}, {y, -60, 60},
    PlotStyle → LightYellow, PlotLabel → "Newton Hyperbola 336941"],
  ContourPlot[f == 0, {x, -60, 60}, {y, -60, 60}], ListLinePlot[{{{24, 0}, {55, 0}, {24, -1}},
      {{-60, -7}, {-50, -4}, {-60, -1}}, {{60, 10}, {50, 7}, {60, 4}}},  PlotStyle → Dashed],
  ListPlot[{Labeled[{1, -60}, "A", Right], Labeled[{-2, -60}, "B", Left],
      Labeled[{-59.5, -34}, "C", Bottom], Labeled[{-60, -7}, "D", Bottom],
      Labeled[{-50, -4}, "b", Right], Labeled[{48, 7}, "a", Left],
      Labeled[{-60, -1}, "E", Top], Labeled[{60, 10}, "D", Top],
      Labeled[{60, 4}, "E", Top], Labeled[{-1, 60}, "A", Left],
      Labeled[{2, 60}, "B", Right], Labeled[{35, 60}, "C", Right] }], ImageSize -> 400 ]
```

Out[52]=

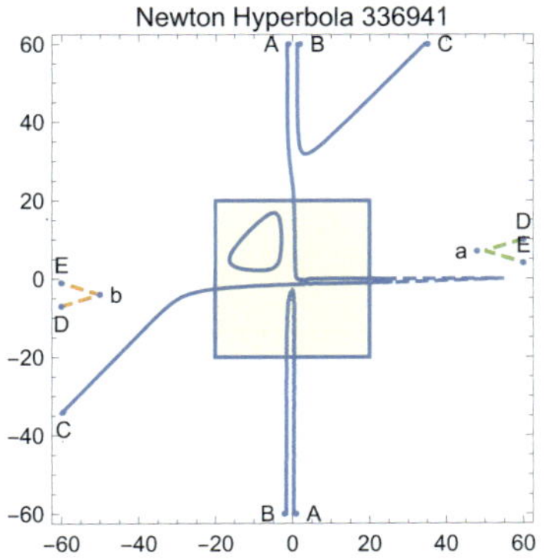

We have answered the second problem posed in Chapter 1 that asked for the various ways to complete the plot in the yellow square. Such a finite plot does not give complete information even though theoretically, interpolating on 20 random enough points, there would be enough to reconstruct the equation. The diamond diagrams given showed hypothetical ways a plot like this could be completed, but it should be noted that discovery of the final graph and plot did not depend on those diagrams, only on analyzing the contour plots near the origin and the infinite points. We did not need path tracing or even calculation of the critical points, but using the projective plane and FLT to go back and forward from the affine curve to the curve near infinite points was crucial.

10.3 Conclusion

In the course of this book, we have learned several ways to analyze and represent plane algebraic real curves. In the course of this journey, we have also learned some of the basic properties of these curves, which have classically been developed using algebraic techniques.

Perhaps the most important thing we have learned is that analyzing and plotting a projective plane curve is a finite problem. As I quoted F. S. Macualay (1916) in my introduction, merely proving a problem is finite is not sufficient if the number of operations is too numerous to be carried out in practice. Even 20 years ago, this would have been the case, but working with actual code, we have seen that, at least in our limited examples up to degree 6, that now, especially with the powerful Wolfram Language, that this is in fact feasible.

Of course, this book is only a first look at these methods and there is still a lot of work to be done automating my techniques and using them with curves of much higher degree. If nothing else comes of this book, I hope it can motivate such work in the future.

More to Explore

For some basic information, check out the Wolfram MathWorld entry on hyperbolas (wolfr.am/v6yyd9gY)

Check out my work in progress on my website for more on these curves (barryhdayton.space)

Appendix 1: Topics in Numerical Linear Algebra

In this appendix, we review some familiar topics in numerical linear algebra that are used in this text and introduce some topics that are not common knowledge. The material here is not necessary to understand the book (assuming the reader will accept certain algorithms, specifically intersection multiplicity, without knowing the details). It is assumed the reader of this appendix will have some prior knowledge of linear algebra and matrices. The expert in numerical linear algebra via Mathematica can jump to A.3. The recommended source for further reading is the wonderful linear algebra text by Gilbert Strang (1976).

As an aside, in my younger days, I was a very abstract, pure-mathematics-only person. My interest in numerical mathematics originated in teaching linear algebra from Strang's textbook.

A.1 Basic Real Linear Algebra

Like classical algebraic geometry, classical linear algebra lives in the complex domain. A main tool is the eigenvalue-eigenvector decomposition, which requires complex numbers—even for simple real matrices. However, modern real numerical linear algebra is based on the singular value decomposition, which gives real singular values even for complex matrices. First, we review Mathematica's linear algebra implementation.

A real numerical matrix is a $m \times n$ array, which is a list of m lists of n machine numbers.

A is a 3 × 4 matrix:

```
In[1]:= A = RandomReal[{-9, 9}, {3, 4}]

Out[1]= {{2.82402, 6.22772, 3.0389, -1.20122},
         {5.273, -3.76647, 7.43988, 5.00426}, {4.89119, 6.2753, 7.73591, -3.11044}}
```

In[2]:= **A // MatrixForm**

Out[2]//MatrixForm=
$$\begin{pmatrix} 2.82402 & 6.22772 & 3.0389 & -1.20122 \\ 5.273 & -3.76647 & 7.43988 & 5.00426 \\ 4.89119 & 6.2753 & 7.73591 & -3.11044 \end{pmatrix}$$

Columns and rows can be extracted. To do so, I prefer Part or Extract rather than Take.

Extract the second column:

In[3]:= **c2 = Extract[A, {All, 2}]**

Out[3]= {6.22772, −3.76647, 6.2753}

In[4]:= **A[[All, 2]]**

Out[4]= {6.22772, −3.76647, 6.2753}

In[5]:= **Take[A, All, {2}] // MatrixForm**

Out[5]//MatrixForm=
$$\begin{pmatrix} 6.22772 \\ -3.76647 \\ 6.2753 \end{pmatrix}$$

Now extract the second row. Note that in the last try, Mathematica views this as a column vector:

In[6]:= **Extract[A, 2] // MatrixForm**

Out[6]= {5.273, −3.76647, 7.43988, 5.00426}

In[7]:= **r2 = A[[2]]**

Out[7]= {5.273, −3.76647, 7.43988, 5.00426}

In[8]:= **Extract[A, 2]**

Out[8]//MatrixForm=
$$\begin{pmatrix} 5.273 \\ -3.76647 \\ 7.43988 \\ 5.00426 \end{pmatrix}$$

While the proper notation for a column is a list of one-element lists (as above), for multiplication purposes, the Wolfram Language will interpret a list of n numbers as either a row or column, depending on the context.

As seen here:

In[9]:= **c2.A**

Out[9]= {28.4202, 92.3501, 39.4484, −45.8482}

Next, we review Strang's four fundamental subspaces (1976) associated with a real $m \times n$ matrix A:

1. The "row space" is the the subspace of $\mathbb{R}^n$ generated by the rows of A (that is, the set of all linear combinations of rows of A).
2. The "column space" is the subspace of $\mathbb{R}^m$ generated by all columns of A.
3. The "null space" is the subspace of $\mathbb{R}^n$ generated by all vectors v with $A.v = 0$.
4. The "left null space" is the subspace of $\mathbb{R}^m$ generated by all vectors w with $w.A = 0$.

Two real vectors of the same length (v and w) are called "orthogonal" if $v.w = 0$. I emphasize that this definition only works in the real case, not in the complex case. The orthogonal complement of a subspace V of $\mathbb{R}^n$ is the subspace of $\mathbb{R}^n$ consisting of all vectors w, which satisfy $v.w = 0$ for all v in V. The orthogonal complement of V is denoted $V^\perp$. Note $(V^\perp)^\perp = V$.

Strang's fundamental theorem of real linear algebra is, "The orthogonal complement of the row space is the null space and conversely. The orthogonal complement of the column space is the left null space and conversely" (1976).

Again, this only works for real linear algebra, not in the complex setting (similarly for the next definition).

A square real matrix A is orthogonal if all rows (and equivalently all columns) are orthonormal—that is, $v.w = 0$ when v and w are different rows (columns) and $v.v = 1$ for all rows (columns). Alternatively, A is orthogonal when the inverse matrix to A is just the transposition of A, $A^{-1} = A^T$.

For example, rotation matrices are orthogonal:

```
In[10]:= R = RotationMatrix[RandomReal[{0, 2 Pi}], RandomReal[{-1, 1}, 3]]

Out[10]= {{0.929277, 0.369373, -0.00276111},
          {0.369245, -0.928697, 0.0343626}, {0.0101284, -0.0329519, -0.999406}}

In[11]:= Transpose[R]

Out[11]= {{0.929277, 0.369245, 0.0101284},
          {0.369373, -0.928697, -0.0329519}, {-0.00276111, 0.0343626, -0.999406}}

In[12]:= Inverse[R]

Out[12]= {{0.929277, 0.369245, 0.0101284},
          {0.369373, -0.928697, -0.0329519}, {-0.00276111, 0.0343626, -0.999406}}
```

A.2 The Singular Value Decomposition

The key idea in real numerical linear algebra is the singular value decomposition (SVD), which gives real results for real matrices (as opposed to the eigenvalue-eigenvector decomposition, which often gives complex results even for real matrices).

Given a real $m \times n$ matrix A, there exists an orthogonal $m \times m$ matrix U, a diagonal $m \times n$ matrix S with decreasing non-negative entries and an orthogonal $n \times n$ matrix U with $A = U.S.V^{\top}$.

An example of an SVD:

```
In[13]:= A = RandomReal[{-4, 4}, {3, 4}];
         A // MatrixForm
```

Out[14]//MatrixForm=
$$\begin{pmatrix} -3.97247 & -3.76528 & 2.91312 & -3.48532 \\ 0.0743706 & 0.751709 & 3.75585 & 1.46374 \\ -1.13094 & 2.93559 & -2.36027 & 3.33028 \end{pmatrix}$$

```
In[15]:= {U, S, V} = SingularValueDecomposition[A];
         U // MatrixForm
```

Out[16]//MatrixForm=
$$\begin{pmatrix} 0.844415 & -0.068226 & 0.531327 \\ 0.0660174 & 0.997549 & 0.0231736 \\ -0.531606 & 0.0155087 & 0.84685 \end{pmatrix}$$

```
In[17]:= S // MatrixForm
```

Out[17]//MatrixForm=
$$\begin{pmatrix} 8.15904 & 0. & 0. & 0. \\ 0. & 4.07491 & 0. & 0. \\ 0. & 0. & 3.2855 & 0. \end{pmatrix}$$

```
In[18]:= V // MatrixForm
```

Out[18]//MatrixForm=
$$\begin{pmatrix} -0.336839 & 0.0804127 & -0.933401 & 0.0940009 \\ -0.574873 & 0.258235 & 0.153045 & -0.761192 \\ 0.485666 & 0.861685 & -0.110771 & -0.0967325 \\ -0.565853 & 0.429357 & 0.305074 & 0.634345 \end{pmatrix}$$

U and *V* are orthogonal:

```
In[19]:= Chop[U.Transpose[U]]

Out[19]= {{1., 0, 0}, {0, 1., 0}, {0, 0, 1.}}
```

```
In[20]:= Chop[V.Transpose[V]]

Out[20]= {{1., 0, 0, 0}, {0, 1., 0, 0}, {0, 0, 1., 0}, {0, 0, 0, 1.}}
```

```
In[21]:= U.S.Transpose[V] // MatrixForm
```

Out[21]//MatrixForm=
$$\begin{pmatrix} -3.97247 & -3.76528 & 2.91312 & -3.48532 \\ 0.0743706 & 0.751709 & 3.75585 & 1.46374 \\ -1.13094 & 2.93559 & -2.36027 & 3.33028 \end{pmatrix}$$

The entries of S are called the singular values. One can do this with exact matrices, and if A is square and invertible, all singular values will be nonzero. If A is not invertible, some of the singular values will be zero. In fact, for an exact matrix, the number of nonzero singular values is the dimension of the row space (which is equal to the dimension of the column space). This number is called the "rank". Further, for any $m \times n$ exact matrix A, the dimension of the null space is $n - r$ and the dimension of the left null space is $m - r$.

For numerical matrices, there may be difficulties in determining the dimensions of the subspaces. The SVD allows us to resolve this problem. We pick a tolerance $\epsilon > 0$ (which is generally quite small but larger than the machine precision) and declare that any singular value $s < \epsilon$ is equal to zero. The number of singular values greater than ϵ is the rank r. The null space is then generated by the last $n - r$ columns of V while the left null space is generated by the last $m - r$ columns of U. The row space is generated by the first r columns of V while the column space is generated by the first r columns of U.

As an example, we set $\epsilon = 10^{-10}$, which is actually a larger-than-normal tolerance.

We modify *A* above to:

```
In[22]:= B = {{-1.2125910411564504`, -2.843903596715471`, -2.618517514835883`,
           -2.7219449070534747`}, {-1.001767330326654`, -1.4814251186573628`,
           -3.266788014281124`, 3.384583512352174`}, {0.8773941003711174`,
           2.2586088618735145`, 1.6393429153186834`, 3.272955439991057`}};
         B // MatrixForm
```

Out[23]//MatrixForm=

$$\begin{pmatrix} -1.21259 & -2.8439 & -2.61852 & -2.72194 \\ -1.00177 & -1.48143 & -3.26679 & 3.38458 \\ 0.877394 & 2.25861 & 1.63934 & 3.27296 \end{pmatrix}$$

Now do an SVD:

```
In[24]:= {U, S, V} = SingularValueDecomposition[B];
         U // MatrixForm
```

Out[25]//MatrixForm=

$$\begin{pmatrix} 0.746559 & -0.0010111 & 0.665318 \\ 0.14866 & -0.974463 & -0.168294 \\ -0.648498 & -0.224548 & 0.727344 \end{pmatrix}$$

```
In[26]:= S // MatrixForm
```

Out[26]//MatrixForm=

$$\begin{pmatrix} 6.53794 & 0. & 0. & 0. \\ 0. & 5.06709 & 0. & 0. \\ 0. & 0. & 1.89358 \times 10^{-11} & 0. \end{pmatrix}$$

```
In[27]:= V // MatrixForm
```

Out[27]//MatrixForm=

$$\begin{pmatrix} -0.248271 & 0.154012 & 0.0861294 & -0.952483 \\ -0.582458 & 0.185373 & 0.751016 & 0.249707 \\ -0.535892 & 0.556117 & -0.610859 & 0.174368 \\ -0.558501 & -0.795393 & -0.235388 & -0.00431952 \end{pmatrix}$$

Note that the third singular value is smaller than ϵ, so we set it to 0:

```
In[28]:= S[[3, 3]] = 0;
         S // MatrixForm
```

Out[29]//MatrixForm=

$$\begin{pmatrix} 6.53794 & 0. & 0. & 0. \\ 0. & 5.06709 & 0. & 0. \\ 0. & 0. & 0 & 0. \end{pmatrix}$$

```
In[30]:= U.S.Transpose[V] // MatrixForm
```

Out[30]//MatrixForm=

$$\begin{pmatrix} -1.21259 & -2.8439 & -2.61852 & -2.72194 \\ -1.00177 & -1.48143 & -3.26679 & 3.38458 \\ 0.877394 & 2.25861 & 1.63934 & 3.27296 \end{pmatrix}$$

The display version of this matrix (with the small singular value set to 0) does not vary from the original B, but the rank of this matrix is 2. The null space of this matrix is two dimensional and generated by the last two columns of V, as is the left null space, which is generated by the last column of U. The column space is generated by the first two columns of U, and the row space is generated by the first two rows of V^T.

Example of the computing basis for null space:

```
In[31]:= NS = Take[V, All, -2];
         NS // MatrixForm
         err1 = B.NS
```

Out[32]//MatrixForm=

$$\begin{pmatrix} 0.0861294 & -0.952483 \\ 0.751016 & 0.249707 \\ -0.610859 & 0.174368 \\ -0.235388 & -0.00431952 \end{pmatrix}$$

Out[33]= $\{\{1.25984\times 10^{-11}, -6.29705\times 10^{-16}\}, \{-3.18756\times 10^{-12}, -8.41341\times 10^{-16}\}, \{1.37726\times 10^{-11}, -1.35308\times 10^{-16}\}\}$

Left null space basis:

```
In[34]:= u0 = Extract[U, {All, 3}]
```

Out[34]= {0.665318, −0.168294, 0.727344}

```
In[35]:= err2 = u0.B
```

Out[35]= $\{1.63158\times 10^{-12}, 1.42217\times 10^{-11}, -1.15667\times 10^{-11}, -4.45644\times 10^{-12}\}$

Column space basis:

```
In[36]:= CS = Take[U, All, 2];
         CS // MatrixForm
```

Out[37]//MatrixForm=

$$\begin{pmatrix} 0.746559 & -0.0010111 \\ 0.14866 & -0.974463 \\ -0.648498 & -0.224548 \end{pmatrix}$$

Row space basis:

```
In[38]:= Take[Transpose[V], 2] // MatrixForm

Out[38]//MatrixForm=
( -0.248271  -0.582458  -0.535892  -0.558501 )
(  0.154012   0.185373   0.556117  -0.795393 )
```

One standard application is the pseudoinverse. Given $A = U.S.V^\top$, the pseudoinverse is obtained by killing diagonal entries in S less than the tolerance, transposing and replacing the remaining nonzero values by their reciprocals (call this ST). Then the pseudoinverse is $P = V.ST.U^\top$.

Alternatively, the Wolfram Language will calculate the pseudoinverse using the same method:

```
In[39]:= P = PseudoInverse[B, Tolerance → 10^-10];
         P // MatrixForm

Out[40]//MatrixForm=
( -0.0283805  -0.0352636  0.017801  )
( -0.0665472  -0.0488935  0.0495592 )
( -0.0613039  -0.119133   0.0285108 )
( -0.0636159   0.140264   0.0906455 )
```

Note:

```
In[41]:= B1 = B.P;
         B1 // MatrixForm

Out[42]//MatrixForm=
(  0.557352  0.111969  -0.483915 )
(  0.111969  0.971677   0.122408 )
( -0.483915  0.122408   0.470971 )
```

Is not the identity matrix, but rather a projection matrix, with B1.B1 = B1 projecting $\mathbb{R}^3$ to the column space of B.B1 acts as the identity on the column space:

```
In[43]:= B1.B1

Out[43]= {{0.557352, 0.111969, -0.483915},
          {0.111969, 0.971677, 0.122408}, {-0.483915, 0.122408, 0.470971}}
```

```
In[44]:= B1.CS // MatrixForm

Out[44]//MatrixForm=
(  0.746559  -0.0010111 )
(  0.14866   -0.974463  )
( -0.648498  -0.224548  )
```

This is used in least squares computations. For example, the Gauss–Newton method refers to a method for finding least squares solutions of over-determined systems.

A special case with three equations in two variables is the following modification of the two-variable version of the Gauss–Newton method, replacing the matrix Inverse with the PseudoInverse.

If you only have two equations, you can use $h = 0$:

```
In[45]:= gaussNewton3[f_, g_, h_, p0_, x_, y_] := Module[{Q, Q0, p1, p2, J0, n},
    n = 30;
    p1 = p0;
    Catch[Do[Q = {f, g, h} /. Thread[{x, y} → p1];
        J0 = D[{f, g, h}, {{x, y}}] /. Thread[{x, y} → p1];
        p2 = p1 - PseudoInverse[J0].Q;
        If[Norm[p1 - p2] < 1.*^-7, Throw[p2]];
        p1 = p2, {n}]];
    p2]
```

Suppose we want to find a singular point of $g = 0$ using the classical definition that a singular point is where g, $D[g, x]$ and $D[g, y]$ all vanish, starting from an approximation taken from a plot.

A bad initial point may converge to a point, which is not a common solution:

```
In[46]:= g = -0.22495692374518167` - 1.1013674560719149` x + 8.256356096694816` x^2 -
        6.3764001275126425` x^3 + 1.3683563251360158` y - 4.696428009684947` x y +
        7.522388070354437` x^2 y - 0.0035046560139825322` y^2 -
        3.591849285015912` x y^2 + 0.7352364129554713` y^3;
    p = {1.58, 3.58};
    {g, D[g, x], D[g, y]} /. Thread[{x, y} → p]

Out[48]= {0.0128948, -0.513936, 0.337176}
```

```
In[49]:= q = gaussNewton3[g, D[g, x], D[g, y], p, x, y]
    {g, D[g, x], D[g, y]} /. Thread[{x, y} → q]

Out[49]= {1.50552, 3.38784}

Out[50]= {-3.55271 × 10^-15, -2.84217 × 10^-14, 7.10543 × 10^-15}
```

This is close enough to meeting the classical definition of a singular point! Much of the time, we will only be interested in the numerical rank of a matrix.

In this case, instead of generating the whole singular value setup, we can just calculate the singular values by:

```
In[51]:= SingularValueList[B]

Out[51]= {6.53794, 5.06709, 1.89358 × 10^-11}
```

The Wolfram Language uses a default tolerance of $s_1 * 100 * 10^{-p}$, where p = $MachinePrecision and s_1 is the largest singular value and smaller singular values are listed as 0. This can be reset by the user, but if you are finding the rank and it is possible that a matrix is the zero matrix, you will get a list of zeros.

It is best to find the number of singular values larger than your tolerance by:

```
In[52]:= M2 = {{1, 0, 0, 0}, {1, 1, 0, 0}, {2, 3, 0, 1.*^-13}};
         Select[SingularValueList[M2], # > 1.*^-12 &]

Out[53]= {3.90898, 0.848463}
```

Programming note: If M is an exact matrix, SingularValueList *and* SingularValueDecomposition *will attempt an exact calculation of these quantities. For a large matrix (such as the matrices in the next two sections), this can take a large amount of time and not give useful results. Thus it is probably a good idea to use these in the form* SingularValueList[N[M]] *or* SingularValueDecomposition[N[M]] *unless you are sure you have a numerical matrix, that is, at least one entry contains a decimal point.*

A.3 Sylvester Matrices

The material here works for polynomials of many variables. We will restrict it to the two-variable case to keep it simple.

This returns a list of rules (as it did in Chapter 9). However, perhaps a more convenient method would be to use Association.

For example, to find the total degree of a bivariate polynomial, we use:

```
tDeg[f_, x_, y_] :=
    Max[Table[Total[p], {p, Keys[Association[CoefficientRules[f, {x, y}]]]}]];
```

Recall the iterative function to get exponent lists:

```
In[55]:= pExps[d_] := Module[{pex},
           pex[0] = {{0, 0}};
           pex[n_] := pex[n] = Join[pex[n - 1], Table[{n - j, j}, {j, 0, n}]];
           pex[d]]

In[56]:= pExps[3]

Out[56]= {{0, 0}, {1, 0}, {0, 1}, {2, 0}, {1, 1}, {0, 2}, {3, 0}, {2, 1}, {1, 2}, {0, 3}}
```

Given a bivariate polynomial $f(x, y)$, we construct an association of coefficients by the following procedure. Since we often need the degree, we can save time by returning the degree as well as the association.

Associations link keys with values, making a more efficient mechanism for finding and updating:

```
In[57]:= fAssoc[f_, x_, y_] := Module[{A, B, d, FA},
             A = Association[CoefficientRules[f, {x, y}]];
             d = Max[Table[Total[p], {p, Keys[A]}]];
             FA = Association[Table[If[MissingQ[A[p]], p → 0, p → A[p]], {p, pExps[d]}]];
             {FA, d}];
```

Rather than actually multiply *f* by monomials, it is easier to work with this association:

```
In[58]:= shiftFA[FA_, q_, n_] := Module[{S, K, d},
             K = Keys[FA];
             S = Association[Table[p + q → FA[p], {p, K}]];
             Association[Table[If[MissingQ[S[p]], p → 0, p → S[p]], {p, pExps[n]}]]];
```

A Sylvester matrix of degree *n* for one polynomial is:

```
In[59]:= sylD1[f_, n_, x_, y_] := Module[{FA, d},
           {FA, d} = fAssoc[f, x, y];
           If[d > n, Print["Must have n≥deg(f)"];
             Abort[]];
           Table[Values[shiftFA[FA, q, n]], {q, pExps[n - d]}]]
```

Finally, the general Sylvester matrix for two polynomials is given by the following (note that it would be easy to modify this procedure to handle an arbitrary number of polynomials, but that won't be necessary for us):

```
sylvesterMatrix[f_, g_, n_, x_, y_] := Join[sylD1[f, n, x, y], sylD1[g, n, x, y]];
```

For example:

```
In[61]:= f = x - y + x^4;
         g = x - y + y^4;
         S = sylvesterMatrix[f, g, 5, x, y];
         S // MatrixForm
```

Out[64]//MatrixForm=

$$\begin{pmatrix} 0 & 1 & -1 & 0 & 0 & 0 & 0 & 0 & 0 & 0 & 1 & 0 & 0 & 0 & 0 & 0 & 0 & 0 & 0 & 0 & 0 \\ 0 & 0 & 0 & 1 & -1 & 0 & 0 & 0 & 0 & 0 & 0 & 0 & 0 & 0 & 0 & 1 & 0 & 0 & 0 & 0 & 0 \\ 0 & 0 & 0 & 0 & 1 & -1 & 0 & 0 & 0 & 0 & 0 & 0 & 0 & 0 & 0 & 0 & 1 & 0 & 0 & 0 & 0 \\ 0 & 1 & -1 & 0 & 0 & 0 & 0 & 0 & 0 & 0 & 0 & 0 & 0 & 0 & 1 & 0 & 0 & 0 & 0 & 0 & 0 \\ 0 & 0 & 0 & 1 & -1 & 0 & 0 & 0 & 0 & 0 & 0 & 0 & 0 & 0 & 0 & 0 & 0 & 0 & 0 & 1 & 0 \\ 0 & 0 & 0 & 0 & 1 & -1 & 0 & 0 & 0 & 0 & 0 & 0 & 0 & 0 & 0 & 0 & 0 & 0 & 0 & 0 & 1 \end{pmatrix}$$

One application of sylD1 is polynomial multiplication. Note that sylD1 $(f, d, x, y)[\![1]\!]$ can be viewed as an alternate notation for a polynomial of degree d.

We can recover the polynomial from sylD1$(f, d, x, y)[\![1]\!]$ using the following function:

```
In[65]:= invSyl[S_, d_, x_, y_] :=
           FromCoefficientRules[Table[pExps[d][[i]] → S[[i]], {i, Length[S]}], {x, y}];
```

```
In[66]:= f = 1 + x^2 - x y + y^3;
         S = sylD1[f, 3, x, y][[1]]

Out[67]= {1, 0, 0, 1, -1, 0, 0, 0, 0, 1}
```

```
In[68]:= f1 = invSyl[S, 3, x, y]
```

Out[68]= $1 + x^2 - x y + y^3$

We have the following identity: given polynomials f, g of degree, $d1$, $d2$ multiplication equation:

```
sylD1[f, d1, x, y].sylD1[g, d1 + d2, x, y] = sylD1[f g, d1 + d2, x, y]
```

For example, for *f* above:

```
In[69]:= g = 2 - x + x^2 - 3 x y;
         sylD1[f, 3, x, y].sylD1[g, 5, x, y]
Out[70]= {{2, -1, 0, 3, -5, 0, -1, 1, 0, 2, 1, -4, 3, -1, 0, 0, 0, 0, 1, -3, 0}}

In[71]:= Sfg = sylD1[f * g, 5, x, y]
Out[71]= {{2, -1, 0, 3, -5, 0, -1, 1, 0, 2, 1, -4, 3, -1, 0, 0, 0, 0, 1, -3, 0}}

In[72]:= invSyl[Sfg[[1]], 5, x, y]
```

Out[72]= $2-x+3x^2-x^3+x^4-5xy+x^2y-4x^3y+3x^2y^2+2y^3-xy^3+x^2y^3-3xy^4$

```
In[73]:= h = Expand[f * g]
```

Out[73]= $2-x+3x^2-x^3+x^4-5xy+x^2y-4x^3y+3x^2y^2+2y^3-xy^3+x^2y^3-3xy^4$

We can use the identity to perform numerical division. In general, one cannot divide one bivariate polynomial by another; the division algorithm does not directly apply to polynomials of several variables. However, if we know that some polynomial (say g) divides numerical polynomial h, then we can find another factor f using the multiplication equation above.

See here:

```
In[74]:= P = PseudoInverse[N[sylD1[g, 5, x, y]]];
         S = Chop[sylD1[h, 5, x, y].P];
         invSyl[S[[1]], 3, x, y]
```

Out[76]= $1.+1.\,x^2-1.\,xy+1.\,y^3$

For another example (more to the point of this book), consider the following numerical curve:

```
In[77]:= h2 = -3.329102349826519` x - 2.383334088473527` x^2 - 0.42613781708474535` x^3 +
            2.2174484310530316` y - 3.1536576003667216` x y - 1.3649471149541397` x^2 y -
            2.616665911526473` y^2 - 1.2169484594921673` x y^2 - 0.19250584655083047` y^3
```

Out[77]= $-3.3291\,x-2.38333\,x^2-0.426138\,x^3+2.21745\,y-$
$3.15366\,xy-1.36495\,x^2y-2.61667\,y^2-1.21695\,xy^2-0.192506\,y^3$

```
In[78]:= ContourPlot[h2 == 0, {x, -8, 2}, {y, -2, 8}]
```

Out[78]=

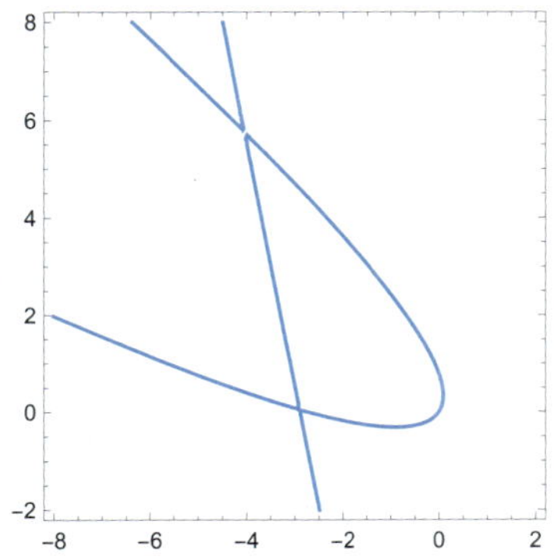

Visually, it appears that this cubic is the union of a line and a quadric.

Find two points on the line:

```
In[79]:= sol2 = {x, y} /. NSolve[{y^2-64, h2}, {x, y}, Reals]

Out[79]= {{-1.2813, -8.}, {-20.3498, 8.}, {-6.37957, 8.}, {-4.48806, 8.}}
```

From our plot, these will be solutions 1 and 4, so using our Chapter 1 line function, the line should be:

```
In[80]:= g2 = line[sol2[[1]], sol2[[4]], x, y]

Out[80]= -2.09476 - 0.72617 x - 0.145541 y
```

Using our numerical division:

```
In[81]:= P = PseudoInverse[N[sylD1[g2, 3, x, y]]];
         S = Chop[sylD1[h2, 3, x, y].P];
         f2 = invSyl[S[[1]], 3, x, y]
```

Out[83]= $1.58925\,x + 0.58683\,x^2 - 1.05857\,y + 1.76204\,x\,y + 1.32269\,y^2$

We get our conic *f*:

```
In[84]:= ContourPlot[{h2 == 0, f2 == 0}, {x, -8, 2}, {y, -2, 8}]
```

Out[84]=

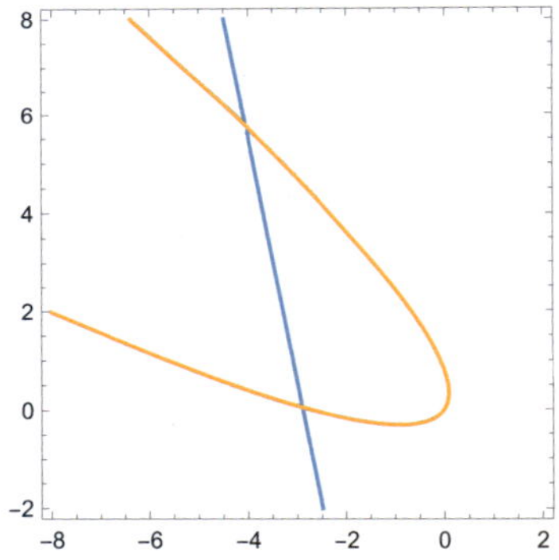

We have the general function:

```
In[85]:= nDivide[h_, g_, x_, y_, tol_] := Module[{d1, d2, P, S, f},
           d1 = tDeg[g, x, y];
           d2 = tDeg[h, x, y];
           If [d1 > d2, Print["Does Not Divide"]; Return[Fail]];
           P = PseudoInverse[N[sylD1[g, d2, x, y]], Tolerance → tol];
           S = Chop[sylD1[h, d2, x, y].P];
           f = invSyl[S[[1]], d2 - d1, x, y];
           If[Norm[Flatten[sylD1[f * g - h, d2, x, y]]] > d2 * tol,
             Print[Style["Warning: g Does not divide h at this tolerance", Orange]];
             Return[Fail]];
           f];
```

Our main goal of this subsection is our numerical greatest common divisor (GCD) algorithm. Suppose we have bivariate polynomials f and h with GCD g. Of course, g is only defined up to scalar multiple. If f and h have no common divisor of degree greater than 0, then g is a constant. Otherwise, we have bivariate polynomials u and v with $f = g\,u$ and $h = g\,v$. Multiplying f by v and h by u gives $f\,v = g\,u\,v = h\,u$ or $f\,v - h\,u = 0$. Using the previous multiplication idea, some linear combination of the rows of certain large enough Sylvester matrices will be zero; hence there will be a nontrivial vector in the left null space. Since $h = v$ and $f = u$ satisfy the equation $f\,v - h\,u = 0$ when the degree is tDeg(*f*) + tDeg(*h*), we need go no further than that. But if there is a nontrivial GCD, it will appear at a lower degree than that, so the following finitely terminating procedure will give the numerical GCD. Note that the simple While loop does the main work in this procedure.

Numerical GCDs ff and hh are bivariate polynomials in x, y. tol is the tolerance. The default is 1.0×10^{-11}, but the larger the tolerance, the more likely it is to get a GCD.

With a big tolerance, a check is in order:

```
In[86]:= nGCD[ff_, hh_, x_, y_, tol_] :=
  Module[{S, Sk, U, V, u, v, su, d1, d2, dv, f, h, n, m, k, W, lnull},
   lnull[M_] := Dimensions[M][[1]] - Length[SingularValueList[N[M], Tolerance → tol]];
   d1 = tDeg[ff, x, y];
   d2 = tDeg[hh, x, y];
   If[d1 ≥ d2, f = ff; h = hh; m = d1; n = d2,
    f = hh; h = ff; m = d2; n = d1];
   k = m - 1;
   Sk = {{0}};
   While[k < m + n && lnull[Sk] == 0,
    k++;              (* Main loop *)
    Sk = sylvesterMatrix[f, h, k, x, y]];
   {U, S, V} = SingularValueDecomposition[N[Sk], Tolerance → tol];
   W = Flatten[Take[U, All, -1]];
   dv = k - m;
   If[dv ≤ 0, Return[h]];
   su = Chop[Take[W, Binomial[dv + 2, 2]], tol];
   u = invSyl[su, dv, x, y];
   nDivide[h, u, x, y, tol]]
```

```
In[87]:= g = randomRealPolynomial[2, x, y]
```

Out[87]= $-0.208937 - 0.232811\,x - 0.0305156\,x^2 - 0.242897\,y - 0.0312451\,x\,y - 0.023855\,y^2$

```
In[88]:= h = Expand[g * randomRealPolynomial[3, x, y]]
```

Out[88]= $-0.145131 - 0.179657\,x - 0.0449851\,x^2 - 0.00684899\,x^3 - 0.000553344\,x^4 + 1.3341 \times 10^{-7}\,x^5 - 0.195602\,y - 0.0701004\,x\,y - 0.00639279\,x^2\,y + 0.00194622\,x^3\,y + 0.000283274\,x^4\,y - 0.0605495\,y^2 - 0.0192549\,x\,y^2 - 0.00169428\,x^2\,y^2 + 0.0000256853\,x^3\,y^2 - 0.0146773\,y^3 - 0.000177492\,x\,y^3 + 0.00041658\,x^2\,y^3 + 0.00225502\,y^4 + 0.000270392\,x\,y^4 + 0.000364197\,y^5$

```
In[89]:= f = Expand[g*randomRealPolynomial[4, x, y]]

Out[89]= -0.0986273 - 0.0650524 x + 0.0331804 x^2 + 0.00647492 x^3 + 0.00265585 x^4 + 0.000519442 x^5 +
  0.0000186722 x^6 - 0.093354 y + 0.0704431 x y + 0.0165452 x^2 y + 0.00279478 x^3 y +
  0.000156916 x^4 y - 0.0000134155 x^5 y + 0.0051807 y^2 + 0.0107682 x y^2 - 0.000693988 x^2 y^2 -
  0.000606977 x^3 y^2 - 0.000102888 x^4 y^2 - 0.00957927 y^3 - 0.00163492 x y^3 -
  0.000871742 x^2 y^3 - 0.0000978573 x^3 y^3 - 0.00371132 y^4 - 0.000190502 x y^4 -
  0.000058843 x^2 y^4 - 0.000323109 y^5 + 3.45439 × 10^-6 x y^5 - 5.57536 × 10^-6 y^6

In[90]:= gout = nGCD[f, h, x, y, 1.*^-10]

Out[90]= -0.186213 - 0.20749 x - 0.0271968 x^2 - 0.21648 y - 0.0278469 x y - 0.0212606 y^2
```

A constant multiplied by the output gives the input. To check, the following should give 0:

```
In[91]:= Expand[g - gout/gout[[1]]*g[[1]]]

Out[91]= 2.77556 × 10^-17 - 6.10623 × 10^-16 x + 2.84495 × 10^-16 x^2 -
  8.60423 × 10^-16 y - 1.07206 × 10^-15 x y - 4.92661 × 10^-16 y^2
```

This algorithm was developed in 2003 independently by E. Kaltofen and me. My colleague Z. Zeng added a preprocessing stage to make the algorithm faster and a postprocessing stage to make it more accurate. His algorithm (Zeng and Dayton, 2004) is still used as a benchmark to test other GCD algorithms. For our purposes, the present version is fast and accurate enough.

A relevant use of the GCD algorithm in this book is to test the hypotheses of Bézout's theorem, especially if the solver detects infinitely many solutions. Another application is to test to see if a bivariate polynomial is square-free and, if not, to give a square-free polynomial defining the same curve. Here, the theory says that if polynomial f is not square-free, then almost any partial derivative $h = a\,D[f, x] + b\,D[f, y]$ will share a common divisor with f. Dividing f by the greatest common divisor will give a square-free polynomial defining the same curve.

The following example illustrates:

```
In[92]:= f4 = -39.72740279218701` x^2 - 6.576432679235679` x^4 - 0.27216394569483804` x^6 +
  20.230248062255377` y + 52.41558252827816` x^2 y + 4.199817230888195` x^4 y -
  24.986060223213187` y^2 - 23.79488900809628` x^2 y^2 - 0.5443278913896761` x^4 y^2 +
  11.063866289626182` y^3 + 4.338410336604424` x^2 y^3 - 2.0680831293809843` y^4 -
  0.27216394569483804` x^2 y^4 + 0.13859310571622874` y^5;
```

In[93]:= **ContourPlot[f4 == 0, {x, -3, 3}, {y, -1, 6}]**

Out[93]=

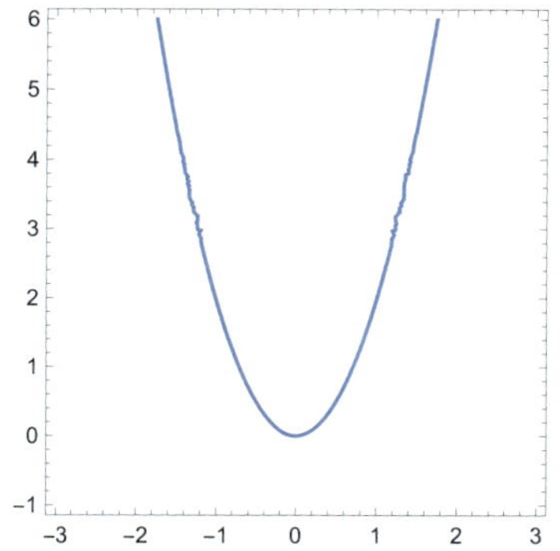

This fifth-degree curve looks like a parabola with some fuzz. Since ContourPlot works by changing the sign of f, any algebraic component that is a square will generally not show up but could cause some noise.

So possibly this curve is not square-free:

In[94]:= **h4 = Expand[{D[f4, x], D[f4, y]}.RandomReal[{-1, 1}, 2]]**

Out[94]= $-18.2055 - 18.6548\,x - 47.1695\,x^2 - 6.17619\,x^3 - 3.77948\,x^4 - 0.3834\,x^5 + 44.9706\,y + 24.6128\,x\,y + 42.8267\,x^2\,y + 3.94422\,x^3\,y + 0.979697\,x^4\,y - 29.8696\,y^2 - 11.1734\,x\,y^2 - 11.7126\,x^2\,y^2 - 0.5112\,x^3\,y^2 + 7.44439\,y^3 + 2.03719\,x\,y^3 + 0.979697\,x^2\,y^3 - 0.62361\,y^4 - 0.1278\,x\,y^4$

In[95]:= **g4 = nGCD[f4, h4, x, y, 1.*^-10]**

Out[95]= $84.5524 + 6.99836\,x^2 - 52.2147\,y + 6.99836\,y^2$

There is a nontrivial GCD that indicates a squared component of h4. Dividing this out will give a square-free curve:

In[96]:= **ff = nDivide[f4, g4, x, y, 1.*^-12]**

Out[96]= $-0.469855\,x^2 - 0.0388897\,x^4 + 0.239263\,y + 0.309959\,x^2\,y - 0.147755\,y^2 - 0.0388897\,x^2\,y^2 + 0.0198037\,y^3$

In[97]:= **RegionPlot[ff > 0, {x, -3, 3}, {y, -1, 6}]**

Out[97]=

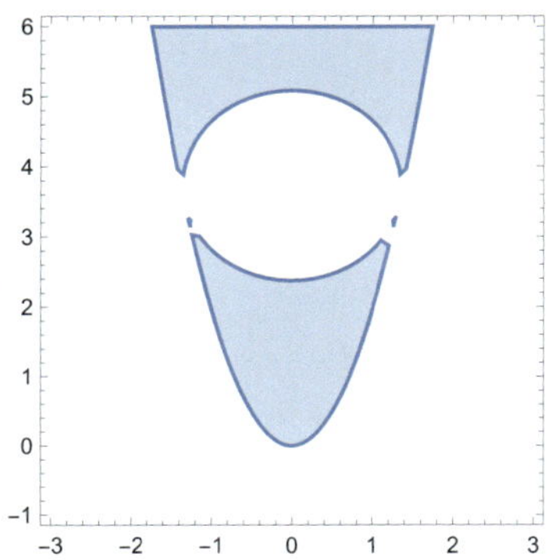

This still is not a good picture. We should graph the components separately:

In[98]:=
```
fp = nDivide[f4, g4^2, x, y, 1.*^-12]
```

Out[98]= $-0.00555697\, x^2 + 0.00282976\, y$

In[99]:=
```
ContourPlot[{fp == 0, g4 == 0}, {x, -3, 3}, {y, -1, 6}]
```

Out[99]=
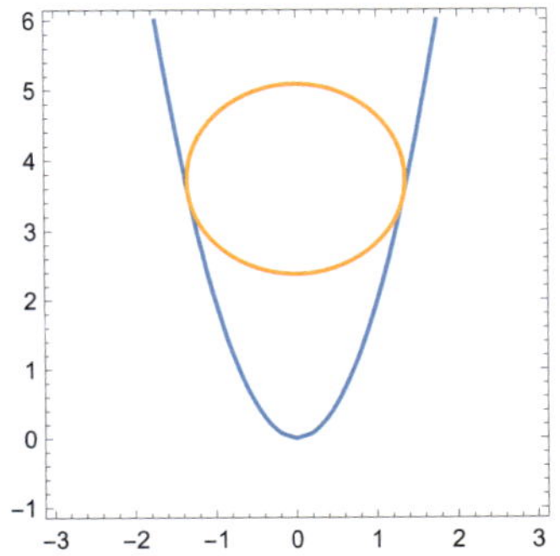

This solves our problem. However, in the algorithm, the last step was a division, but we found *ff* by another division, which just returned what we had originally, so it is better to have the square-free reduction as a separate procedure and save two divisions, as well as saving some clerical steps (because we know degree *f* is greater than degree *h*).

For example:

In[100]:=
```
sqFree[f_, x_, y_, tol_] := Module[{S, Sk, U, V, v, su, dv, h, n, m, k, W},
    h = Expand[{D[f, x], D[f, y]}.RandomReal[{-1, 1}, 2]];
    lnull[M_] :=
      Dimensions[M][[1]] - Length[SingularValueList[N[M], Tolerance → tol]];
    m = tDeg[f, x, y];
    n = tDeg[h, x, y];
    k = m - 1;
    Sk = {{0}};
    While[k < m + n && lnull[Sk] == 0,
      k++;
      Sk = sylvesterMatrix[f, h, k, x, y]];
    {U, S, V} = SingularValueDecomposition[N[Sk], Tolerance → tol];
    W = Flatten[Take[U, All, -1]];
    dv = k - n;
    su = Chop[Take[W, -Binomial[dv + 2, 2]], tol];
    invSyl[su, dv, x, y]];
```

For the example above, we get a constant multiple of the square-free reduction from earlier:

In[101]:=
```
sqFree[f4, x, y, 1.*^-12]
```

Out[101]= $0.49279\, x^2 + 0.040788\, x^4 - 0.250942\, y - 0.325089\, x^2\, y + 0.154967\, y^2 + 0.040788\, x^2\, y^2 - 0.0207703\, y^3$

The Wolfram Language also has a square-free algorithm, but it doesn't work on numerical polynomials:

In[102]:=
```
SquareFreeQ[f4, {x, y}]
```

Out[102]= True

Here is an interesting example:

In[103]:=
```
f5 = 4 y + 5 y^2 + 7 x y^2 + 5 x^2 y^2 - 2 x^4 y^2 + 5 y^3 + 8 x y^3;
```

Plotting f = 0 shows a problem:

In[104]:=
```
ContourPlot[f5 == 0, {x, -4, 4}, {y, -4, 4}]
```

Out[104]=

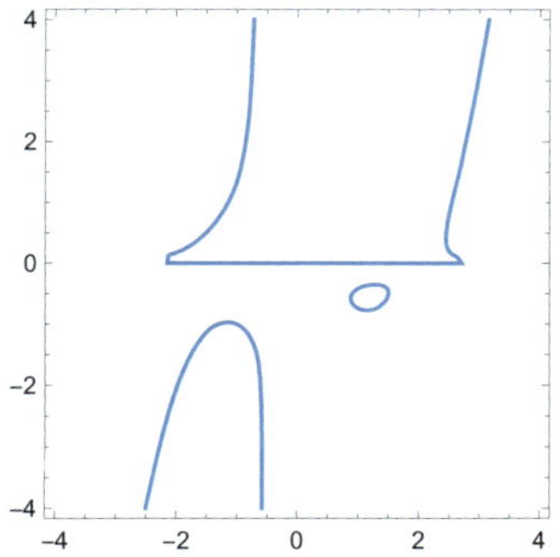

Checking Factor, we see this is reducible but square-free as an exact curve:

In[105]:=
```
Factor[f5]
```

Out[105]= $y(4+5y+7xy+5x^2y-2x^4y+5y^2+8xy^2)$

The component $y = 0$ only partly appears in the plot—it disappears, as in a squared component. Running sqFree with a tight tolerance returns the curve as square-free, but with a very loose tolerance, we get various answers because the algorithm uses randomization.

But one run gives (non-editable graphic):

In[106]:=
```
g = sqFree[f, x, y, .02]
```

Out[106]= $0.153522 + 0.210315\,y + 0.294896\,x\,y + 0.215976\,x^2\,y - 0.0768224\,x^4\,y + 0.198772\,y^2 + 0.31495\,x\,y^2$

In[107]:=
```
ContourPlot[{g == 0, f/y == 0}, {x, -4, 4}, {y, -4, 4}]
```

Out[107]=

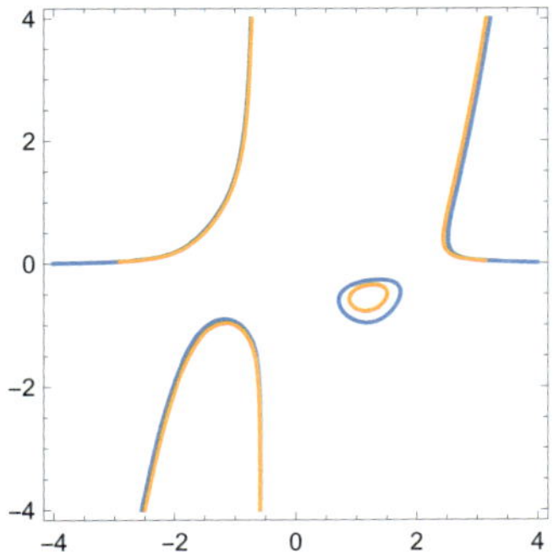

So the almost-square component of f is removed, giving close to the second exact component of f.

A.4 Macaulay Matrices

Sylvester matrices are global objects because we use the whole defining polynomials, but Macaulay matrices are based at a point. Further, we will see that the degree of the polynomials is not used in defining Macaulay matrices (which allows us to extend the notion of intersection multiplicity to curves defined by analytic functions, as well as polynomial functions).

There are several strategies for calculating Macaulay matrices, which tend to be large. Fortunately, we will only work in two dimensions with polynomials of modest degree (where the number of entries in these matrices run in the thousands), but the concept can be expanded to any number of variables. The size of Macaulay matrices grows very quickly as the degree and number of variables increase, and entries are measured in millions. Thus in choosing methods, considering computer time and memory is important.

In choosing methods, I compared two strategies. The first, copying the idea of constructing Sylvester matrices, uses the fact that polynomials are given in terms of coefficients. We can use the `fAssoc` function in A.3 to get an association directly from `CoefficientRules` after using a change of basis calculation

```
fp = Expand[f /. Thread[{x, y} -> {x, y} + p].
```

To calculate a single Macaulay matrix, this is twice as fast and adds about 60% less memory to the session.

However (as in the previous GCD algorithm), in our main intersection multiplicity algorithm that follows, we will again be increasing the degree until we reach a stopping point, but the Macaulay matrix of one degree larger has the previous Macaulay matrix as a submatrix, and one can possibly save time and memory by not recalculating these entries (although sometimes memory lookup can be more time consuming than recalculating). Thus the algorithm below (given the calculation of the previous degree Macaulay matrix) is 25% faster and adds about the same amount of new memory as the Sylvester algorithm. Although we won't need this, it will also work with analytics as well as polynomial functions.

We start with the differential operator `difOp`, which would calculate the coefficient of $x^i\ y^j$ of the transformed function `fp` mentioned above:

```
difOp[f_, pt_, i_, j_] := N[(1/Factorial[i]/Factorial[j] D[f, {x, i}, {y, j}]) /. Thread[{x, y} → pt]]
```

We assume that we will only calculate Macaulay matrices at common points of a pair of curves, so the functions (which can now be analytic as well as polynomials) will be approximately zero at the point `pt`. The recursive function `dAss` will be treated as a global variable, so all associations calculated will be stored for the remainder of the session.

The code for dAss:

```
In[108]:= dAss[f_, pt_, 0] = <|{0, 0} → 0|>;
dAss[f_, pt_, n_] := dAss[f, pt, n] =
    AssociateTo[dAss[f, pt, n - 1], Table[{i, n - i} → difOp[f, pt, i, n - i], {i, 0, n}]];
```

Then we can construct the Macaulay matrix of order *d* by:

```
In[110]:= macaulayD1[f_, p_, n_] := Module[{M},
    Table[Values[shiftFA[dAss[f, p, n - Total[q]], q, n]], {q, pExps[n - 1]}]]

In[111]:= macaulayMatrix[f_, g_, p_, n_] := Join[macaulayD1[f, p, n], macaulayD1[g, p, n]]
```

We suppress the variables *x* and *y*, so the user must be careful to use these variables. Here's an example:

```
In[112]:= f6 = -0.007570349710845024` x^2 + 0.0038550230270411076` y;
g6 = -72.4414468536883` - 5.995940647212547` x^2 +
    44.73564407546775` y - 5.995940647212419` y^2;
pt6 = {-1.3304175490338985`, 3.475880965863065`}

Out[114]= {-1.33042, 3.47588}
```

```
In[115]:= M6 = macaulayMatrix[f6, g6, pt6, 3];
Dimensions[M6]

Out[116]= {12, 10}
```

```
In[117]:= M6 // MatrixForm
```

Out[117]//MatrixForm=

0	0.0201435	0.00385502	−0.00757035	0.	0.	0.	0.	0.	0.
0	0	0	0.0201435	0.00385502	0	−0.00757035	0.	0.	0
0	0	0	0	0.0201435	0.00385502	0	−0.00757035	0.	0.
0	0	0	0	0	0	0.0201435	0.00385502	0	0
0	0	0	0	0	0	0	0.0201435	0.00385502	0
0	0	0	0	0	0	0	0	0.0201435	0.00385502
0	15.9542	3.05329	−5.99594	0.	−5.99594	0.	0.	0.	0.
0	0	0	15.9542	3.05329	0	−5.99594	0.	−5.99594	0
0	0	0	0	15.9542	3.05329	0	−5.99594	0.	−5.99594
0	0	0	0	0	0	15.9542	3.05329	0	0
0	0	0	0	0	0	0	15.9542	3.05329	0
0	0	0	0	0	0	0	0	15.9542	3.05329

```
In[118]:= SingularValueList[M6]

Out[118]= {22.9457, 19.2867, 17.5101, 16.2438, 14.3956, 11.0763, 0.00728866, 0.00669726}
```

We see that the null space will have numerical rank 2:

```
In[119]:= {U, S, V} = SingularValueDecomposition[M6];
NS = Take[V, All, -2];
Chop[NS] // MatrixForm
```

Out[121]//MatrixForm=

0	−1.
−0.187967	0
0.982175	0
0	0
0	0
0	0
0	0
0	0
0	0
0	0

```
In[122]:= diff1 = NS[[1, 2]] * difOp[#, pt6, 0, 0] &;
          diff2 = NS[[2, 1]] * difOp[#, pt6, 1, 0] + NS[[3, 1]] difOp[#, pt6, 0, 1] &;
          {diff1[f6], diff1[g6]}

Out[124]= {3.83374 × 10^-16, 5.68434 × 10^-14}

In[125]:= {diff2[f6], diff2[g6]}

Out[125]= {3.29597 × 10^-17, 2.1716 × 10^-13}
```

We have two differential operators that numerically kill (tolerance 10^{-12}) both f and g. Leaving out the technical details, this is the numerical analyst's definition of intersection multiplicity 2. (See the introduction to [Dayton, Li, and Zeng, 2011] for the details.)

We can now describe the intersection multiplicity algebraically according to F. S. Macaulay (1916). Given two curves $f = 0$, $g = 0$ and a point p on both curves, one computes the Macaulay matrices macaulayMatrix[f, g, p, i] of order $i = 1, 2, 3$ ... calculating via SingularValueList the dimension n of the null space (that is, the number of columns minus the number of singular values larger than your tolerance). The default is 10^{-12}, but this can be changed. As i increases, this sequence, known as the Hilbert function, will be nondecreasing. When the same number appears twice, it will (theoretically) remain constant. This is the intersection multiplicity. The following procedure gives the multiplicity, optionally printing out the Hilbert function. It will assume the equations of the curves involve variables x and y, and no other variables appear. By default (to avoid an infinite loop), the procedure will stop at multiplicity maxm = 20, but this is easy to change.

If the two equations have a nontrivial common divisor and the point p is on the common curve, then the multiplicity is infinity (that is, the procedure will not terminate). Intersection multiplicity of curves f, g are in variables x, y only at point p with tolerance tol.

Running with option HF → True gives the depth, the number of iterations (minus 2), and the Hilbert function:

```
In[126]:= Options[intersectionMultiplicity] = {HF → False};
          intersectionMultiplicity[f_, g_, p_, tol_, OptionsPattern[]] :=
            Module[{M, c, r, m, n, del, maxm, i, L, x, y},
              maxm = 20;
              If[Abs[f /. Thread[{x, y} → p]] > tol, Print["p not zero of f"]; Abort[]];
              If[Abs[g /. Thread[{x, y} → p]] > tol, Print["p not zero of g"]; Abort[]];
              del = 1;
              m = 1;
              L = {1};
              For[i = 1, i < maxm && del > 0, i++,
                M = macaulayMatrix[f, g, p, i];
                c = Dimensions[M][[2]];
                r = Length[Select[SingularValueList[M], # > tol &]];
                n = c - r;
                del = n - m;
                m = n;
```

```
    AppendTo[L, m]];
  If[m == maxm, Print["Multiplicity Greater or equal to ", maxm]];
  If[OptionValue[HF], Print["depth = ", i-2, " HF = ", L]];
  m]
```

Some examples follow.

Normal crossings have multiplicity 1:

```
In[128]:= dTol = 1.*^-9;
          intersectionMultiplicity[x, y, {0, 0}, dTol]
Out[129]= 1
```

```
In[130]:= intersectionMultiplicity[y-x^2, x-y, {1, 1}, dTol]
Out[130]= 1
```

Simple tangents have multiplicity 2:

```
In[131]:= intersectionMultiplicity[y, y-x^2, {0, 0}, dTol]
Out[131]= 2
```

A nice family of examples with high multiplicity are $f = x - y - x^n$, $g = x - y - y^n$. The multiplicity is $2n - 1$:

```
In[132]:= intersectionMultiplicity[x-y+x^7, x-y + y^7, {0, 0}, dTol, HF → On]
Out[132]= 13
```

Another large multiplicity:

```
In[133]:= intersectionMultiplicity[x^3, x^2y+y^4, {0, 0}, dTol, HF → On]
Out[133]= 12
```

This algorithm was originally proposed by F. S. Macaulay in 1916 but was largely forgotten because it required finding the rank of large matrices. My colleague Z. Zeng and I implemented his algorithm numerically in 2004 (a discussion on the topic can be read in [Dayton, Li, and Zeng, 2011]). A problem that could arise is when an intersection has high multiplicity, functions such as NSolve may not be able to calculate the point accurately, and then intersectionMultiplicity may not give the correct multiplicity. Fortunately, at approximately the same time, our neighbors A. Leykin, J. Verschelde, and A. Zhao (2006) proposed a deflation algorithm that would replace the system in the non-quadratically converging Newton's method with a quadratically converging system. These two algorithms are closely related and, ideally, should be used together. Unfortunately, in the case that the multiple intersection point is not known exactly, the deflation algorithm may be too complicated for normal use.

As an example of the relationship of deflation and our multiplicity algorithm, consider the following numerical system:

```
In[134]:= f8 = -4+12x-27x^2+27x^3-3y;
          g8 = 5+3x+33y+54y^2+27y^3;
          p8 = {x, y} /. NSolve[{f8, g8}, {x, y}, Reals][[1]]
Out[136]= {0.333703, -0.666297}
```

Using the default in the global intersectionMultiplicity:

```
In[137]:= intersectionMultiplicity[f8, g8, p8, 1.*^-9]

Out[137]= 2
```

A first step in deflation might be to add the Jacobian determinate to the system, as it must vanish at a multiple zero. This increases the depth and multiplicity:

```
In[138]:= J8 = Det[D[{f8, g8}, {{x, y}}]]

Out[138]= 405 - 1782 x + 2673 x^2 + 1296 y - 5832 x y + 8748 x^2 y + 972 y^2 - 4374 x y^2 + 6561 x^2 y^2

In[139]:= p8a = gaussNewton3[f8, g8, J8, p8, x, y]

Out[139]= {0.333377, -0.666623}

In[140]:= intersectionMultiplicity[f8, g8, p8a, 1.*^-9]

Out[140]= 3
```

If we could continue deflation, we would find that the intersection point is:

```
In[141]:= p8e = N[{1/3, -2/3}]
          intersectionMultiplicity[f8, g8, p8e, 1.*^-9, HF → On]

Out[141]= {0.333333, -0.666667}

Out[142]= 5
```

An easier workaround (if we just want the multiplicity) is to use a much looser tolerance:

```
In[143]:= intersectionMultiplicity[f8, g8, p8, .02]

Out[143]= 5
```

A.5 Finding Algebraic Equations of Curves Parameterized by Polynomials

In Chapter 1, we mentioned that curves parameterized by polynomial or rational parameterizations are contained in algebraic curves. In Chapter 2, we suggested finding these parameterizations by choosing an appropriate number of random points on the curve and using interpolation to find the algebraic curve.

Here, we show that there is another approach for polynomial parameterizations that does not require finding points. This is especially advantageous if the polynomials are exact (integer coefficients), as the algebraic curve is exact. Moreover, we can explain why it is always possible to find an algebraic equation of a curve containing the parameterized curve. Finally, we will sketch (but not give details) on how to extend this to rational parameterizations.

The main tool is a matrix I will call the p2aMatrix, which is somewhat analogous to the Sylvester matrix. For convenience, we will index the rows by monomials of total degree d or less and the columns by powers of t. Assuming the parameterized curve is given by $F = \{u, v\}$ where u, v are polynomials in t, then the entry in the row $x^i\, y^j$ and column t^k is the coefficient of t^k in the expansion of $a^i\, b^j$. Since the row indices have total degree d or less, $a^i\, b^j$ will have as a polynomial in t degree no larger than $m\, d$, where m is the larger of the degrees of a, b. Thus the p2aMatrix will have dimensions $\frac{(d+2)\,(d+1)}{2} \times (d\, m + 1)$. To get a result, we will need $d \geq m$.

Consider the easy parameterized curve $F = \{x, y\} = \{t^2, 2\,t - 1\}$:

```
In[144]:= F1 = {t^2, 2 t - 1};
```

The maximum degree is $m = 2$, so we try $d = 2$. Then upon substitution of the parameters into the monomials, and if the constant remains 1, x becomes t^2; y becomes $2\,t - 1$; x^2 becomes t^4; $x\,y$ becomes $t^2(2\,t - 1) = 2\,t^3 - t^2$; and y^2 becomes $(2\,t - 1)^2 = 4\,t^2 - 4\,t + 1$.

So our p2aMatrix will be:

	1	t	t^2	t^3	t^4
1	1	0	0	0	0
x	0	0	1	0	0
y	−1	2	0	0	0
x^2	0	0	0	0	1
x y	0	0	−1	2	0
y^2	1	−4	4	0	0

Thus the p2aMatrix is 6×5, and by elementary linear algebra, the rows must be dependent, that is, some linear combination of the rows gives the zero row. By inspection, we see that if we add row 1, -4 times row x, twice row y, and row y^2, we get the zero row. In other words, we could write $1 - 4\,x + 2\,y + y^2 = 0$.

Or in the Wolfram Language:

```
In[145]:= Expand[(1 - 4 x + 2 y + y^2) /. Thread[{x, y} → F1]]
Out[145]= 0
```

That means every point on the parameterized curve lies in the curve $1 - 4\,x + 2\,y + y^2 = 0$.

By this approach, we can find a curve that contains our parameterized curve. In the next example, the curve is a parabola and every point in the parabola can be written in the form $F(t)$ for some t. In general, the parameterized curve may lie in some topological component of the algebraic curve and need not give the entire curve.

See below:

In[146]:=
```
Show[ParametricPlot[F1, {t, -3, 3}, Axes → False, Frame → True, PlotStyle → Cyan],
  ContourPlot[1 - 4 x + 2 y + y^2 == 0, {x, -1, 9}, {y, -7, 5}, ContourStyle → Dashed],
  ImageSize → Medium]
```

Out[146]=

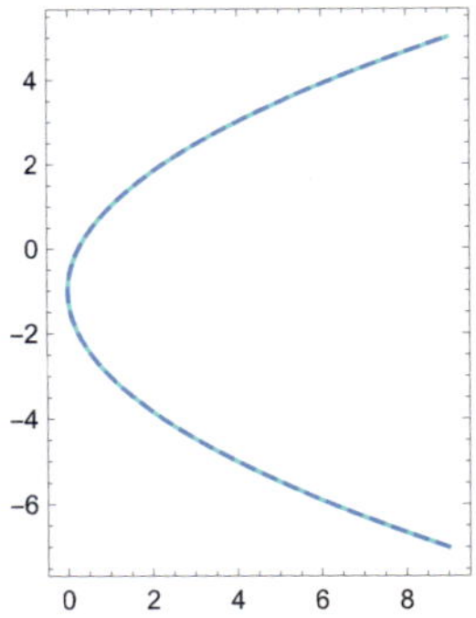

The general code to find the p2aMatrix is:

In[147]:=
```
p2aMatrix[F_, d_, t_, x_, y_] := Module[{m, n, M, c, ca, ra},
    m = Max[Exponent[F[[1]], t], Exponent[F[[2]], t]];
    n = m * d;
    M = {};
    Do[c = CoefficientRules[Expand[(x^p[[1]] y^p[[2]]) /. Thread[{x, y} → F]], t];
      ca = Association[c];
      ra = Association[Table[If[MissingQ[ca[{k}]], {k} → 0, {k} → ca[{k}]], {k, 0, n}]];
      AppendTo[M, Values[ra]], {p, pExps[d]}];
    M];
```

From our example:

In[148]:=
```
p2aMatrix[F1, 2, t, x, y] // MatrixForm
```

Out[148]//MatrixForm=

$$\begin{pmatrix} 1 & 0 & 0 & 0 & 0 \\ 0 & 0 & 1 & 0 & 0 \\ -1 & 2 & 0 & 0 & 0 \\ 0 & 0 & 0 & 0 & 1 \\ 0 & 0 & -1 & 2 & 0 \\ 1 & -4 & 4 & 0 & 0 \end{pmatrix}$$

The theory is that since m is fixed by the parameterization, the number of rows is a quadratic function of d, but the number of columns is only a linear function of d. Thus for some d, we must have more rows than columns. There will be a linear combination of the rows adding to zero, and our parameterized curve will be contained in the corresponding algebraic curve. We will see that we might get such a linear combination even before this happens. Using the smallest value of d will make it likely that the algebraic curve will not be reducible with algebraic components (which are not needed).

It is common sense that our algebraic equation should have degree $s = m$ since the parameterization has a t^m in it. We will see that this isn't actually true, but counter-examples are unlikely (and it will be apparent if we have one). We calculate p2aMatrix for $d = m$ and look for a nonzero vector u in the left null space. How we do it will depend on whether we want an exact or numerical answer. In either case, we can use the following procedure to find the equation.

p2aEq converts a sequence of coefficients to an equation. The typical use is to apply to a row of the left null space of p2aMatrix:

```
In[149]:= p2aEq[u_, x_, y_] := Module[{P, d},
    If[Length[Dimensions[u]] > 1, Print["u must be a vector"]; Abort[]];
    d = Switch[Length[u], 3, 1, 6, 2, 10, 3, 15, 4, 21, 5, 28, 6, 36, 7, 45, 8, 55, 9];
    If[Length[d] > 1, Abort[]];
    P = pExps[d];
    FromCoefficientRules[Table[P[[k]] → u[[k]], {k, (d+2) (d+1)/2}], {x, y}]];
```

If our parametric function has integer coefficients for both component polynomials, then p2aMatrix will be an integer matrix and Mathematica will treat it as such. Calculating the left null space of PM = p2aMatrix (F, d, t, x, y) by NullSpace[Transpose[PM]] will result in (at worst) rational numbers, which can be converted to integers by multiplying by the greatest common denominator.

We demonstrate by finding the equation of the degree 5 Chebyshev curve, as given parametrically in Chapter 1:

```
In[150]:= c5 = {ChebyshevT[4, t], ChebyshevT[5, t]}
```

Out[150]= $\{1-8t^2+8t^4, 5t-20t^3+16t^5\}$

Since $m = 5$, we do:

```
In[151]:= PM = p2aMatrix[c5, 5, t, x, y];
```

```
In[152]:= NS = NullSpace[Transpose[PM]]
```

```
Out[152]= {{-1, 5, 0, 0, 0, 8, -20, 0, 0, 0, 0, 0, 0, 0, -8, 16, 0, 0, 0, 0, 0}}
```

```
In[153]:= f = p2aEq[NS[[1]], x, y]
```

Out[153]= $-1+5x-20x^3+16x^5+8y^2-8y^4$

Check:

```
In[154]:= Expand[f /. Thread[{x, y} → c5]]
```

```
Out[154]= 0
```

In[155]:=
```
Show[ParametricPlot[c5, {t, -2, 2}, Axes → False, Frame → True, PlotRange → 2,
    PlotStyle → Cyan], ContourPlot[f == 0, {x, -2, 2}, {y, -2, 2}, ContourStyle → Dashed]]
```

Out[155]=

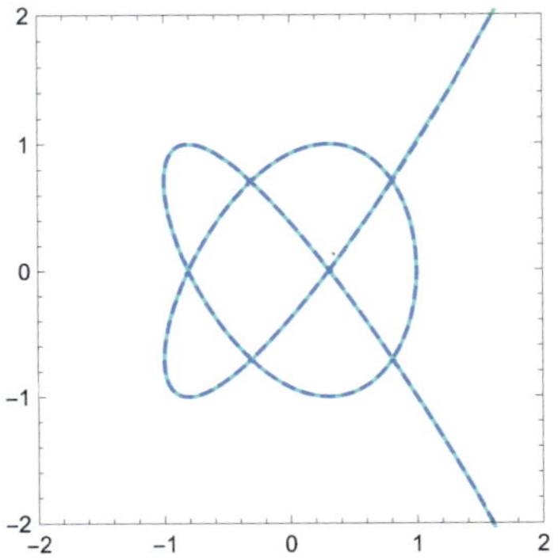

Here is a simple example of a bad case, explaining why we don't have one automatic procedure to do the whole problem:

In[156]:=
```
G1 = {t^2-1, t^2+1};
PG1 = p2aMatrix[G1, 2, t, x, y];
```

In[158]:=
```
NS = NullSpace[Transpose[PG1]]
```

Out[158]= {{−4, −4, 0, −1, 0, 1}, {0, −2, 0, −1, 1, 0}, {−2, −1, 1, 0, 0, 0}}

So our null space has three vectors, which gives three answers:

In[159]:=
```
g1a = p2aEq[NS[[1]], x, y]
g1b = p2aEq[NS[[2]], x, y]
g1c = p2aEq[NS[[3]], x, y]
```

Out[159]= $-4-4x-x^2+y^2$

Out[160]= $-2x-x^2+xy$

Out[161]= $-2-x+y$

In[162]:=
```
{ContourPlot[g1a == 0, {x, -4, 4}, {y, -4, 4}],
  ContourPlot[g1b == 0, {x, -4, 4}, {y, -4, 4}], ContourPlot[g1c == 0, {x, -4, 4}, {y, -4, 4}],
  ParametricPlot[G1, {t, -2, 2}, Axes → False, Frame → True, PlotRange → 4]}
```

Out[162]=

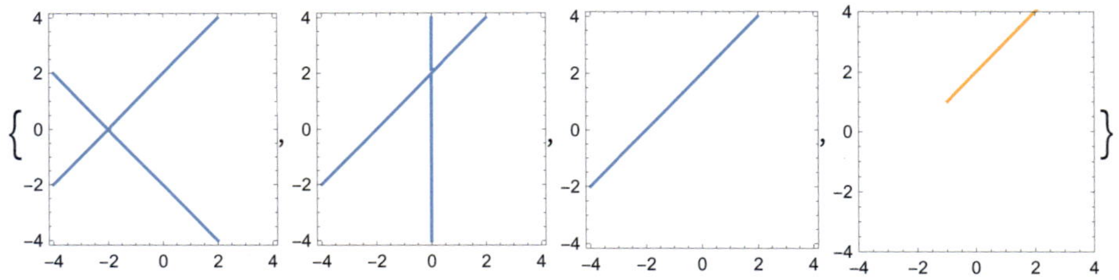

In this case, the parametric curve is half the line given by g1c. The other two curves have an extra algebraic component, as well as the extra half line. The curve has degree 1 while $m = 2$.

Further, note:

```
In[163]:= PG1a = p2aMatrix[G1, 1, t, x, y];
          NS1 = NullSpace[Transpose[PG1a]]
Out[164]= {{-2, -1, 1}}
```

```
In[165]:= p2aEq[NS1[[1]], x, y]
Out[165]= -2 - x + y
```

If the parametrizing function has machine number coefficients, we should use singular value decomposition to get the left null space:

```
In[166]:= F = N[{CubeRoot[2] t^2, t^3}]
Out[166]= {1.25992 t^2, t^3}
```

```
In[167]:= PF = p2aMatrix[F, 3, t, x, y];
          Dimensions[PF]
Out[168]= {10, 10}
```

```
In[169]:= {u, s, v} = SingularValueDecomposition[PF];
          Diagonal[s]
Out[170]= {2.23607, 1.5874, 1.5874, 1.25992, 1.25992, 1.25992, 1., 1., 1., 0.}
```

The last zero shows that the left null space is nontrivial:

```
In[171]:= ns = Flatten[Take[u, All, -1]]
Out[171]= {0., 0., 0., 0., 0., -0.894427, 0.447214, 0., 0., 0.}
```

```
In[172]:= f = p2aEq[ns, x, y]
Out[172]= 0. + 0.447214 x^3 - 0.894427 y^2
```

```
In[173]:= Expand[f/Coefficient[f, x^3]]
Out[173]= 0. + 1. x^3 - 2. y^2
```

The previous examples show that while integer polynomial parameterized curves are contained in exact curves, exact curves most often do not have integer polynomial parameterizations but may have numerical parameterizations.

Finally, this method can be extended to deal with rational parameterizations. We can assume the two component functions have the same denominator, i.e. $F = \left\{\frac{a}{c}, \frac{b}{c}\right\}$ for polynomials a, b, c in t. Then by homogenizing, we get a polynomial parameterization $H = \{a, b, c\}$. Extending the procedures p2aMatrix, p2aEq to three variables and adding a random second equation $r_1 x + r_2 y + r_3 z = 1$, we can find a homogeneous equation $h[x, y, z] = 0$ whose projective solution contains the image of H. Setting $z = 1$, for example, gives us an affine equation.

A.6 Factoring Highly Singular Curves

In Chapter 5, we gave the famous classical result that an irreducible curve of degree d has, at most, $\frac{(d-1)(d-2)}{2}$ singularities. Rather than the usual indirect proof, we sketched the direct proof of the contrapositive: a curve of degree d with more than $\frac{(d-1)(d-2)}{2}$ singularities is reducible. I promised that we could actually make this into an algorithm implemented in Mathematica. We do this here.

Assume that at least $\frac{(d-1)(d-2)}{2}+1$ is affine (possibly complex).

The following algorithm will produce (with proper choice of tolerance) factors based on the algorithm described in Chapter 5:

```
In[174]:= singularFactor[f_, x_, y_, tol_] := Module[{h, csp, d, sp, M, M1, cr, c0, rhs, rr, XX, h1},
   csp = Chop[complexSingularPoints[f, x, y]];
   d = tDeg[f, x, y];
   If[d < 3, Print["degree must be at least 3"]; Abort[]];
   If[Length[csp] < ((d - 1) (d - 2)/2 + 1), Print["Not enough affine singular points"];
    Abort[]];
   csp = RandomSample[csp, (d - 1) (d - 2)/2 + 1];
   sp = If[d == 3, csp, Join[csp, realRegularPoints[f, x, y, d - 3]]];
   XX = pExps[d - 2];
   M = Table[Which[e == {0, 0}, 1, e[[1]] == 0, p[[2]] ^ e[[2]], e[[2]] == 0,
       p[[1]] ^ e[[1]], True, p[[1]] ^ e[[1]] * p[[2]] ^ e[[2]]], {p, sp}, {e, XX}];
   rr = Table[0, {Length[XX]}];
   rr[[-1]] = 1;
   M1 = Append[M, RandomReal[{-1, 1}, Length[XX]]];
   rhs = Table[0, {Length[XX]}];
   rhs[[-1]] = 1;
   cr = LeastSquares[M1, rhs];
   c0 = Min[Select[Abs[cr], # > .1 &]];
   h = FromCoefficientRules[Table[XX[[i]] → cr[[i]], {i, Length[XX]}], {x, y}];
   h = ComplexExpand[Re[h]];
   h1 = nGCD[f, h, x, y, tol];
   If[tDeg[h1, x, y] == 0,
    Print[Style["Warning: GCD not found, Returning h", Orange]];
    Return[{h, 1}]];
   Chop[{h1, nDivide[f, h1, x, y, tol]}, 10 * tol]];
```

For our first example, consider the quartic curve of the example from Section 5.4 with four complex singular points:

```
In[175]:= g7 = 18 - 24 x + 12 x^2 - 8 x^3 + 2 x^4 - 24 y + 32 x y - 8 x^2 y + 9 y^2 - 8 x y^2 + 3 x^2 y^2 - 4 y^3 + y^4;

In[176]:= complexSingularPoints[g7, x, y]

Out[176]= {{6.88879 - 4.73523 ⅈ, 6.76932 + 6.83955 ⅈ}, {6.88879 + 4.73523 ⅈ, 6.76932 - 6.83955 ⅈ},
  {1.11121 + 0.517534 ⅈ, 1.23068 + 0.747525 ⅈ}, {1.11121 - 0.517534 ⅈ, 1.23068 - 0.747525 ⅈ}}
```

In[177]:= **{h, k} = singularFactor[g7, x, y, 1.*^-8]**

Out[177]= {−11.6984 + 15.5979 x − 3.89948 x^2 − 1.94974 y^2,
−1.53867 − 0.512889 x^2 + 2.05156 y − 0.512889 y^2}

In[178]:= **ContourPlot[{h == 0, k == 0}, {x, −2, 4}, {y, −2, 4}]**

Out[178]=

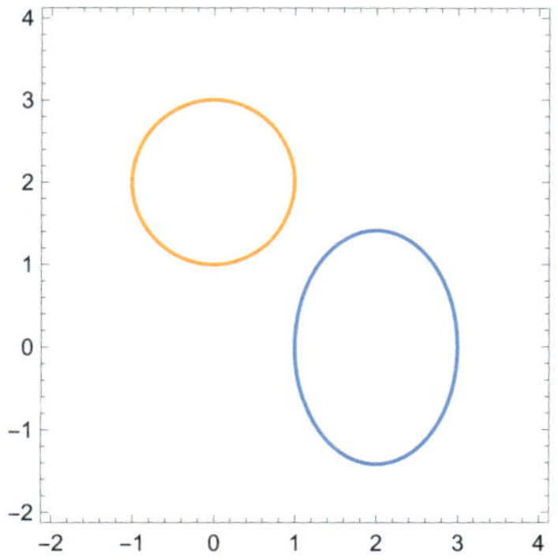

As another example, consider the curve of degree 6:

In[179]:= **f7 = x^2 + 5 x^3 + 8 x^4 + 4 x^5 − y^2 − 4 x y^2 − 4 x^2 y^2 + x^2 y^3 + x^3 y^3 − y^5;**
rsp = Chop[realSingularPoints[f7, x, y]]

Out[180]= {{1.61803, −2.61803}, {−0.618034, −0.381966}, {−0.41135, −0.315602}, {0, 0}}

In[181]:= **ContourPlot[f7 == 0, {x, −2, 2}, {y, −3, 3}, Epilog → {PointSize[Medium], Red, Point[rsp]}]**

Out[181]=

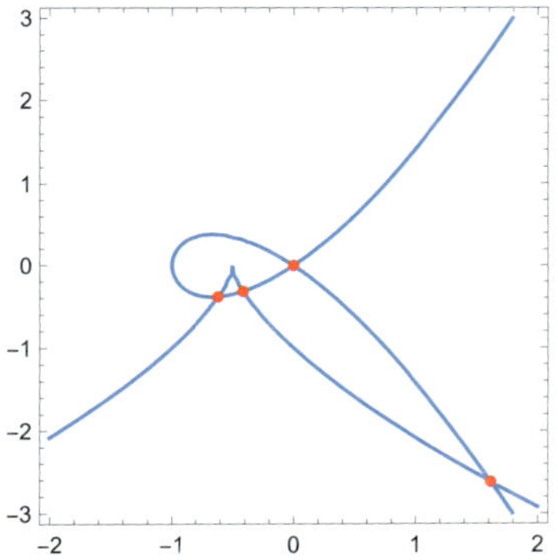

In[182]:= **Length[complexSingularPoints[f7, x, y]]**

Out[182]= 11

Since $\frac{(6-1)(6-2)}{2} + 1 = 11$, f should factor. However, the curve through fourteen points is of degree 4, and there is no factor of degree 4 or complementary degree 2.

We can find this degree 4 curve by using a ridiculously tight tolerance. singularFactor does not give the factorization, but it does return the curve *h* through 14 points:

In[183]:=
```
{h, k} = singularFactor[f7, x, y, 1.*^-17]
```

Warning: g Does not divide h at this tolerance

Warning: GCD not found, Returning h

Out[183]= $\{5.07263\times10^{-15}+0.150044\,x+0.600174\,x^2+0.600174\,x^3-4.50666\times10^{-11}\,x^4+0.0863285\,y+0.345314\,x\,y+0.345314\,x^2\,y-3.18197\times10^{-11}\,x^3\,y+7.25152\times10^{-11}\,y^2+7.52819\times10^{-11}\,x\,y^2+5.41022\times10^{-11}\,x^2\,y^2-4.41168\times10^{-11}\,y^3+0.150044\,x\,y^3+0.0863285\,y^4,\ 1\}$

In[184]:=
```
ContourPlot[{f7 == 0, h == 0}, {x, -3, 3}, {y, -5, 5}]
```

Out[184]=

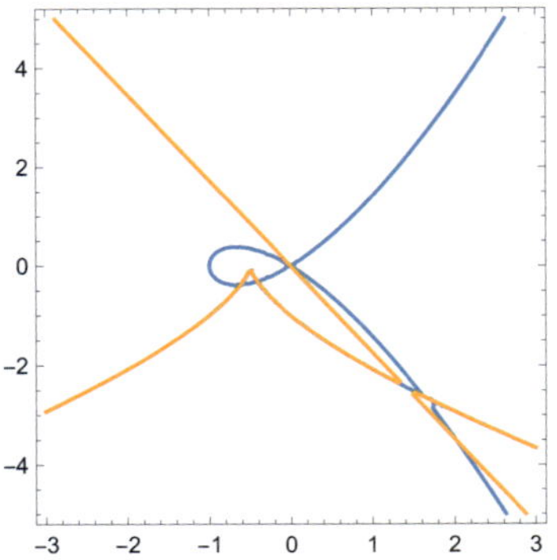

h contains an extra line. This is why singularFactor needs the nGCD function.

Several tries may be necessary with different tolerances, but we can get our desired factorization into two cubics:

In[185]:=
```
{h, k} = Chop[singularFactor[f7, x, y, 1.*^-7], 1.*^-7]
```

Out[185]= $\{-1.73616-6.94464\,x-6.94464\,x^2-1.73616\,y^3,\ -0.575983\,x^2-0.575983\,x^3+0.575983\,y^2\}$

In[186]:=
```
ContourPlot[{f7 == 0, h == 0}, {x, -3, 3}, {y, -5, 5}]
```

Out[186]=

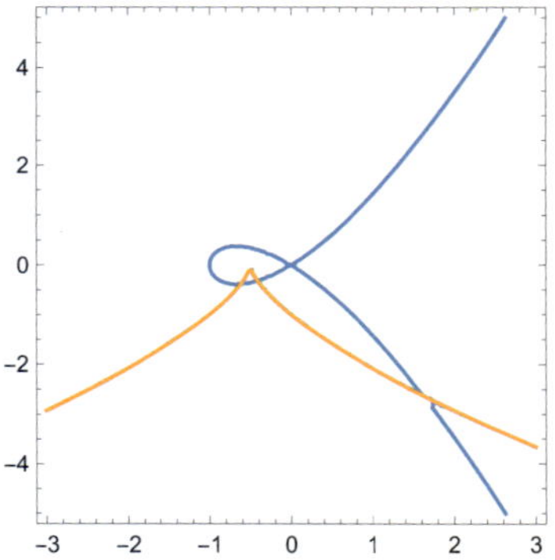

Some of the singularities will be infinite, so we use our trick from Chapter 6 to get our general factoring algorithm. Perhaps you may need to run this several times with different tolerances:

```
In[187]:= singularFactorInf[f_, x_, y_, tol_] := Module[{d, fh, f1, z, A, h, k},
      d = tDeg[f, x, y];
      A = iTransform[RandomReal[{-1, 1}, 3].{x, y, 1}, x, y];
      f1 = FLT[f, A, x, y];
      {h, k} = singularFactor[f1, x, y, tol];
      Chop[{FLT[h, Inverse[A], x, y], FLT[k, Inverse[A], x, y]}, 10*tol]
    ];
```

As a simple example, consider the degree 4 curve:

```
In[188]:= g9 = 5 - 2 x + 2 y - 6 x y + 2 x^2 y - 2 x y^2 + x^2 y^2;

In[189]:= Length[complexSingularPoints[g9, x, y]]

Out[189]= 2
```

Since the only term of degree 4 is $x^2 y^2$, there are two obvious infinite singular points {0, 1, 0}, {1, 0, 0}. So:

```
In[190]:= {h, k} = singularFactorInf[g9, x, y, 1.*^-7]

    degree must be at least 3

Out[190]= $Aborted

In[191]:= ContourPlot[{h == 0, k == 0}, {x, -5, 5}, {y, -5, 5}]
```

Out[191]=
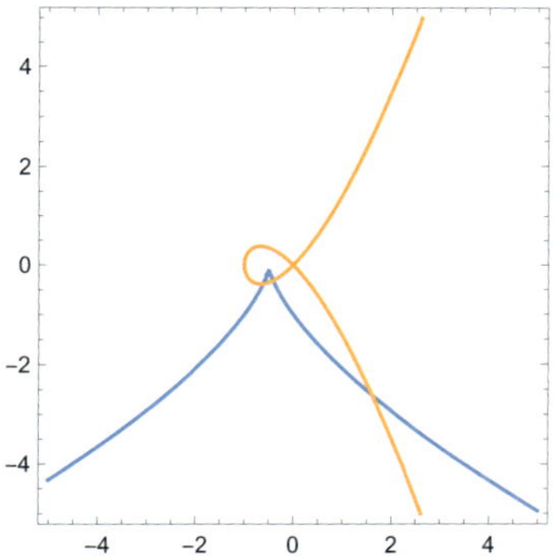

This is not a general factoring algorithm as there are many reducible curves of degree $d > 3$ with fewer than $\frac{(d-1)(d-2)}{2}$ singular points.

In fact, it seems that the only examples of reducible curves with more than this many singular points are curves such that each factor has almost the maximum number of singular points. For example, reducible curves with factors of degree 1 and 2, singular curves of degree 3, and Chebyshev curves.

Tech Notes

- Mathematica has a built-in function called SquareFreeQ, which will give True if an integer bivariate polynomial (variables optional) is square-free. It does not work on numerical polynomials.
- Some comments on intersection multiplicity: In the multivariable case, there are some invariants other than simply the multiplicity for multiple solutions. The number *i* returned by my procedure gives the order of the first Macaulay matrix with the correct null space. I call this number the *depth* of the intersection. The second entry in the list minus the first entry (always one) I call the "breadth". Other terms are used by algebraic geometers. For example, the breadth is called the *imbedding dimension* and the depth is the *regularity index*.

Appendix 2: Global Functions

These functions were coded by me in Mathematica 11.3 as simple illustrations of algorithms in *A Numerical Approach to Real Algebraic Curves with the Wolfram Language* only.

This appendix, also periodically called the global functions notebook, contains examples of most of the functions I have written for this book. You can check the integrity of these functions on your Mathematica version by running this sheet with **Evaluate Notebook**. There should be no error messages produced.

Chapter 1

(1.1) line: Gives a line through two points, finite or infinite. You can enter points as either pairs or triples. Note that there is randomization; rerunning the same data in a session will give the same answer, but it will vary by a constant multiple in different sessions.

```
In[1]:= rrle3 = RandomReal[{-1, 1}, 3];
        line[p_, q_, x_, y_] := Module[{e1, e2, a, b, c, M},
            e1 = Switch[Length[p], 2, Append[p, 1], 3, p];
            e2 = Switch[Length[q], 2, Append[q, 1], 3, q];
            M = {e1, e2, rrle3};
            ({a, b, c} /. (Solve[M.{a, b, c} == {0, 0, 1}][[1]])).{x, y, 1}
          ];

In[3]:= line[{2, 3}, {1, 4, 0}, x, y]

Out[3]= -1.35002 + 1.08001 x - 0.270004 y
```

(1.3) newtonHyperbolas: $1 \leq n \leq 7$ degree 1, $8 \leq n \leq 63$ degree 2, $64 \leq n \leq 1023$ degree 3, $1024 \leq n \leq 32\,767$ degree 4, $32\,768 \leq n \leq 2\,097\,151$ degree 5, $2\,097\,152 \leq n < 2^{28}$ degree 6.

```
In[4]:= newtonHyperbola[n_, x_, y_] := Module[{v, c, k, n2dig, s},
          If[(n > 2^28) || (n < 0), Print["n out of range"]; Return[Fail]];
          v = {{0, 0}, {1, 0}, {0, 1}, {2, 0}, {1, 1}, {0, 2}, {3, 0}, {2, 1},
               {1, 2}, {0, 3}, {4, 0}, {3, 1}, {2, 2}, {1, 3}, {0, 4}, {5, 0}, {4, 1}, {3, 2},
               {2, 3}, {1, 4}, {0, 5}, {6, 0}, {5, 1}, {4, 2}, {3, 3}, {2, 4}, {1, 5}, {0, 6}};
        c = {1, 1, 1, .2, 1, .2, .008, .2, .2, .008, .000064,
               .008, .2, .008, .000064, .2^10, .2^6, .008, .008,
```

```
    .2^6, .2^10, .2^15, .2^10, .2^6, .2^3, .2^6, .2^10, .2^15};
  k = Which[n < 8, 3, n < 64, 6, n < 1024, 10, n < 32 768, 15, n < 2^21, 21, True, 28];
  n2dig = IntegerDigits[n, 2, k];
  s = Reverse[n2dig];
  FromCoefficientRules[Table[v[[i]] → (-1)^s[[i]] * c[[i]], {i, k}], {x, y}]]
```

```
In[5]:= newtonHyperbola[213, x, y]
```

Out[5]= $-1+x+0.2\,x^2-0.008\,x^3-y-x\,y-0.2\,x^2\,y+0.2\,y^2+0.2\,x\,y^2+0.008\,y^3$

(1.4) gaussCurve: p is a polynomial in z with possibly complex coefficients. ComplexExpand assumes that variables x, y are real.

```
In[6]:= gaussCurve[p_, x_, y_, z_] := ComplexExpand[Re[p /. {z → x + I * y}]];
```

```
In[7]:= gaussCurve[z^2 - 2 z + 3, x, y, z]
```

Out[7]= $3-2\,x+x^2-y^2$

(1.5) sqFree: sqFree[g, x, y, tol] gives the square free decomposition of polynomial $g(x, y)$ to tolerance tol. See Appendix (A.3) below for code and example.

Chapter 2

(2.2) newton1: One-dimensional Newton method.

```
In[8]:= newton1[f_, p0_, z_] := Module[{p1, p2},
  p1 = N[p0];
  Chop[Catch[Do[p2 = p1 - (f / D[f, z]) /. {z → p1};
  If[Abs[p1 - p2] < 1.*^-14, Throw[p2]];
  p1 = p2, {30}]]]
  ]
```

```
In[9]:= newton1[x^3 - 3 x - 1, 4.712, x]
```

```
Out[9]= 1.87939
```

(2.2) gaussNewton2: Two-dimensional Newton method for two functions. In practice, use the built-in FindRoot instead.

```
In[10]:= gaussNewton2[f_, g_, p0_, x_, y_] := Module[{Q, Q0, p1, p2, J0, n},
    n = 20;
    p1 = N[p0];
    Catch[Do[Q = {f, g} /. Thread[{x, y} → p1];
      J0 = D[{f, g}, {{x, y}}] /. Thread[{x, y} → p1];
      p2 = p1 - Inverse[J0].Q;
      If[Norm[p1 - p2] < 1.*^-13, Throw[p2]];
      p1 = p2, {n}]]]
```

```
In[11]:= gaussNewton2[x - y + x y, x^2 - y^2 - 3, {2, 3}, x, y]
```

```
Out[11]= {-1.84968, -0.649083}
```

(2.4) tDeg: The total degree of a bivariate polynomial.

```
In[12]:= tDeg[f_, x_, y_] := Max[Table[Total[t], {t, Keys[CoefficientRules[f, {x, y}]]}]];

In[13]:= tDeg[x + y^2 + x y^3, x, y]

Out[13]= 4
```

(2.4) pExps: A list of exponent pairs in a bivariate polynomial of degree d can be constructed recursively.

```
In[14]:= pExps[d_] := Module[{pex},
           pex[0] = {{0, 0}};
           pex[n_] := pex[n] = Join[pex[n-1], Table[{n-j, j}, {j, 0, n}]];
           pex[d]]

In[15]:= pExps[3]

Out[15]= {{0, 0}, {1, 0}, {0, 1}, {2, 0}, {1, 1}, {0, 2}, {3, 0}, {2, 1}, {1, 2}, {0, 3}}
```

(2.4) randomRealPolynomial: A procedure to construct random numerical bivariate polynomials of degree d.

```
In[16]:= randomRealPolynomial[d_, x_, y_] := FromCoefficientRules[
           Table[e → RandomReal[{-1, 1}/(4^Total[e])], {e, pExps[d]}], {x, y}]

In[17]:= randomRealPolynomial[3, x, y]

Out[17]= -0.164264 + 0.138124 x + 0.0104697 x^2 - 0.00764866 x^3 - 0.110536 y -
           0.0299281 x y + 0.0107359 x^2 y + 0.0235272 y^2 - 0.00256342 x y^2 + 0.00384543 y^3
```

(2.4) randomIntegerPolynomial: The function to produce a random integer polynomial has a new variable kk, which is the maximum number of nonzero terms which must be a positive integer. At least one term will have total degree d.

```
In[18]:= randomIntegerPolynomial[d_, kk_, x_, y_] := Module[{k, S, b, j, m, t},
           b = Binomial[d + 2, 2];
           If[kk < 1 || kk > b, k = b, k = kk];
           S = RandomSample[pExps[d], k];
           m = Max[Table[Total[t], {t, S}]];
           If[m < d, j = RandomInteger[{0, d}]; S = Append[Take[S, k - 1], {j, d - j}]];
           FromCoefficientRules[Table[e → RandomInteger[{-9, 9}], {e, S}], {x, y}]]

In[19]:= randomIntegerPolynomial[3, 4, x, y]

Out[19]= x - 8 x^2 y + 4 y^2 - 4 y^3
```

(2.4) aCurve: Gives a curve of degree d through $\binom{d+2}{2} - 1$ points provided that these points are sufficiently random.

```
In[20]:= aCurve[pts_, x_, y_] := Module[{d, P, M, B, n, c, pow},
           pow[a_, n_] := If[n == 0, 1, a^n];
           d = Switch[Length[pts], 2, 1, 5, 2, 9, 3, 14, 4, 20, 5,
               27, 6, _, Return["number of points must be 2,5,9,14,20,27"]];
           P = pExps[d];
```

```
        n = Length[P];
        M = Table[
            If[Length[p] == 2, pow[p[[1]], e[[1]]] * pow[p[[2]], e[[2]]], pow[p[[1]], e[[1]]] *
                pow[p[[2]], e[[2]]] * pow[p[[3]], d - e[[1]] - e[[2]]]], {p, pts}, {e, P}];
        AppendTo[M, RandomReal[{-1, 1}, n]];
        B = Append[Table[0, {n - 1}], 1];
        c = LinearSolve[M, B];
        FromCoefficientRules[Table[P[[i]] → c[[i]], {i, n}], {x, y}]
    ];
```

```
In[21]:= f = aCurve[{{1, 1}, {2, 2}, {3, 5}, {4, 10}, {5, 17}}, x, y];
         Expand[f / -Coefficient[f, y]]

Out[22]= 2. - 2. x + 1. x^2 - 1. y
```

Chapter 3

(3.1) tLine: Tangent line at a point.

```
In[23]:= tLine[f_, p_, x, y] := line[p, {D[f, y], -D[f, x], 0} /. Thread[{x, y} → p], x, y];
```

```
In[24]:= f = 2 - 2 x + x^2 - y;
         g = tLine[f, {3, 5}, x, y];
         Expand[g / -Coefficient[g, y]]

Out[26]= -7. + 4. x - 1. y
```

(3.1) tangentRealPoints: Points of $f = 0$ with tangents parallel to line k.

```
In[27]:= tangentRealPoints[f_, k_, x_, y_] := Module[{J, p, q, sol},
             J = Det[D[{f, k}, {{x, y}}]];
             sol = NSolve[{f, J}, {x, y}, Reals];
             If[Length[sol] > 0, {x, y} /. sol, {}] ];
```

```
In[28]:= f = 2 - 2 x + x^2 - y;
         tangentRealPoints[f, 4 x - y, x, y]

Out[29]= {{3., 5.}}
```

(3.2) realSingularPoints: Real affine singular points. Note that we have a similar function for complex projective singular points in (6.4).

```
In[30]:= realSingularPoints[f_, x_, y_] := Module[{eps},
             eps = 1.*^-6;
             Flatten[
               Take[Select[Tally[tangentRealPoints[f, RandomReal[{-1, 1}, 2].{x, y}, x, y],
                    Norm[#1 - #2] < eps &], #[[2]] > 1 &], All, 1], 1]];
```

```
In[31]:= f = (x - 1)^3 - (y - 2)^2;
         realSingularPoints[f, x, y]

Out[32]= {{1., 2.}}
```

(3.3) form: Homogeneous form consisting of all monomials of degree k in polynomial h.

```
In[33]:= form[h_, k_, x_, y_] :=
           FromCoefficientRules[Select[CoefficientRules[h, {x, y}], Total[#[[1]]] == k &], {x, y}];

In[34]:= f = (x - 1)^3 - (y - 2)^2;
         form[f, 2, x, y]

Out[35]= -3 x^2 - y^2
```

(3.3) maxForm: Maximal form of polynomial h.

```
In[36]:= maxForm[h_, x_, y_] := form[h, tDeg[h, x, y], x, y];

In[37]:= f = (x - 1)^3 - (y - 2)^2;
         maxForm[f, x, y]

Out[38]= x^3
```

(3.3) infiniteRealPoints: Note that we have a similar function for complex projective singular points in Chapter 6.

```
In[39]:= infiniteRealPoints[f_, x_, y_] := Module[{fm, rnl, sol},
             fm = maxForm[f, x, y];
             rnl = RandomReal[{-1, 1}, 2].{x, y} - 1;
             sol = NSolve[{fm, rnl}, {x, y}, Reals];
             If[Length[sol] == 0, Return[{}], sol = {x, y} /. sol];
             sol = DeleteDuplicates[sol, Norm[#1 - #2] < .001 &];
             Table[Append[sol[[i]], 0], {i, Length[sol]}]
           ];

In[40]:= f = (x - 1)^3 - (y - 2)^2;
         infiniteRealPoints[f, x, y]

Out[41]= {{0, -1.10232, 0}}
```

(3.4) pointMinForm: Minimum form at a point not necessarily 0. The tolerance tol can be changed.

```
In[42]:= pointMinForm[f_, p_, x_, y_] := Module[{A, h, j, tol},
             tol = 1.*^-9;
             h = Chop[Expand[f /. Thread[{x, y} → {x, y} + p]], tol];
             A = <| CoefficientRules[h, {x, y}] |>;
             j = Min[Table[Total[k], {k, Keys[A]}]];
             form[h, j, x, y]];

In[43]:= f = (x - 1)^3 - (y - 2)^2;
         pointMinForm[f, {1, 2}, x, y]

Out[44]= -y^2
```

(3.5) criticalPoints:

```
In[45]:= criticalPoints[f_, x_, y_] := DeleteDuplicates[
             {x, y} /. NSolve[{f, y D[f, x] - x D[f, y]}, {x, y}, Reals], Norm[#1 - #2] < 1.*^-6 &];

In[46]:= f = (x - 1)^3 - (y - 2)^2;
         cp = criticalPoints[f, x, y]

Out[47]= {{1.1593, 1.93642}, {1., 2.}, {1.75284, 1.34679}}
```

(3.5) closetPoint: Closest point on a curve to q. If several are equidistant, only one is given.

```
In[48]:= closestPoint[f_, q_, x_, y_] := Module[{g, p, cp},
           g = Expand[f /. Thread[{x, y} → {x + q[[1]], y + q[[2]]}]];
           cp = criticalPoints[g, x, y];
           p = MinimalBy[cp, Norm[#] &][[1]];
           p + q];

In[49]:= f = (x - 1)^3 - (y - 2)^2;
         closestPoint[f, {5, 3}, x, y]

Out[50]= {2.6856, 4.18842}
```

Chapter 4

(4.2) gaussPlot: Utility function to plot a curve inside a disk, showing critical and/or boundary points, if given. f is the curve, r is the radius of the disk, and pts is the points, possibly an empty set.

```
In[51]:= gaussPlot[f_, r_, pts_] := Module[{plta, pltb, pltc},
           plta := RegionPlot[x^2 + y^2 < r^2, {x, -1.1 r, 1.1 r},
             {y, -1.1 r, 1.1 r}, PlotStyle → LightOrange, BoundaryStyle → Orange];
           pltb := RegionPlot[{f > 0 && x^2 + y^2 < r^2}, {x, -r, r},
             {y, -r, r}, BoundaryStyle → Blue];
           pltc := ListPlot[Table[Labeled[pts[[n]], n], {n, Length[pts]}],
             PlotStyle → Red, PlotMarkers → Automatic];
           Show[plta, pltb, pltc]];

In[52]:= g = x^3 - x + 1 - y^2;
         bp = {x, y} /. NSolve[{x^2 + y^2 - 16, g}, {x, y}, Reals];
         gaussPlot[g, 4, bp]
```

Out[54]=

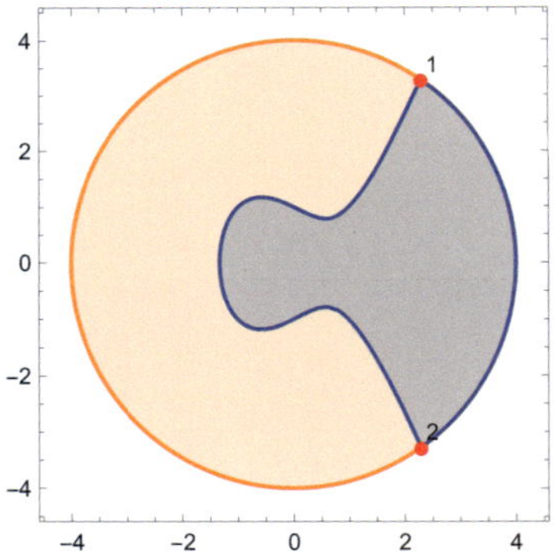

(4.3) gTvec: Directed tangent vector clockwise about components of $f > 0$.

```
In[55]:= gTvec[f_, p_, x_, y_] :=
           Cross[Normalize[Chop[{D[f, x], D[f, y]} /. Thread[{x, y} → p], 1.*^-6]]]

In[56]:= f = (x - 1)^3 - (y - 2)^2;
         gTvec[f, cp[[3]], x, y]

Out[57]= {-0.609271, 0.792962}
```

(4.3) gPoints: gPoints of the curve f inside a disk of radius r.

```
In[58]:= gPoints[f_, r_, x_, y_] := Module[{bp, cp, sol},
           sol = NSolve[{f, x^2+y^2-r^2}, {x, y}, Reals];
           If[Length[sol] > 0, bp = {x, y} /. sol, bp = {}];
           bp = Reverse[SortBy[bp, ArcTan @@ # &]];
           cp = DeleteDuplicates[ Select[{x, y} /. NSolve[{f, y D[f, x] - x D[f, y]}, {x, y}, Reals],
                Norm[#] < r &], Norm[#1 - #2] < .1 &];
           Join[
             bp,
             cp]];

In[59]:= g = x^3 - x + 1 - y^2;
         gPoints[g, 4, x, y]

Out[60]= {{2.29191, 3.27828}, {2.29191, -3.27828}, {-1., -1.},
           {0.333333, 0.83887}, {-1.32472, 0.}, {0.333333, -0.83887}, {-1., 1.}}
```

(4.4) pathFinder: pathFinder uses naive tracing using the closest point. f is the curve, p is the initial point, q is the final point, and s is the step size.

```
In[61]:= pathFinder[f_, p_, q_, s_, x_, y_] := Module[{L, p0, p1, u, maxit, k},
           maxit = 30;
           p0 = p;
           L = Reap[Sow[p];
                k = 0;
                While[Norm[q - p0] > 2 s && k < maxit,
                  u = p0 + s * gTvec[f, p0, x, y];
                  p1 = closestPoint[f, u, x, y];
                  Sow[p1];
                  p0 = p1;
                  k++];
                Sow[q]][[2, 1]];
           If[k ≥ maxit, Print["Warning: iteration limit reached at", p1]];
           L];

In[62]:= g = x^3 - x + 1 - y^2;
         pathFinder[g, {-1, -1}, {-1, 1}, .4, x, y]

Out[63]= {{-1, -1}, {-1.20137, -0.683701}, {-1.30229, -0.306044},
           {-1.3229, 0.0879482}, {-1.26819, 0.478065}, {-1, 1}}
```

(4.4) pathFinderT: pathFinderT traces using the tangent vector given by gTvec. It has the same arguments as pathFinder. It will be faster, especially in higher degrees, but may not work as well in some cases. An option can be used to change the maximum number of iterations.

```
In[64]:= Options[pathFinderT] := {maxit → 20};
         pathFinderT[f_, p_, q_, s_, x_, y_, OptionsPattern[]] :=
           Module[{L, p0, p1, tv, u, l, maxit2, k},
             maxit2 = OptionValue[maxit];
             p0 = p;
             L = Reap[Sow[p];
                  k = 0;
                  While[Norm[q - p0] > 2 s && k < maxit2,
                    tv = gTvec[f, p0, x, y];
```

```
            u = p0 + s * tv;
            l = line[u, {tv[[2]], -tv[[1]], 0}, x, y];
            p1 = {x, y} /. FindRoot[{f, l}, {x, u[[1]]}, {y, u[[2]]}];
            Sow[p1];
            p0 = p1;
            k++];
          Sow[q]][[2, 1]];
      If[k ≥ maxit2, Print["Warning: iteration limit reached at", p1]];
      L];
```

```
In[66]:= g = x^3 - x + 1 - y^2;
         pathFinderT[g, {-1, -1}, {-1, 1}, .4, x, y]

Out[67]= {{-1, -1}, {-1.21495, -0.649267}, {-1.30898, -0.257151},
           {-1.31979, 0.14458}, {-1.25119, 0.540821}, {-1, 1}}
```

(4.5) plotPath: Curve plotting using several points. *f* is the curve, *G* is the graph, *V* is the actual points of the curve used as vertices, and *s* is the step size.

```
In[68]:= plotPath[f_, G_, V_, x_, y_, s_] := Module[{n, L},
            n = Length[G];
            L = Flatten[Table[pathFinder[f, V[[G[[i, 1]]]], V[[G[[i, 2]]]], s, x, y], {i, n}], 1]];

In[69]:= g = x^3 - x + 1 - y^2;
         G = {1 → 2, 2 → 3, 3 → 4};
         V = {{0, -1}, {-1, -1}, {-1, 1}, {0, 1}};
         Timing[plotPath[g, G, V, x, y, .3]]

Out[72]= {0.168681, {{0, -1}, {-0.273452, -1.11938}, {-0.558065, -1.17655}, {-1, -1},
            {-1, -1}, {-1.16487, -0.764342}, {-1.26583, -0.487397}, {-1.31577, -0.194503},
            {-1.32222, 0.103181}, {-1.28611, 0.398494}, {-1.20192, 0.682358},
            {-1, 1}, {-1, 1}, {-0.766341, 1.14729}, {-0.49002, 1.17148}, {0, 1}}}
```

In many cases, we can use the faster version.

```
In[73]:= plotPathT[f_, G_, V_, x_, y_, s_] := Module[{n, L},
            n = Length[G];
            L = Flatten[Table[pathFinderT[f, V[[G[[i, 1]]]], V[[G[[i, 2]]]], s, x, y], {i, n}], 1]];

In[74]:= g = x^3 - x + 1 - y^2;
         G = {1 → 2, 2 → 3, 3 → 4};
         V = {{0, -1}, {-1, -1}, {-1, 1}, {0, 1}};
         Timing[plotPathT[g, G, V, x, y, .3]]

Out[77]= {0.008668, {{0, -1}, {-0.275387, -1.12005}, {-0.573151, -1.1768}, {-1, -1},
            {-1, -1}, {-1.17266, -0.748396}, {-1.27182, -0.463257}, {-1.31823, -0.165864},
            {-1.32043, 0.134891}, {-1.27885, 0.432848}, {-1.18601, 0.719552},
            {-1, 1}, {-1, 1}, {-0.732968, 1.15723}, {-0.423626, 1.16086}, {0, 1}}}
```

(4.6) deTrace: Curve tracing using NDSolveValue. *f* is the curve, *p* is the initial point, *s* is the maximal distance from the initial point, and *r* is the maximum distance from the origin. deTrace returns a pair of interpolating functions, i.e. *x*, *y* parameters of the curve. Interpolating function approximates the curve but does not necessarily give actual points on the curve.

```
In[78]:= deTrace[f_, p_, x_, y_, s_, r_] := Module[{X, Y, F, u, t, tv, ttv, XS, YS},
          F = f /. Thread[{x, y} → {X[t], Y[t]}];
          tv = gTvec[f, p, x, y];
```

```
        u = Normalize[RandomReal[{.8, 1.2}, 2]];
        ttv = u.tv;
        NDSolveValue[{u.{X'[t], Y'[t]} == ttv, D[F, t] == 0, {X[0], Y[0]} == p,
            WhenEvent[X[t]^2 + Y[t]^2 > r^2 ||
                u[[1]] * X'[t] + u[[2]] * Y'[t] == 0 || (X[t] - p[[1]])^2 + (Y[t] - p[[2]])^2 > s^2,
              "StopIntegration"]}, {X, Y}, {t, 0, 20}]
    ]
```

Note: NDSolveValue::ndsz is not an error message but an information message that may or may not appear here depending on the random value used.

```
In[79]:= g = x^3 - x + 1 - y^2;
         {X, Y} = deTrace[g, {-1, -1}, x, y, 2, 3]

Out[80]= {InterpolatingFunction[ Domain: {{0., 20.}} Output: scalar ],

          InterpolatingFunction[ Domain: {{0., 20.}} Output: scalar ]}
```

(4.6) deEnd: deEnd uses end point and maximum value for the domain of the interpolating function pair from deTrace.

```
In[81]:= deEnd[{X_, Y_}] := Module[{t1},
            t1 = X[[1, 1, 2]];
            {{X[t1], Y[t1]}, t1}]

In[82]:= (* using output {X,Y} from example above *)
         deEnd[{X, Y}]

Out[82]= {{-1.10242, 0.873283}, 20.}
```

(4.6) pathFinderDE: Here, r is the maximum distance from the origin and eps is the tolerance. When the distance from a calculated point to q is determined, the algorithm stops.

```
In[83]:= pathFinderDE[f_, p_, q_, r_, x_, y_, eps_] := Module[{i, e, X, Y, s},
              Off[NDSolveValue::ndsz];
              e = p;
              s = Norm[p - q];
              i = 1;
              Reap[Sow[p];
                  While[s > eps && Norm[e] < 2 * r && i < 60,
                    {X, Y} = deTrace[f, e, x, y, s * eps, r];
                    e = deEnd[{X, Y}][[1]];
                    Sow[e];
                    s = Norm[e - q];
                    i++];
                  Sow[q]][[2, 1]]];

In[84]:= g = x^3 - x + 1 - y^2;
         pathFinderDE[g, {-1, -1}, {-1, 1}, 3, x, y, .5]

Out[85]= {{-1, -1}, {-1.25993, -0.509796}, {-1.3093, 0.254615}, {-1.21576, 0.647129}, {-1, 1}}
```

(4.6) plotPathDE: Has the same syntax as plotPath.

```
In[86]:= plotPathDE[f_, G_, V_, x_, y_, s_] := Module[{r, n, L, LL, PF},
           n = Length[G];
           r = Max[Norm /@ V];
           L = Flatten[Table[pathFinderDE[f, V[[G[[i, 1]]]], V[[G[[i, 2]]]], r, x, y, s], {i, n}], 1];
           Append[L, V[[G[[n, 2]]]]]]

In[87]:= g = x^3 - x + 1 - y^2;
         G = {1 → 2, 2 → 3, 3 → 4};
         V = {{0, -1}, {-1, -1}, {-1, 1}, {0, 1}};
         Timing[plotPathDE[g, G, V, x, y, .5]]

Out[90]= {0.030813, {{0, -1}, {-0.470601, -1.16892},
           {-0.747995, -1.15304}, {-1, -1}, {-1, -1}, {-1.01165, -0.988069},
           {-1.32425, -0.0444485}, {-1.26284, 0.498904}, {-1.16488, 0.764327},
           {-1, 1}, {-1, 1}, {-0.531747, 1.17533}, {-0.258566, 1.11413}, {0, 1}, {0, 1}}}
```

(4.6) endFinder: Routine to attempt to find the next point of a set of points CP on a curve reached by tracing from the nth point of CP. The goal is met when the trace is within eps of a point in CP.

```
In[91]:= endFinder[f_, n_, CP_, x_, y_, eps_] := Module[{r, i, k, e, X, Y, s, CPD},
           Off[NDSolveValue::ndsz];
           r = Max[Norm /@ CP];
           e = CP[[n]];
           CPD = CP;
           CPD[[n]] = Nothing;
           s = Norm[e - First[Nearest[CPD, e]]];
           i = 1;
           e = Catch[
             While[s > eps && Norm[e] < 1.01 * r && i < 60,
               {X, Y} = deTrace[f, e, x, y, s, r];
               e = deEnd[{X, Y}][[1]];
               If[Norm[e] > r, Throw[e]];
               s = Norm[e - First[Nearest[CPD, e]]];
               i++];
             Throw[e]];
           If[Norm[e] > r, Return[0]];
           k = 1;
           While[k ≤ Length[CP] && Norm[e - CP[[k]]] > s, k++];
           If[k <= Length[CP], k, Fail]];

In[92]:= g = x^3 - x + 1 - y^2;
         V = {{0, -1}, {-1, -1}, {-1, 1}, {0, 1}};
         endFinder[g, 2, V, x, y, .4]

Out[94]= 3
```

(4.6) nearPoints: Are points nearby a singularity *pt* of *f* but with gTvec at those points pointing away from the singularity. *r* is the radius of a bounding circle and *ep* is the distance of the nearPoints from the singularity. These are necessary because one can trace into a singularity but cannot trace out of a singularity.

```
In[95]:= nearPoints[f_, r_, x_, y_, pt_, ep_] := Module[{np, stnpts},
           np = {x, y} /. NSolve[{f, (x - pt[[1]])^2 + (y - pt[[2]])^2 - ep^2}, {x, y}, Reals];
           np = Select[np, Norm[# - pt + ep/2 * Normalize[gTvec[f, #, x, y]]] > ep &];
           Select[np, Norm[#] < r &]];

In[96]:= f = (x - 1)^3 - (y - 2)^2;
         nearPoints[f, 4, x, y, {1, 2}, .4]

Out[97]= {{1.34492, 2.20257}}
```

(4.6) gaussGraphNS: An attempt at a black box function that traces a curve *f* inside a disk of gPoints CP and returns a plot of the curve with gPoints labeled and a graph of the curve inside the disk. eps is used in nearPoints and endFinder. If run without, it will return the gPoints and graph rules. gaussGraphNS is simpler and works in the nonsingular case. gaussGraph works best for Gauss curves, possibly even with singularities.

```
In[98]:= gaussGraphNS[f_, CP_, x_, y_, eps_] := Module[{n, G},
           n = Length[CP];
           G = Select[Table[k → endFinder[f, k, CP, x, y, eps],
               {k, n}], #[[2]] > 0 &];
           Graph[G, VertexLabels → "Name", ImageSize → Small]];

In[99]:= g = x^3 - x + 1 - y^2;
         V = {{0, -1}, {-1, -1}, {-1, 1}, {0, 1}};
         gaussGraphNS[g, V, x, y, .4]
```

Out[101]=

```
In[102]:= gaussGraph[f_, gp_, x_, y_, eps_] := Module[{n, k, e, G, np, CP, r, S},
        CP = gp;
        r = Max[Norm[#] &/@ gp];
        n = Length[CP];
        G = {};
        k = 1;
        S = Reap[While[k ≤ n,
              If[Norm[gTvec[f, CP[[k]], x, y]] < eps/2,
                np = nearPoints[f, r, x, y, CP[[k]], 1.25 * eps];
                Do[Sow[k → n + j], {j, Length[np]}];
                CP = Join[CP, np];
                n = Length[CP];
                e = 0,
                e = endFinder[f, k, CP, x, y, eps]];
              If[e > 0, Sow[k → e]];
              k++]];
        G = S[[2, 1]];
        Echo[{Show[ContourPlot[f == 0, {x, -1.2 r, 1.2 r}, {y, -1.2 r, 1.2 r}],
              ListPlot[Table[Labeled[CP[[j]], j], {j, Length[CP]}]]],
            Graph[G, VertexLabels → "Name"]}];
        {CP, G}]
```

```
In[103]:= f = (x - 1)^3 - (y - 2)^2;
          Vf = {{2, 1}, {1, 2}, {2, 3}};
          gaussGraph[f, Vf, x, y, .4]
```

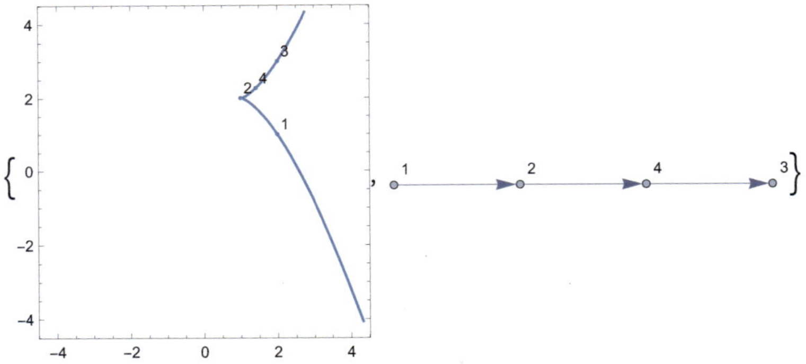


```
Out[105]= {{{2, 1}, {1, 2}, {2, 3}, {1.41964, 2.27184}}, {1 → 2, 2 → 4, 4 → 3}}
```

(4.6) plotPathS and plotPathDES: Two routines to plot singular curves from graphs that include user near points. The first is without the DE path tracing and the second is with.

```
In[106]:= plotPathS[f_, G_, V_, x_, y_, s_] := Module[{n, L, r},
            n = Length[G];
            r = Max[Norm /@ V];
            L = Flatten[
                Table[If[Norm[gTvec[f, V[[G[[i, 1]]]], x, y]] < s/2, {V[[G[[i, 1]]]], V[[G[[i, 2]]]]},
                    pathFinder[f, V[[G[[i, 1]]]], V[[G[[i, 2]]]], s, x, y]],
                  {i, n}], 1];
            AppendTo[L, V[[G[[n, 2]]]]];
            L];

In[107]:= plotPathDES[f_, G_, V_, x_, y_, eps_] := Module[{r, n, L},
            n = Length[G];
            r = Max[Norm[#] & /@ V];
            L = Flatten[Table[
                  If[Norm[gTvec[f, V[[G[[i, 1]]]], x, y]] < eps/2, {V[[G[[i, 1]]]], V[[G[[i, 2]]]]},
                    pathFinderDE[f, V[[G[[i, 1]]]], V[[G[[i, 2]]]], r, x, y, eps]],
                  {i, n}], 1];
            AppendTo[L, V[[G[[n, 2]]]]];
            L];

In[108]:= f = (x - 1)^3 - (y - 2)^2;
          Vsf = {{2, 1}, {1, 2}, {1.4196433776070834`, 2.271844506346037`}, {2, 3}};
          Gsf = {1 → 2, 2 → 3, 3 → 4};
          Timing[plotPathS[f, Gsf, Vsf, x, y, .3]]
          Timing[plotPathDES[f, Gsf, Vsf, x, y, .3]]

Out[111]= {0.080483, {{2, 1}, {1.8285, 1.24589}, {1.64463, 1.48243},
            {1.44395, 1.7042}, {1, 2}, {1, 2}, {1.41964, 2.27184}, {1.41964, 2.27184},
            {1.62258, 2.49124}, {1.80808, 2.72641}, {2, 3}, {2, 3}}}

Out[112]= {0.02738, {{2, 1}, {1.75381, 1.34553}, {1.61314, 1.5199},
            {1.45563, 1.69244}, {1.33563, 1.80556}, {1.24509, 1.87866}, {1, 2},
            {1, 2}, {1.41964, 2.27184}, {1.41964, 2.27184}, {1.60993, 2.47635},
            {1.73357, 2.62829}, {1.8164, 2.73766}, {1.87281, 2.81542}, {2, 3}, {2, 3}}}
```

Chapter 5

(5.1) pLine: Projective line through two points (triples) of the real or complex projective plane.

```
In[113]:= pLine[p_, q_, x_, y_, z_] := Module[{a, b, c, M},
            M = {p, q, RandomReal[{-1, 1}, 3]};
            ({a, b, c} /. (Solve[M.{a, b, c} == {0, 0, 1}][[1]])).{x, y, z}
          ];

In[114]:= pl = pLine[{2, 3, 4}, {-1, 3, 0}, x, y, z]

Out[114]= -0.620794 x - 0.206931 y + 0.465596 z

In[115]:= Expand[pl/Coefficient[pl, x]]

Out[115]= 1. x + 0.333333 y - 0.75 z

In[116]:= pLine[{2 + ⅈ, 3, 1}, {1, ⅈ, 0}, x, y, z]

Out[116]= (-0.155507 + 0.314609 ⅈ) x - (0.314609 + 0.155507 ⅈ) y + (1.56945 - 0.00719169 ⅈ) z
```

(5.2) homog: Homogenization of affine curve f.

```
In[117]:= homog[f_, x_, y_, z_] := Expand[z^tDeg[f, x, y] (f /. Thread[{x → x/z, y → y/z}])];

In[118]:= homog[x y^2 - x^2 + y^2 - xy - x + 1, x, y, z]

Out[118]= x y^2 - x^2 z + y^2 z - x z^2 + z^3 - xy z^3
```

(5.3) dTol: Default tolerance (constant).

```
In[119]:= dTol := 1.*^-12;
```

(5.3) intersectionMultiplicity[f, g, p]: Gives the intersection multiplicity of curves f, g at intersection point p. See (A.4).

```
In[120]:= f = x - y + x^3;
          g = x - y + y^3;
          intersectionMultiplicity[f, g, {0, 0}, dTol]

Out[122]= 5
```

(5.3) complexSingularPoints: Finds complex affine singular points.

```
In[123]:= complexSingularPoints[f_, u_, v_] := Module[{a, b, sol1, sol2, solz},
            {a, b} = RandomComplex[{-1 - I, 1 + I}, 2];
            sol1 = {u, v} /. NSolve[{f, a D[f, u] + b D[f, v]}];
            {a, b} = RandomComplex[{-1 - I, 1 + I}, 2];
            sol2 = {u, v} /. NSolve[{f, a D[f, u] + b D[f, v]}];
            solz = Join[sol1, sol2];
            Flatten[Take[Select[Tally[solz, Norm[#1 - #2] < .001 &], #[[2]] > 1 &], All, 1], 1]]
```

```
In[124]:= h = 9 - 10 x^2 + x^4 + 6 y^2 + 2 x^2 y^2 + y^4;
          complexSingularPoints[h, x, y]
Out[125]= {{0., 0. + 1.73205 ⅈ}, {0., 0. - 1.73205 ⅈ}}
```

(5.3) realRegularPoints: Finds n random real regular points for curve f of degree > 1.

```
In[126]:= realRegularPoints[f_, x_, y_, n_] := Module[{a, b, c, d, k},
            If[tDeg[f, x, y] < 2, Echo["Degree must be > 1"]; Return[{}]];
            k = 0;
           Reap[While[k < n,
                 {a, b} = RandomReal[{-10, 10}, 2];
                 {c, d} = closestPoint[f, {a, b}, x, y];
                 If[
                  Abs[Det[D[{f, a x - b y}, {{x, y}}]] /. Thread[{x, y} → {c, d}]] > .01, Sow[{c, d}];
                  k++]]][[2, 1]]];

In[127]:= f = x^4 - 4 x y + y^4;
          realRegularPoints[f, x, y, 4]
Out[128]= {{0.504896, 1.25311}, {-1.18374, -1.50861}, {1.42692, 0.797046}, {-1.48082, -1.31193}}
```

Chapter 6

(6.2) Fractional linear transformations: FLT

flt - Transforms affine points—always returns a pair.

flti - Transforms possibly infinite points—may return a triple.

flts - Transforms symbolic points—returns a pair of possibly rational symbolic expressions.

flth - Transforms projective point to projective point.

FLT - Transforms curve equations—gives an affine equation.

FLTh - Transforms homogeneous curve equation results in a homogeneous equation.

```
In[129]:= flt[{x_, y_}, A_] := Module[{u, v, w},
            {u, v, w} = A.{x, y, 1};
            If[Abs[w] < 1.*^-6, {}, {u/w, v/w}]];
          flti[p_, A_]  := Module[{u, v, w},
            If[Length[p] == 2, {u, v, w} = A.Append[p, 1], {u, v, w} = A.p];
            If[Abs[w] < 1.*^-6, {u, v, 0}, {u/w, v/w}]];
          flts[{x_, y_}, A_] := Module[{u, v, w},
            {u, v, w} = A.{x, y, 1};
            {u/w, v/w}];
          FLT[f_, A_, x_, y_] := Module[{fh, gh, AI, B, d, z},
            d = tDeg[f, x, y];
            fh = Expand[z^d (f /. {x → x/z, y → y/z})];
            AI = Inverse[A];
            B = AI.{x, y, z};
            gh = Expand[fh /. Thread[{x, y, z} → B]];
            Chop[gh /. {z → 1}, 1*^-10]
           ];

          flth[{x_, y_, z_}, A_] := A.{x, y, z};

          FLTh[h_, A_, x_, y_, z_] := Expand[h /. Thread[{x, y, z} -> Inverse[A].{x, y, z}]];
```

```
In[135]:= A = {{-2, 2, 7}, {-1, -6, 0}, {3, 3, 8}};
          flt[{2, 1}, A]
```

Out[136]= $\left\{\frac{5}{17}, -\frac{8}{17}\right\}$

```
In[137]:= flt[{-3, 1/3}, A]
          (* answer {} expected *)
```

Out[137]= {}

```
In[138]:= flti[{-3, 1/3}, A]
```

Out[138]= $\left\{\frac{41}{3}, 1, 0\right\}$

```
In[139]:= flts[{x, y}, A]
```

Out[139]= $\left\{\frac{7-2x+2y}{8+3x+3y}, \frac{-x-6y}{8+3x+3y}\right\}$

```
In[140]:= FLT[y - x^2, N[A], x, y]
```

Out[140]= $-0.0395421+0.0857737\,x-0.0463803\,x^2-0.0217036\,y+0.000445964\,x\,y-0.00995986\,y^2$

```
In[141]:= flth[{-9, 1, 3}, A]
```

Out[141]= {41, 3, 0}

```
In[142]:= FLTh[y z - x^2, A, x, y, z]
```

Out[142]= $-\frac{312\,x^2}{6727}+\frac{3\,x\,y}{6727}-\frac{67\,y^2}{6727}+\frac{577\,x\,z}{6727}-\frac{146\,y\,z}{6727}-\frac{38\,z^2}{961}$

(6.3) pqTranslation: Translation taking point p to point q. An example of a Euclidean transformation as FLT.

```
In[143]:= pqTranslation[p_, q_] := {{1, 0, q[[1]] - p[[1]]}, {0, 1, q[[2]] - p[[2]]}, {0, 0, 1}};

In[144]:= flt[{0, 1}, pqTranslation[{0, 1}, {2, 4}]]
```

Out[144]= {2, 4}

(6.3) klRotation: Rotation taking line k to line l. An example of a Euclidean transformation as FLT.

```
In[145]:= klRotation[k_, l_, x_, y_] := Module[{p, a1, a2, b1, b2, T, R, A, B, m},
             a1 = Coefficient[k, x];
             b1 = Coefficient[k, y];
             a2 = Coefficient[l, x];
             b2 = Coefficient[l, y];
             If[Abs[a1 * b2 - a2 * b1] < 1.*^-9, Print["No affine center"];
               Return[Fail]];
             p = {x, y} /. NSolve[{k, l}][[1]];
             T = {{1, 0, -p[[1]]}, {0, 1, -p[[2]]}, {0, 0, 1}};
             R = N[RotationMatrix[{{a1, b1}, {a2, b2}}]];
```

```
        R = Join[Append[R, {0, 0}], {{0}, {0}, {1}}, 2];
        Inverse[T].R.T
    ]
```

In[146]:=
```
flt[{1, 0}, klRotation[x, y, x, y]]
```

Out[146]= {0., 1.}

(6.3) kRelfection: Reflection in line k. An example of a Euclidean transformation as FLT.

In[147]:=
```
kReflection[k_, x_, y_] := Module[{a, b, c, T, RX, RO},
        RX = {{1, 0, 0}, {0, -1, 0}, {0, 0, 1}};
        a = Coefficient[k, x];
        b = Coefficient[k, y];
        c = k /. Thread[{x, y} → {0, 0}];
        If[a == 0,
          T = pqTranslation[{0, 0}, {0, -c / b}];
          Return[T.RX.Inverse[T]],
          RO = klRotation[y, k, x, y];
          Return[RO.RX.Inverse[RO]];
        ]
    ];
```

In[148]:=
```
flt[{1, 0}, kReflection[x, x, y]]
```

Out[148]= {-1., 0.}

(6.3) kShear: A standard shear moving line k to the x-axis. An example of affine transformations as FLT.

In[149]:=
```
kShear[k_, x_, y_] := Module[{a, b, c, d, e},
        c = Coefficient[k, x];
        d = Coefficient[k, y];
        e = k /. {x → 0, y → 0};
        a = -e / d;
        b = -c / d;
        {{1, 0, 0}, {-b, 1, -a}, {0, 0, 1}}
    ];
```

In[150]:=
```
FLT[x - y, kShear[x - y, x, y], x, y]
```

Out[150]= -y

(6.3) homothety: Homothety where r, s are positive reals. If one of r, s is negative, this adds in a reflection about the given axis. Both negative adds in a half-turn. An example of affine transformations as FLT.

In[151]:=
```
homothety[r_, s_] := {{r, 0, 0}, {0, s, 0}, {0, 0, 1}};
```

In[152]:=
```
flt[{1, 1}, homothety[2, -3]]
```

Out[152]= {2, -3}

(6.3) ip2z: Take infinite point *ip* to the origin. *x*, *y* are dummy variables. An example of projective transformation.

```
In[153]:= ip2z[ip_, x_, y_] := Module[{p, ral, A, B},
      If[Length[ip] ≠ 3, Print["not infinite point"];
        Abort[], If[Abs[ip[[3]]] > 1.*^-6, Print["not infinite point"];
          Abort[], p = Chop[ip, 1.*^-6]]];
      B = {{1.414213562373095`, 0.`, 0.`}, {0.`, 0.`, 1.414213562373095`},
          {0.`, -0.7071067811865475`, -0.7071067811865475`}};
      If[Abs[p[[1]]] < 1.*^-6, Return[B]];
      ral = line[p, {0, 0}, x, y];
      B.klRotation[ral, x, x, y]]

In[154]:= flti[{0, 1, 0}, ip2z[{0, 1, 0}, x, y]]

Out[154]= {0., 0.}
```

(6.3) iTransform: Takes line *l* to the infinite line, moving the original infinite line to the *x*-axis. There is a fixed random vector r1 that could be replaced by another at the user's discretion.

```
In[155]:= iTransform[l_, x_, y_] := Module[{ph, r1},
      ph = Normalize[N[{Coefficient[l, x], Coefficient[l, y], l /. Thread[{x, y} → {0, 0}]}]];
      r1 = {-0.7812879011987071`, -0.3168196688992504`, 0.9191639531555853`};
      {r1, {0, 0, 1}, ph}]

In[156]:= flti[{5, 7}, iTransform[x - y, x, y]]

Out[156]= {3.6805, -0.707107}

In[157]:= flti[{3.2, 3.2}, iTransform[x - y, x, y]]

Out[157]= {-2.59478, 1., 0}

In[158]:= flti[{2, 5, 0}, iTransform[3 x - 2 y + 5, x, y]]

Out[158]= {4.84935, 0.}
```

(6.4) complexProjectiveSingularPoints: Here we often use a very loose tolerance. It is recommended that this function be run several times with different tolerances, but even running it twice with the same tolerance will give different results. A weak tolerance, e.g. .001, will identify all possible singular points. A tight tolerance will eliminate some that are not singular but may miss some that are.

```
In[159]:= complexProjectiveSingularPoints[h_, u_, v_, eps_] :=
     Module[{f, w, A, a, b, sol1, sol2, solz, sing, sing1, sing2},
       A = {RandomReal[{-1, 1}, 3], {0, 0, RandomReal[{1, 2}]}, RandomReal[{-1, 1}, 3]};
       f = FLT[h, A, u, v];
       {a, b} = RandomComplex[{-1 - I, 1 + I}, 2];
       sol1 = {u, v} /. NSolve[{f, a D[f, u] + b D[f, v]}];
       {a, b} = RandomComplex[{-1 - I, 1 + I}, 2];
       sol2 = {u, v} /. NSolve[{f, a D[f, u] + b D[f, v]}];
       solz = Join[sol1, sol2];
       sing =
         Flatten[Take[Select[Tally[solz, Norm[#1 - #2] < eps &], #[[2]] > 1 &], All, 1], 1];
       sing1 = Table[{sing[[j, 1]], If[Abs[sing[[j, 2]]] > eps, sing[[j, 2]], 0]},
```

```
{j, Length[sing]}];
sing2 = flti[#, Inverse[A]] &/@sing1;
sing2 = Table[Which[Length[p] == 2, p, Abs[p[[1]]] > eps, p/p[[1]],
    Abs[p[[2]]] > eps, p/p[[2]], True, {0, 0, p[[3]]}], {p, sing2}];
Chop[sing2, eps]];
```

```
In[160]:= h = 9 - 10 x^2 + x^4 + 6 y^2 + 2 x^2 y^2 + y^4;
complexProjectiveSingularPoints[h, x, y, 1.*^-7]
Out[161]= {{0, 0. - 1.73205 i}, {0, 0. + 1.73205 i}, {1., 0. + 1. i, 0}, {1., 0. - 1. i, 0}}
```

(6.4) complexProjectiveIntersectionpoints:

```
In[162]:= complexProjectiveIntersectionpoints[f_, g_, x_, y_] := Module[{f1, g1, A, sol, sol2},
  A = {RandomReal[{-1, 1}, 3], {0, 0, RandomReal[{1, 2}]}, RandomReal[{-1, 1}, 3]};
  f1 = FLT[f, A, x, y];
  g1 = FLT[g, A, x, y];
  sol = {x, y} /. NSolve[{f1, g1}];
  flti[#, Inverse[A]] &/@sol]

In[163]:= complexProjectiveIntersectionpoints[x, x y - 1, x, y]

Out[163]= {{0., -27.0532, 0}, {0., -27.0532, 0}}
```

Chapter 7

(7.1) asymptotes: See infiniteRealPoints in (3.3) of this appendix for infinite points.

```
In[164]:= asymptotes[f_, p_, x_, y_] := Module[{g, gm, A, sol, rnd, k},
    A = ip2z[p, x, y];
    g = FLT[f, A, x, y];
    gm = pointMinForm[g, {0, 0}, x, y];
    rnd = RandomReal[{-1, 1}, 2].{x, y} - 1;
    sol = DeleteDuplicates[
        {x, y} /. NSolve[{gm, rnd}, {x, y}, Reals], Norm[#1 - #2] < .001 &];
    k = Length[sol];
    If[k == 0, Return[{}]];
    Table[FLT[sol[[i, 2]] x - sol[[i, 1]] y, Inverse[A], x, y], {i, k}]
  ];
```

Example: rational function $y = \frac{x^3+2x^2}{x^2-1}$.

```
In[165]:= fa = y (x^2 - 1) - x^3 - 2 x^2;
ipfa = infiniteRealPoints[fa, x, y]
Out[166]= {{0., 2.13159, 0}, {1.01751, 1.01751, 0}}

In[167]:= asymptotes[fa, {0, 1, 0}, x, y]

Out[167]= {-1.09518 + 1.09518 x, -4.00069 - 4.00069 x}

In[168]:= asymptotes[fa, {1, 1, 0}, x, y]

Out[168]= {2.46152 + 1.23076 x - 1.23076 y}
```

(7.2) conicWithFoci: Here we find the equation of a conic given foci and a "good" point q on the conic. The two points p_1, p_2 define the major axis. By "good", q must not be on the axis, but the closest point on the axis to q must lie between p_1, p_2 and no farther from the axis than Norm[p_1 - p_2]/2. This function has options, the default is "ellipse" and the other is "hyperbola"; either may be abbreviated to three or more letters.

```
In[169]:= Options[conicWithFoci] = {conic → "ell"};
conicWithFoci[p1_, p2_, q_, x_, y_, OptionsPattern[]] :=
  Module[{c, p0, l, qc, R, T, M, qq, a, b, aa, bb, cc, g, pp0},
    p0 = N[(p1 + p2)/2];
    c = Norm[p1 - p0];
    l = line[p1, p2, x, y];
    qc = closestPoint[l, q, x, y];
    If[Norm[qc - q] > c || Abs[ Norm[p1 - qc] + Norm[p2 - qc] - 2 c] > 1.*^-10,
      Echo["Bad Point q "];
      Abort[]];
    If[p1[[1]] == 0 && p2[[1]] == 0, R = IdentityMatrix[3], R = klRotation[l, x, x, y]];
    pp0 = flt[p0, R];
    T = pqTranslation[pp0, {0, 0}];
    M = T.R;
    qq = flt[q, M];
    Switch[StringTake[OptionValue[conic], 3], "ell",
      sol = NSolve[{c^2 - b^2 + a^2,
            qq[[1]]^2*b^2 + qq[[2]]^2*a^2 - a^2 b^2}, {a, b}, Reals], "hyp",
      sol = NSolve[{c^2 - b^2 - a^2, -qq[[1]]^2*b^2 + qq[[2]]^2*a^2 - a^2 b^2},
          {a, b}, Reals], _, Echo[Null, "options ell, hyp only"];
      Abort[]];
    If[Length[sol] == 0, Echo[Null, "no solution"]; Abort[]];
    {aa, bb} = {a, b} /. sol[[1]];
    Switch[StringTake[OptionValue[conic], 3], "ell",
      g = x^2/aa^2 + y^2/bb^2 - 1, "hyp", g = -x^2/aa^2 + y^2/bb^2 - 1];
    FLT[g, Inverse[M], x, y]
  ]

In[171]:= p1 = {1, 4};
p2 = {3, 1};
q = {2, 1};
conicWithFoci[p1, p2, q, x, y, conic → "hyperbola"]

Out[174]= -4.66623 + 3.36084 x - 0.069417 x^2 + 0.244313 y - 1.23327 x y + 0.444444 y^2
```

(7.3) cTransform: Places p at infinite point {0, 1, 0} with the infinite line as the tangent line to the transform of f. This is deterministic; no randomization is used.

```
In[175]:= cTransform[f_, p_, x_, y_] := Module[{fh, ph, nh, t, cs, A, d},
      d = tDeg[f, x, y];
      fh = Expand[t^d*(f /. Thread[{x, y} → {x/t, y/t}])];
      ph = If[Length[p] == 2, N[Append[p, 1]], N[p]];
      ph = ph/Norm[ph];
      nh = {D[fh, x], D[fh, y], D[fh, t]} /. Thread[{x, y, t} → ph];
      nh = nh/Norm[nh];
```

```
        cs = Cross[nh, ph];
        cs = cs/Norm[cs];
        A = {cs, ph, nh}];
```

```
In[176]:= circ1 = x^2+y^2-1;
          FLT[circ1, cTransform[circ1, {0, 1}, x, y], x, y]

Out[177]= 1. x^2 + 2. y
```

(7.3) p2cTransform: The FLT P2C takes the parabola $y = x^2$ to the circle $x^2 + y^2 = 1$. In particular, the infinite point {0, 1, 0} goes to {0, 1} and the vertex {0, 0} goes to {0, − 1}. Note that this is a global constant rather than a function.

```
In[178]:= p2cTransform := pqTranslation[{0, -1}, {0, 0}].ip2z[{0, 1, 0}, x, y];

In[179]:= FLT[y-x^2, p2cTransform, x, y]

Out[179]= 0.5 - 0.5 x^2 - 0.5 y^2
```

(7.3) flts: See (6.2) in this appendix.

(7.4) affineInflectionPoints: Affine real inflection points of curve $f = 0$.

```
In[180]:= affineInflectionPoints[f_, x_, y_] := Module[{H, fh, d, z, sol},
            d = tDeg[f, x, y];
            fh = homog[f, x, y, z];
            H = Det[D[fh, {{x, y, z}, 2}]];
            sol = NSolve[{fh, H, z-1}, {x, y, z}, Reals];
            ips = If[Length[sol] == 0, Return[{}], {x, y} /. sol];
            Chop[DeleteDuplicates[ips, Norm[#1-#2] < 1.*^-10 &]]
          ];

In[181]:= f = (x^2 + y^2)^2 - 4 x^3 y^2;
          aip = affineInflectionPoints[f, x, y]

Out[182]= {{1.23241, 1.96227}, {2.43426, 0.882676}, {1.23241, -1.96227}, {2.43426, -0.882676}, {0, 0}}
```

(7.4) allInflectionPoints: Finds all inflection points including infinite real inflection points.

```
In[183]:= allInflectionPoints[f_, x_, y_] := Module[{H, fh, d, z, l, ips, sol},
            fh = homog[f, x, y, z];
            H = Det[D[fh, {{x, y, z}, 2}]];
            l = RandomReal[{-10, 10}, 3].{x, y, z};
            sol = NSolve[{fh, H, l-1}, {x, y, z}, Reals];
            If[Length[sol] == 0, Return[{}]];
            ips = {x, y, z} /. sol;
            ips = DeleteDuplicates[ips, Norm[#1-#2] < 1.*^-10 &];
            Which[Abs[#[[3]]] > 1.*^-9, Take[#/#[[3]], 2],
                 Abs[#[[2]]] > 1.*^-9, #/#[[2]], True, #/#[[1]]] &/@ips
          ];

In[184]:= f = (x^2 + y^2)^2 - 4 x^3 y^2;
          allInflectionPoints[f, x, y]

Out[185]= {{2.43426, 0.882676}, {1.23241, 1.96227},
           {2.43426, -0.882676}, {0., 1., 0.}, {1.23241, -1.96227}}
```

(7.5) weirstrassNormalForm: Weierstrass normal form of a cubic f given a particular inflection point p. Also returns FLT normalizing f and the j-invariant.

```
In[186]:= weierstrassNormalForm[f_, ip_, x_, y_] :=
   Module[{B1, B2, B3, B4, B, f1, f2, f3, f4, j, k, c4, c6, cy2},
    B1 = cTransform[f, ip, x, y];
    f1 = Chop[FLT[f, B1, x, y], 1.*^-9];
    cy2 = Coefficient[f1, y^2];
    If[Length[cy2] > 0, Print["Sorry, this example requires special handling"];
     Abort[]];
    k = Expand[Coefficient[f1, y]/cy2/2+y];
    B2 = {{1, 0, 0}, {Coefficient[k, x], 1, k[[1]]}, {0, 0, 1}};
    f2 = FLT[f1, B2, x, y];
    B3 = homothety[CubeRoot[-(Coefficient[f2, x^3]/Coefficient[f2, y^2])], 1];
    f3 = FLT[f2, B3, x, y];
    f3 = -Expand[f3/Coefficient[f3, y^2]];
    B4 = {{1, 0, Coefficient[f3, x^2]/3}, {0, 1, 0}, {0, 0, 1}};
    f4 = FLT[f3, B4, x, y];
    B = B4.B3.B2.B1;
    c4 = Coefficient[f4, x]/27;
    c6 = f4[[1]]/54;
    j = 1728 c4^3/(c4^3-c6^2);
    {f4, B, j}];

In[187]:= nh210 = newtonHyperbola[210, x, y];
   aip = affineInflectionPoints[nh210, x, y]

Out[188]= {{3.38372, -0.0850327}, {0.0850327, -3.38372}}

In[189]:= weierstrassNormalForm[nh210, aip[[1]], x, y]

Out[189]= {0.588221 - 1.1183 x + 1. x^3 - 1. y^2, {{0.395727, -0.598182, -1.38989},
   {0.855051, -0.320078, 0.608946}, {0.0558517, -0.96104, -0.270706}}, 647.198}
```

Chapter 8

(8.3) ghsMap: Map from projective plane to Gauss hemisphere. p is the point and r is the radius of the inner circle. The entire Gauss hemisphere has radius $3\,r$.

```
In[190]:= ghsMap[p_, r_] := If[Length[p] == 2, N[6*r/Pi*ArcTan[Norm[p]/6/r*Pi]]*Normalize[p],
     3*r*N[Normalize[Take[p, 2]]]];

In[191]:= ghsMap[{1, 1, 0}, 3]

Out[191]= {6.36396, 6.36396}
```

(8.3) inverseGHSmap: Inverse map from Gauss hemisphere to projective plane. Note that infinite points will appear as affine points with a very large Norm. The reason is that we want this function to work with symbolic values of q.

```
In[192]:= inverseGHSmap[q_, r_] := Tan[Norm[q]/6/r*Pi]*6*r/Pi*Normalize[q];
```

In[193]:=
```
inverseGHSmap[{x, y}, 5]
```

Out[193]=

$$\left\{\frac{30\,x\,\mathrm{Tan}\left[\frac{1}{30}\pi\sqrt{\mathrm{Abs}[x]^2+\mathrm{Abs}[y]^2}\right]}{\pi\sqrt{\mathrm{Abs}[x]^2+\mathrm{Abs}[y]^2}},\ \frac{30\,y\,\mathrm{Tan}\left[\frac{1}{30}\pi\sqrt{\mathrm{Abs}[x]^2+\mathrm{Abs}[y]^2}\right]}{\pi\sqrt{\mathrm{Abs}[x]^2+\mathrm{Abs}[y]^2}}\right\}$$

(8.5) graphSD: Symmetric difference of graphs.

In[194]:=
```
graphSD[G_, H_] := With[{GH = GraphIntersection[G, H]}, FindCycle[
    EdgeList[GraphUnion[GraphDifference[G, GH], GraphDifference[H, GH]]]]][[1]];
```

In[195]:=
```
G = {14 — 11, 11 — 13, 13 — 12, 12 — 15, 15 — 9, 9 — 14};
H = {18 — 11, 11 — 13, 13 — 12, 12 — 19, 19 — 7, 7 — 18};
graphSD[G, H]
```

Out[196]=
```
{7 — 18, 18 — 11, 11 — 14, 14 — 9, 9 — 15, 15 — 12, 12 — 19, 19 — 7}
```

(8.7) moebiusB: Point on a Möbius band.

In[197]:=
```
moebiusB[t_, s_] := Module[{a = N[Sqrt[2] - 1], mb, MB, tt},
    mb[tt_] :=
      {Cos[2 tt + Pi] / (1 + a Cos[tt]), Sin[2 tt + Pi] / (1 + a Cos[tt]), 1 / (1 + a Cos[tt + Pi / 2])};
    s * mb[t] + (1 - s) * mb[t + Pi]];
```

(8.7) moebiusPLMB: Draws a Möbius band with an infinite line. Use with Show.

In[198]:=
```
moebiusPLMB :=
   Show[ParametricPlot3D[moebiusB[t, s], {t, -Pi, Pi}, {s, 0, .5}, Mesh → None,
     PlotStyle → Opacity[0.8]], ParametricPlot3D[moebiusB[t, .5], {t, 0, Pi},
     PlotStyle → {Thick, Blue, Dashed}], Boxed → False, Axes → False];
```

In[199]:=
```
Show[moebiusPLMB]
```

Out[199]=

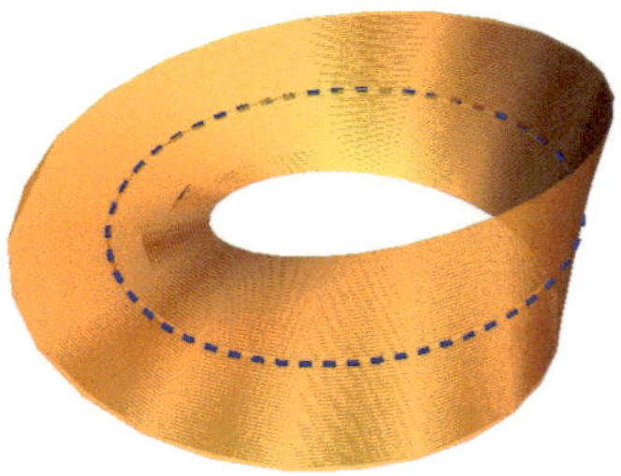

(8.7) moebiusRMPP: Outline of a rectangular hemisphere.

In[200]:=
```
moebiusRMPP := ListLinePlot[{{{-Pi, 0}, {Pi, 0}},
      {{-Pi, .5}, {Pi, .5}}, {{Pi, 0}, {Pi, .5}}, {{-Pi, 0}, {-Pi, .5}}, {{0, 0}, {0, .5}}},
    PlotStyle → {Brown, Orange, Green, Green, Cyan}, Axes → False,
    Frame → True, AspectRatio → 1 / 5];
```

In[201]:=
```
moebiusRMPP
```

Out[201]=

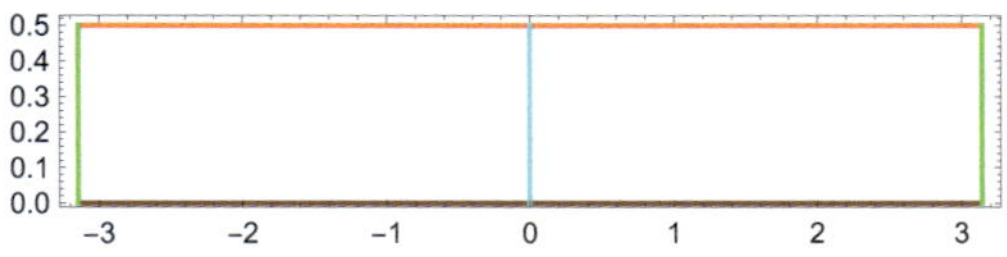

(8.7) moebiusPhi: The function for a projective plane minus the negative *x*-axis to a rectangle hemisphere model of a projective plane. It is assumed projective points will have third coordinate 0 or 1. If any points are on negative *x*-axis, including {1, 0, 0}, then the function will return an error message and {}. As usual, to apply to a list *L* of points, use M = moebiusPhi /@ L. You can find the position of error points in *M* by Position[M, {}]. Then fix these points manually. If they occur as other than the first or last points of *M*, you probably need to split the list before applying MB.

```
In[202]:= moebiusPhi[p_] := If[p[[2]] ≠ 0,
             If[Length[p] == 2,
               {ArcTan @@ N[p], N[ArcTan[Norm[p]] / Pi]}, {ArcTan @@ N[Take[p, 2]], .5}],
             Which[p[[1]] > 0 && Length[p] == 2, {N[Pi/2], N[ArcTan[Norm[p]] / Pi]},
               p[[1]] > 0, {N[Pi/2], .5}, p[[1]] ≤ 0 || Length[p] > 2, Echo[{}, p]]];

In[203]:= L = moebiusPhi /@ {{1, 1}, {2, 1}, {0, 1, 0}, {1, 1}}
          Show[moebiusRMPP, ListLinePlot[L]]

Out[203]= {{0.785398, 0.304087}, {0.463648, 0.36614}, {1.5708, 0.5}, {0.785398, 0.304087}}
```

Out[204]=

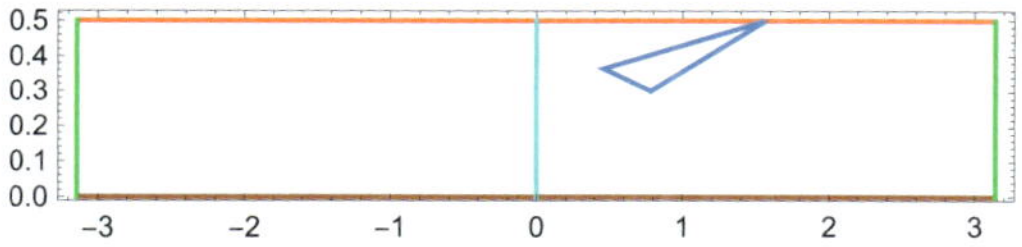

Chapter 9

(9.2) coefficientSigns: Creates an association directly from a polynomial in one or more variables. The polynomial is *f* and *X* is a list of the variables.

```
In[205]:= coefficientSigns[f_, X_] := Module[{A},
            A = Association[CoefficientRules[f, X]];
            Association[Table[p → −(Sign[A[p]] − 1) / 2, {p, Keys[A]}]]];

In[206]:= A = coefficientSigns[x^3 + x^2 − x + 1, {x}]

Out[206]= <| {3} → 0, {2} → 0, {1} → 1, {0} → 0 |>
```

(9.2) descartesDiagram: *A* is an association with rules $\{j\} \to e$ where *j* goes from 0 to the degree and each *e* is 0 or 1 depending on whether the jth coefficient is positive, 0, negative, or 1. See (9.A5) for the code.

```
In[207]:= A = <| {3} → 0, {2} → 0, {1} → 1, {0} → 0 |>;
          descartesDiagram[A]
```

Out[208]=

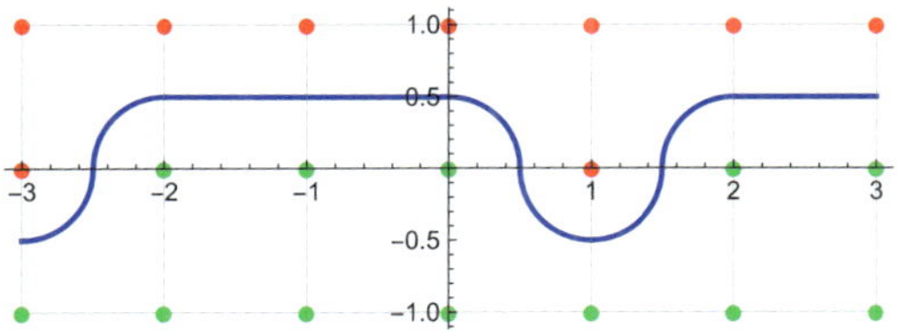

(9.2) descartesSketch: Uses polynomial values rather than signs. d is the degree, and now the association keys include $\{j\}$ for j from $-d$ to d. See 9.A7 for the code.

In[209]:=
```
B = <|{-3} → 1, {-2} → 0, {-1} → 0, {3} → 0, {2} → 0, {1} → 1, {0} → 0|>;
descartesSketch[B, 3]
```

Out[210]=
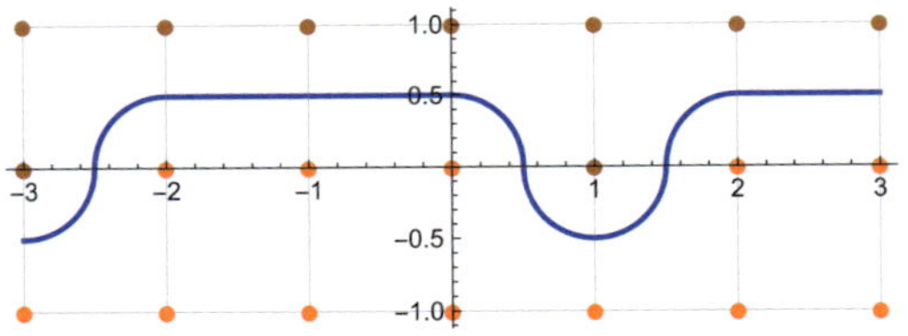

(9.2) dvAssoc: The function dvAssoc creates the association used in the viroDiagram function. For some purposes, you may want to start with this association and modify it. The input is an association with *all* non-negative keys, such as coefficientSigns for a full polynomial above and the output is an association with all keys. This is usually combined with coefficientSigns as viroAssociation. See (9.A10) for the code.

In[211]:=
```
x^2-xy+y^2-x+y-1;
A2 = coefficientSigns[x^2-xy+y^2-x+y-1, {x, y}]
```

Out[212]=
```
<|{2, 0} → 0, {1, 1} → 1, {1, 0} → 1, {0, 2} → 0, {0, 1} → 0, {0, 0} → 1|>
```

In[213]:=
```
dvAssoc[A2]
```

Out[213]=
```
<|{0, 0} → 1, {1, 0} → 1, {0, 1} → 0, {-1, 0} → 0, {0, -1} → 1, {2, 0} → 0, {1, 1} → 1,
  {0, 2} → 0, {-1, 1} → 0, {-2, 0} → 0, {0, -2} → 0, {-1, -1} → 1, {1, -1} → 0|>
```

(9.2) dDiagram: This is the general engine behind all of the two-dimensional diamond diagrams. This has options giving the colors of the dots. In addition, there is a large number of utility functions to enhance these diagrams.

In[214]:=
```
A = <|{0, 0} → 0, {1, 0} → 1, {0, 1} → 0, {-1, 0} → 0, {0, -1} → 1, {2, 0} → 0, {1, 1} → 1,
      {0, 2} → 0, {-1, 1} → 0, {-2, 0} → 0, {0, -2} → 0, {-1, -1} → 1, {1, -1} → 0|>;
DDf = dDiagram[A, pcolr → Orange, ncolr → Brown]
```

Out[215]=
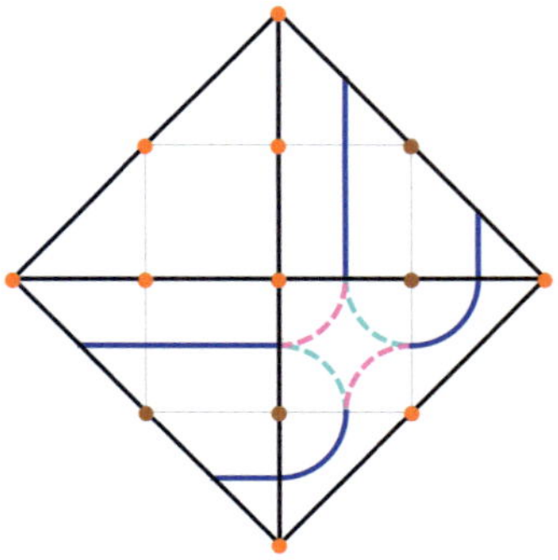

(9.2) viroAssociation and viroDiagram: viroDiagram draws the Viro diagram of a bivariate polynomial.

```
In[216]:= viroAssociation[f_, x_, y_] := dvAssoc[coefficientSigns[f, {x, y}]];
          viroDiagram[f_, x_, y_] :=
            dDiagram[viroAssociation[f, x, y], pcolr → Green, ncolr → Red];
```

```
In[218]:= f = 1 - x + y + x^2 - x y + y^2;
          Af = viroAssociation[f, x, y]
```

```
Out[219]= <|{0, 0} → 0, {1, 0} → 1, {0, 1} → 0, {-1, 0} → 0, {0, -1} → 1, {2, 0} → 0, {1, 1} → 1,
           {0, 2} → 0, {-1, 1} → 0, {-2, 0} → 0, {0, -2} → 0, {-1, -1} → 1, {1, -1} → 0|>
```

```
In[220]:= viroDiagram[f, x, y]
```

Out[220]=

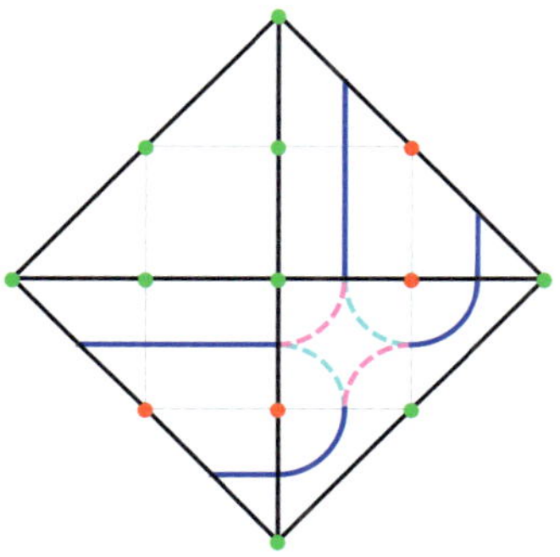

(9.2) pickCyan: Graphics primitives to change ambiguous cells to fixed cells. Here, $\{a, b\}$ is the lower-left corner of the cell.

```
In[221]:= pickCyan[a_, b_] := Graphics[{White, Rectangle[{a + .1, b + .1}, {a + .9, b + .9}], Thick,
                Blue, Circle[{a + 1, b + 1}, .5, {Pi, 3 Pi/2}], Circle[{a, b}, .5, {0, Pi/2}]}];
```

```
In[222]:= f = 1 - x + y + x^2 - x y + y^2;
          Show[viroDiagram[f, x, y], pickCyan[0, -1]]
```

Out[223]=

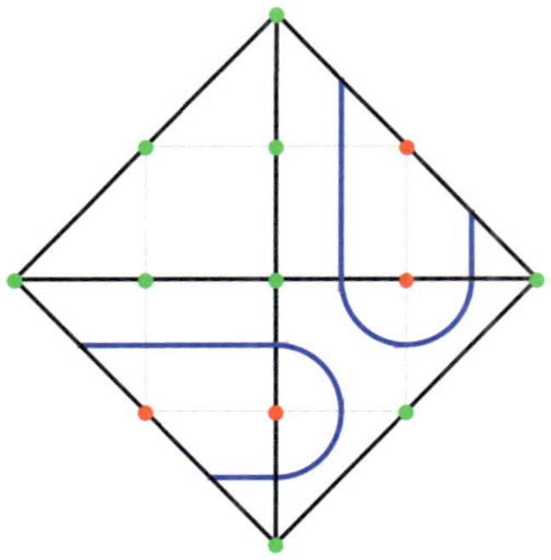

(9.2) pickMagenta:

```
In[224]:= pickMagenta[a_, b_] := Graphics[{White, Rectangle[{a+.1, b+.1}, {a+.9, b+.9}], Thick,
            Blue, Circle[{a+1, b}, .5, {Pi/2, Pi}], Circle[{a, b+1}, .5, {3 Pi/2, 2 Pi}]}];

In[225]:= f = 1 - x + y + x^2 - x y + y^2;
          Show[viroDiagram[f, x, y], pickMagenta[0, -1]]
```

Out[226]=

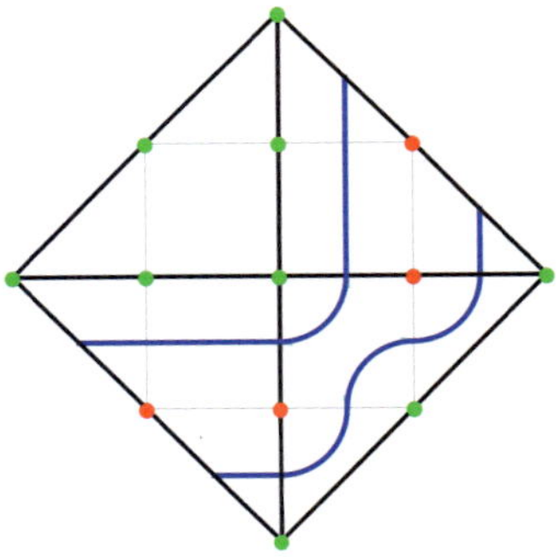

(9.2) pickSingularity:

```
In[227]:= pickSingularity[a_, b_] :=
           Graphics[{White, Rectangle[{a+.1, b+.1}, {a+.9, b+.9}], Thick, Blue,
             Line[{{a, b+.5}, {a+1, b+.5}}], Line[{{a+.5, b}, {a+.5, b+1}}]}]

In[228]:= f = 1 - x + y + x^2 - x y + y^2;
          Show[viroDiagram[f, x, y], pickSingularity[0, -1]]
```

Out[229]=

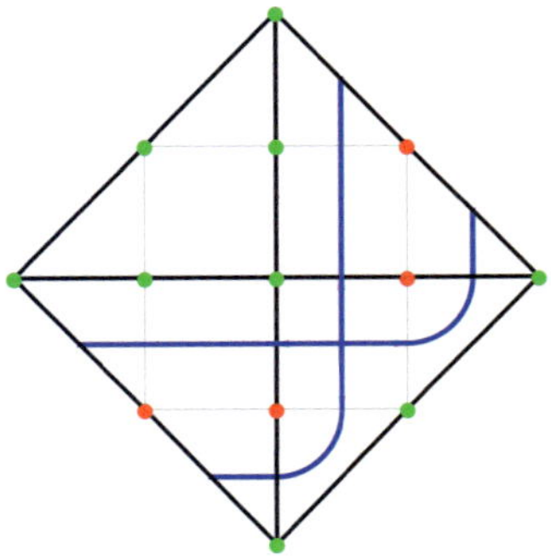

(9.2) labelI, labelII, labelIII, labelIV: Labeling functions for infinite points. The following places labels according to quadrant. For example, in the first quadrant, the label is above and to the right. The coordinates *a*, *b* are absolute, not entirely cell based, and need not be integers. Some trials may be necessary for proper labels.

```
In[230]:= labelI[a_, b_, txt_] := ListPlot[{{a+.5, b+.5}}, PlotStyle → Directive[Transparent, Thin],
            PlotLabels → Placed[txt, After], Axes → None]
```

In[231]:=
```
f = 1 - x + y + x^2 - x y + y^2;
Show[viroDiagram[f, x, y], labelI[-.2, 1.2, "A"]]
```

Out[232]=

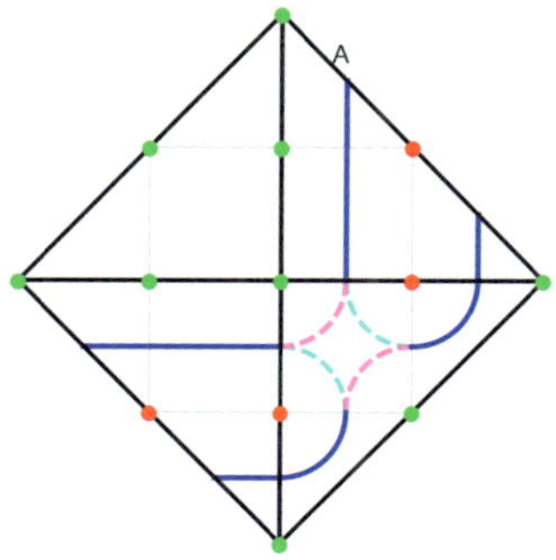

In[233]:=
```
labelII[a_, b_, txt_] := ListPlot[{{a - .5, b + .5}}, PlotStyle → Directive[Transparent, Thin],
    PlotLabels → Placed[txt, Above], Axes → None]
```

In[234]:=
```
labelIII[a_, b_, txt_] := ListPlot[{{a - .5, b - .5}}, PlotStyle → Directive[Transparent, Thin],
    PlotLabels → Placed[txt, Below], Axes → None]
```

In[235]:=
```
labelIV[a_, b_, txt_] := ListPlot[{{a + .5, b - .5}}, PlotStyle → Directive[Transparent, Thin],
    PlotLabels → Placed[txt, After], Axes → None]
```

In[236]:=
```
f = 1 - x + y + x^2 - x y + y^2;
Show[viroDiagram[f, x, y], labelII[-1, .1, "B"], labelIII[-1, .1, "A"]]
```

Out[237]=

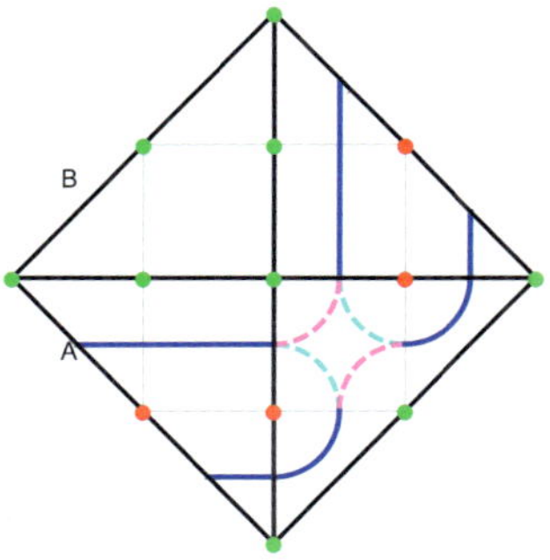

(9.2) label13 and **label24**: When labeling infinite points with a single letter or number, the symmetry of infinite points allows the use of one label for each point. Again, *a*, *b* are absolute coordinates and may need some trials to get proper placement.

In[238]:=
```
label13[a_, b_, txt_] := If[a ≥ 0 && b ≥ 0, ListPlot[
      {{{a + .5, b + .5}}, {{-a - .5, -b - .5}}}, PlotStyle → Directive[Transparent, Thin],
      PlotLabels → {Placed[txt, After], Placed[txt, Below]}, Axes → None],
    Echo["label13 error:must have a,b positive"]];
```

In[239]:=
```
label24[a_, b_, txt_] := If[a ≤ 0 && b ≥ 0, ListPlot[
      {{{a - .5, b + .5}}, {{-a + .5, -b - .5}}}, PlotStyle → Directive[Transparent, Thin],
      PlotLabels → {Placed[txt, Above], Placed[txt, After]}, Axes → None],
    Echo["label24error: must have a≤0,b≥0"]];
```

In[240]:=
```
f = 1 - x + y + x^2 - x y + y^2;
Show[viroDiagram[f, x, y], label13[0, 1, "A"]]
```

Out[241]=

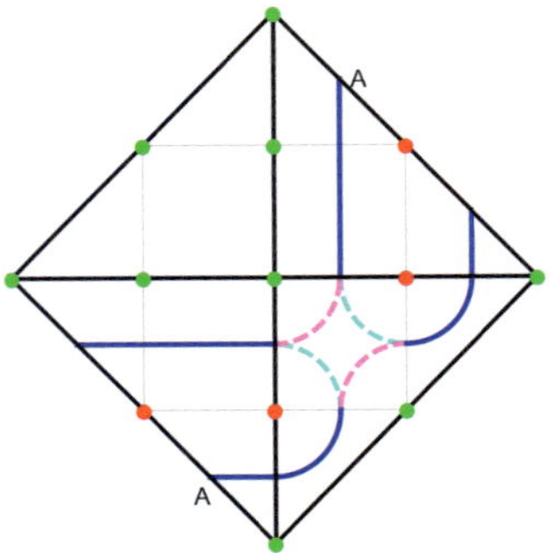

(9.2) toggle: Changes values of sign associations between 0, 1.

In[242]:=
```
toggle[A_, K_] := Module[{B},
    B = A;
    Do[If[B[p] == 0, B[p] = 1, B[p] = 0], {p, K}];
    B]
```

In[243]:=
```
Af = <|{0, 0} → 0, {1, 0} → 1, {0, 1} → 0, {-1, 0} → 0, {0, -1} → 1, {2, 0} → 0, {1, 1} → 1,
      {0, 2} → 0, {-1, 1} → 0, {-2, 0} → 0, {0, -2} → 0, {-1, -1} → 1, {1, -1} → 0|>;
Af1 = toggle[Af, {{0, 0}, {1, 0}}]
```

Out[244]=
```
<|{0, 0} → 1, {1, 0} → 0, {0, 1} → 0, {-1, 0} → 0, {0, -1} → 1, {2, 0} → 0, {1, 1} → 1,
  {0, 2} → 0, {-1, 1} → 0, {-2, 0} → 0, {0, -2} → 0, {-1, -1} → 1, {1, -1} → 0|>
```

In[245]:=
```
{dDiagram[Af], dDiagram[Af1]}
```

Out[245]=

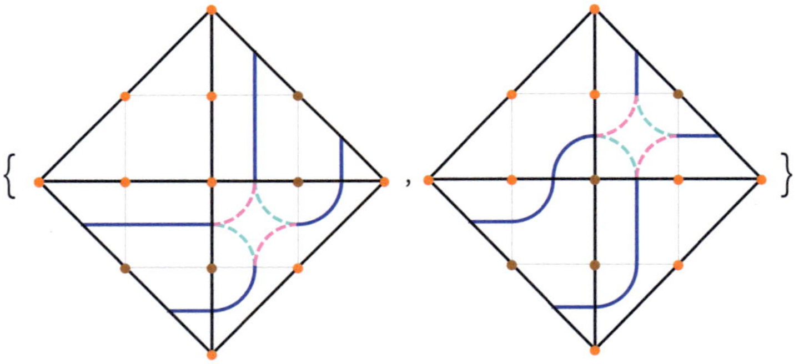

(9.3) axisPoints: Axis points are points on either the *x*- or *y*-axis. Points on the *x*-axis are listed first.

In[246]:=
```
axisPoints[f_, x_, y_] :=
  Quiet[Check[SortBy[{x, y} /. NSolve[{f, x y}, {x, y}, Reals], Abs[#[[2]]] &], {}]]
```

In[247]:=
```
axisPoints[1 - x + y - x y, x, y]
```

Out[247]=
```
{{1., 0.}, {0., -1.}}
```

(9.3) inspectInfPoint: inspectInfPoint puts together several other functions to allow us to quickly see what is happening at an infinite point and to give affine points that directly lead to the infinite points. *ip* is the infinite point and *r* is a positive radius usually less than one. We want a circle that intersects the curve only near the infinite points. Data shown can be changed by options: "plot" just gives plot, "data" just gives data, and the default All is both. Data consists of the transformation, the equation at infinity, the boundary points on the circle around the infinite point, and the affine coordinates of these boundary points.

```
In[248]:= Options[inspectInfPoint] = {output → All};
inspectInfPoint[f_, ip_, r_, x_, y_, OptionsPattern[]] :=
  Module[{A, g, bp, ap, sol, GP},
    A = ip2z[ip, x, y];
    g = FLT[f, A, x, y];
    sol = NSolve[{g, x^2+y^2-r^2}, {x, y}, Reals];
    If[Length[sol] > 0, bp = {x, y} /. sol;
      ap = flti[#, Inverse[A]] &/@bp,
      bp = {}; ap = {}];
    GP := Show[gaussPlot[g, r, bp], ContourPlot[y == 0, {x, -r, r}, {y, -r, r},
          ContourStyle → Directive[Orange, Dashed]], ImageSize → Small];
    Switch[OptionValue[output], "plot", Return[GP], "data",
      Return[{A, g, bp, ap}], Default, Continue];
    Echo[A, "Transformation"];
    Echo[g, " Transformed equation"];
    Echo[bp, "boundary points coordinates relative to infinite point"];
    Echo[ap, "affine coordinates boundry points"];
    GP]
```

```
In[250]:= inspectInfPoint[y (x^2-1)-1, {0, 1, 0}, 1, x, y]
```

» Transformation {{1.41421, 0., 0.}, {0., 0., 1.41421}, {0., −0.707107, −0.707107}}

» Transformed equation $-0.707107\,x^2 - 0.353553\,x^2\,y + 0.707107\,y^2$

» boundary points coordinates relative to infinite point
{{0.648098, 0.761557}, {−0.771135, −0.636672}, {−0.648098, 0.761557}, {0.771135, −0.636672}}

» affine coordinates boundry points {{0.851016, −3.6262}, {1.2112, 2.14134}, {−0.851016, −3.6262}, {−1.2112, 2.14134}}

Out[250]=

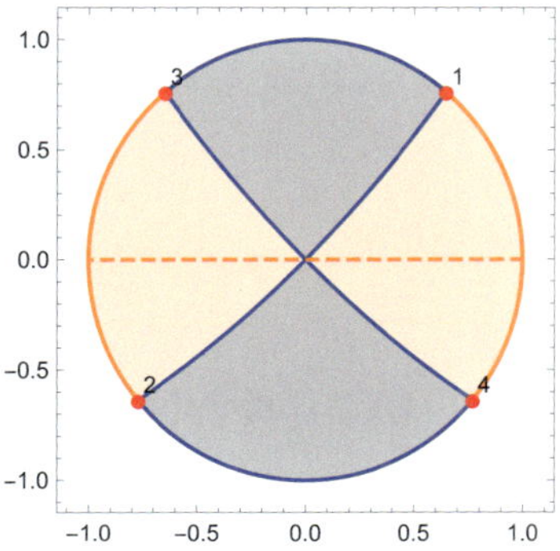

In[251]:=
```
inspectInfPoint[y (x^2-1)-1, {0, 1, 0}, 1, x, y, output → "data"]
```

Out[251]=
```
{{{1.41421, 0., 0.}, {0., 0., 1.41421}, {0., -0.707107, -0.707107}},
  -0.707107 x^2 - 0.353553 x^2 y + 0.707107 y^2, {{0.648098, 0.761557},
   {-0.771135, -0.636672}, {-0.648098, 0.761557}, {0.771135, -0.636672}},
  {{0.851016, -3.6262}, {1.2112, 2.14134}, {-0.851016, -3.6262}, {-1.2112, 2.14134}}}
```

(9.5) plotDiamonds: That is, equations $|x| + |y| = r$, on a region plot given a list *rr* of increasing positive values of *r*.

In[252]:=
```
plotDiamonds[f_, rr_, x_, y_] := Module[{m},
    m = 1.1 * Max[rr];
    If[Length[rr] ≠ tDeg[f, x, y], Echo[" wrong number of diamonds", "WARNING"]];
    Show[RegionPlot[f > 0, {x, -m, m}, {y, -m, m}], ContourPlot[Abs[x] + Abs[y],
        {x, -m, m}, {y, -m, m}, Contours → rr, ContourShading → None]]];
```

In[253]:=
```
fd = y^4 - y^2 - x^4 + 2 x^2;
rrfd = {.2, .4, 1.3, 3};
plotDiamonds[fd, rrfd, x, y]
```

Out[255]=

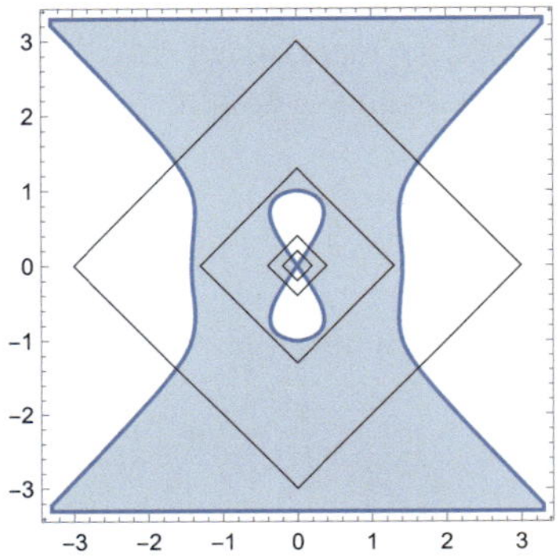

(9.5) gaussAssoc: Creates an association by sampling points on diamonds of radii in the list *rr*. This is useful for sparse exact nonsingular equations symmetric about the *y*-axis. Note that this association considers the function value of 0 to be positive. The resulting diamond diagram is called a Gauss diagram.

In[256]:=
```
gaussAssoc[f_, rr_, x_, y] := Module[{A, r, V, d, sgn},
    sgn = Switch[Sign[#], 1, 0, -1, 1, 0, 0] &;
    d = tDeg[f, x, y];
    If[d ≠ Length[rr], Echo[d, "rr size"]; Abort[]];
    r = Prepend[rr, 0];
    V = ddVerts[d];
    <|
     Table[v → sgn[f /. Thread[{x, y} -> r[[Total[Abs[v]] + 1]] / (Total[Abs[v]] + .001) * v]],
       {v, V}]|>
  ];
```

In[257]:=
```
dDiagram[gaussAssoc[fd, rrfd, x, y]]
```

Out[257]=

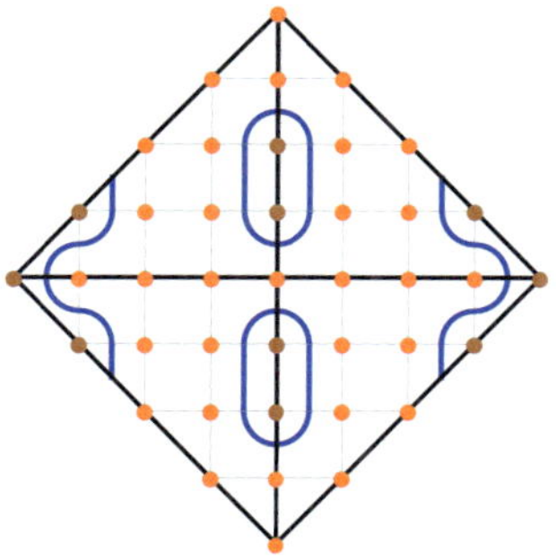

(9.6) n2s: We recall the utility functions changing between sign sequences and index numbers for Newton hyperbolas.

In[258]:=
```
n2s[n_] := Module[{k, n2dig, s},
    k = Which[ n < 8, 3, n < 64, 6, n < 1024 , 10, n < 32 768, 15,
        n < 2 097 152, 21, n < 2 ^ 28, 28, n < 2 ^ 36, 36, n < 2 ^ 45, 45, True, Fail];
    n2dig = IntegerDigits[n, 2, k];
    s = Reverse[n2dig]
  ];
```

In[259]:=
```
n2s[38]
```

Out[259]= {0, 1, 1, 0, 0, 1}

In[260]:=
```
s2n[s_] := Module[{a},
    a = Reverse[s];
    FromDigits[a, 2]
  ];
```

In[261]:=
```
s2n[{0, 1, 1, 0, 0, 1, 0, 0}]
```

Out[261]= 38

Subroutines for Chapter 9

Here are subroutines and code for the Chapter 9 functions listed in this appendix. If the code for Chapter 9 works for all of the examples, then the code below should also be OK, so we are not giving examples since it is unlikely that you will be using any of these directly. However, if there is a problem with the code for Chapter 9, the problem could be here.

(9.A1) ddVerts: Vertices of a diamond diagram of dimension *d*.

In[262]:=
```
ddVerts[d_] := Module[{V},
    V[0] = {{0, 0}};
    V[n_] :=
      V[n] = Join[V[n - 1], Reverse[Table[{j, n - j}, {j, 0, n}]], Table[{-j, n - j}, {j, 1, n}],
          Table[{-j, j - n}, {j, 0, n - 1}], Table[{j, j - n}, {j, 1, n - 1}]];
    V[
      d]];
```

(9.A2) zeroA: Zero diagram association.

```
In[263]:= zeroA[d_] := Association[Table[v → 0, {v, ddVerts[d]}]]
```

Raw plot of diamond diagram association. Variables c1, c2 are the positive and negative colors, e.g. orange, RGBColor[1, .5, 0], etc.

```
In[264]:= plotAssoc[A_, c1_, c2_] := Module[{K, C1, C2},
    K = Keys[A];
    C1 = Select[K, A[#] == 0 &];
    C2 = Select[K, A[#] == 1 &];
    If[Length[C1] == 0, Return[ListPlot[K, Axes → False, PlotStyle → c2]]];
    If[Length[C2] == 0, Return[ListPlot[K, Axes → False, PlotStyle → c1]]];
    ListPlot[{C1, C2}, PlotMarkers → {●, ●}, PlotStyle → {c1, c2}, Axes → False]];
```

(9.A3) sqrSeq: Values bounding a cell.

```
In[265]:= sqrSeq[p_, A_] := {A[p], A[p + {1, 0}], A[p + {0, 1}], A[p + {1, 1}]};
```

sqAssoc: To associate these primitives to sqrSeq we have the following.

```
In[266]:= sqAssoc = <| {0, 0, 0, 0} → ps0, {1, 1, 1, 1} → ps0, {1, 0, 0, 0} → ps1, {0, 1, 1, 1} → ps1,
    {0, 1, 0, 0} → ps2, {1, 0, 1, 1} → ps2, {1, 1, 0, 0} → ps3, {0, 0, 1, 1} → ps3,
    {0, 0, 1, 0} → ps4, {1, 1, 0, 1} → ps4, {1, 0, 1, 0} → ps5, {0, 1, 0, 1} → ps5,
    {1, 1, 1, 0} → ps6, {0, 0, 0, 1} → ps6, {0, 1, 1, 0} → ps7, {1, 0, 0, 1} → ps7 |>;
```

(9.A4) ps0–ps7: Graphics primitives for cells.

```
In[267]:= ps0[a_, b_] :=
   Graphics[{Thin, Gray, Line[{{a, b}, {a + 1, b}, {a + 1, b + 1}, {a, b + 1}, {a, b}}]}];

In[268]:= ps1[a_, b_] := Graphics[{Thick, Blue, Circle[{a, b}, .5, {0, Pi/2}],
     Thin, Gray, Line[{{a, b}, {a + 1, b}, {a + 1, b + 1}, {a, b + 1}, {a, b}}]}];

In[269]:= ps2[a_, b_] := Graphics[{Thick, Blue, Circle[{a + 1, b}, .5, {Pi/2, Pi}],
     Thin, Gray, Line[{{a, b}, {a + 1, b}, {a + 1, b + 1}, {a, b + 1}, {a, b}}]}];

In[270]:= ps3[a_, b_] := Graphics[{Thick, Blue, Line[{{a, b + .5}, {a + 1, b + .5}}],
     Thin, Gray, Line[{{a, b}, {a + 1, b}, {a + 1, b + 1}, {a, b + 1}, {a, b}}]}];

In[271]:= ps4[a_, b_] := Graphics[{Thick, Blue, Circle[{a, b + 1}, .5, {3 Pi/2, 2 Pi}],
     Thin, Gray, Line[{{a, b}, {a + 1, b}, {a + 1, b + 1}, {a, b + 1}, {a, b}}]}];

In[272]:= ps5[a_, b_] := Graphics[{Thick, Blue, Line[{{a + .5, b}, {a + .5, b + 1}}],
     Thin, Gray, Line[{{a, b}, {a + 1, b}, {a + 1, b + 1}, {a, b + 1}, {a, b}}]}];

In[273]:= ps6[a_, b_] := Graphics[{Thick, Blue, Circle[{a + 1, b + 1}, .5, {Pi, 3 Pi/2}],
     Thin, Gray, Line[{{a, b}, {a + 1, b}, {a + 1, b + 1}, {a, b + 1}, {a, b}}]}];

In[274]:= ps7[a_, b_] :=
   Graphics[{Thin, Gray, Line[{{a, b}, {a + 1, b}, {a + 1, b + 1}, {a, b + 1}, {a, b}}], Thick,
     Cyan, Dashed, Circle[{a + 1, b + 1}, .5, {Pi, 3 Pi/2}], Circle[{a, b}, .5, {0, Pi/2}],
     Magenta, Circle[{a + 1, b}, .5, {Pi/2, Pi}], Circle[{a, b + 1}, .5, {3 Pi/2, 2 Pi}]}];
```

On the boundary of the diamond diagram, we have a triangle. The corresponding primitives for triangles are as follows.

```
In[275]:= triseq1[p_, A_] := {A[p], A[p + {1, 0}], A[p + {0, 1}]};
          triseq2[p_, A_] := {A[p], A[p + {-1, 0}], A[p + {0, 1}]};
          triseq3[p_, A_] := {A[p], A[p + {-1, 0}], A[p + {0, -1}]};
          triseq4[p_, A_] := {A[p], A[p + {1, 0}], A[p + {0, -1}]};

In[279]:= pt10[a_, b_] := Graphics[{Thin, Gray, Line[{{a, b}, {a + 1, b}, {a, b + 1}, {a, b}}]}];

In[280]:= pt11[a_, b_] := Graphics[{Thin, Gray, Line[{{a, b}, {a + 1, b}, {a, b + 1}, {a, b}}],
              Thick, Blue, Circle[{a, b}, .5, {0, Pi/2}]}];

In[281]:= pt12[a_, b_] := Graphics[{Thin, Gray, Line[{{a, b}, {a + 1, b}, {a, b + 1}, {a, b}}],
              Thick, Blue, Line[{{a, b + .5}, {a + .5, b + .5}}]}];

In[282]:= pt13[a_, b_] := Graphics[{Thin, Gray, Line[{{a, b}, {a + 1, b}, {a, b + 1}, {a, b}}],
              Thick, Blue, Line[{{a + .5, b}, {a + .5, b + .5}}]}];

In[283]:= t1Assoc = <|{0, 0, 0} → pt10, {1, 1, 1} → pt10, {1, 0, 0} → pt11, {0, 1, 1} → pt11,
            {1, 1, 0} → pt12, {0, 0, 1} → pt12, {0, 1, 0} → pt13, {1, 0, 1} → pt13 |>;

In[284]:= pt20[a_, b_] := Graphics[{Thin, Gray, Line[{{a, b}, {a - 1, b}, {a, b + 1}, {a, b}}]}];
          pt21[a_, b_] := Graphics[{Thin, Gray, Line[{{a, b}, {a - 1, b}, {a, b + 1}, {a, b}}],
              Thick, Blue, Circle[{a, b}, .5, {Pi/2, Pi}]}];
          pt22[a_, b_] := Graphics[{Thin, Gray, Line[{{a, b}, {a - 1, b}, {a, b + 1}, {a, b}}],
              Thick, Blue, Line[{{a, b + .5}, {a - .5, b + .5}}]}];
          pt23[a_, b_] := Graphics[{Thin, Gray, Line[{{a, b}, {a - 1, b}, {a, b + 1}, {a, b}}],
              Thick, Blue, Line[{{a - .5, b}, {a - .5, b + .5}}]}];
          t2Assoc = <|{0, 0, 0} → pt20, {1, 1, 1} → pt20, {1, 0, 0} → pt21, {0, 1, 1} → pt21,
            {1, 1, 0} → pt22, {0, 0, 1} → pt22, {0, 1, 0} → pt23, {1, 0, 1} → pt23 |>;
          pt30[a_, b_] := Graphics[{Thin, Gray, Line[{{a, b}, {a - 1, b}, {a, b - 1}, {a, b}}]}];
          pt31[a_, b_] := Graphics[{Thin, Gray, Line[{{a, b}, {a - 1, b}, {a, b - 1}, {a, b}}],
              Thick, Blue, Circle[{a, b}, .5, {Pi, 3 Pi/2}]}];
          pt32[a_, b_] := Graphics[{Thin, Gray, Line[{{a, b}, {a - 1, b}, {a, b - 1}, {a, b}}],
              Thick, Blue, Line[{{a, b - .5}, {a - .5, b - .5}}]}];
          pt33[a_, b_] := Graphics[{Thin, Gray, Line[{{a, b}, {a - 1, b}, {a, b - 1}, {a, b}}],
              Thick, Blue, Line[{{a - .5, b}, {a - .5, b - .5}}]}];
          t3Assoc = <|{0, 0, 0} → pt30, {1, 1, 1} → pt30, {1, 0, 0} → pt31, {0, 1, 1} → pt31,
            {1, 1, 0} → pt32, {0, 0, 1} → pt32, {0, 1, 0} → pt33, {1, 0, 1} → pt33 |>;
          pt40[a_, b_] := Graphics[{Thin, Gray, Line[{{a, b}, {a + 1, b}, {a, b - 1}, {a, b}}]}];
          pt41[a_, b_] := Graphics[{Thin, Gray, Line[{{a, b}, {a + 1, b}, {a, b - 1}, {a, b}}],
              Thick, Blue, Circle[{a, b}, .5, {3 Pi/2, 2 Pi}]}];
          pt42[a_, b_] := Graphics[{Thin, Gray, Line[{{a, b}, {a + 1, b}, {a, b - 1}, {a, b}}],
              Thick, Blue, Line[{{a, b - .5}, {a + .5, b - .5}}]}];
          pt43[a_, b_] := Graphics[{Thin, Gray, Line[{{a, b}, {a + 1, b}, {a, b - 1}, {a, b}}],
              Thick, Blue, Line[{{a + .5, b}, {a + .5, b - .5}}]}];
          trAssoc = <|{0, 0, 0} → pt40, {1, 1, 1} → pt40, {1, 0, 0} → pt41, {0, 1, 1} → pt41,
            {1, 1, 0} → pt42, {0, 0, 1} → pt42, {0, 1, 0} → pt43, {1, 0, 1} → pt43 |>;
```

(9.A5) dDiagram: The main engine for diamond diagrams is as follows.

```
In[299]:= Options[dDiagram] = {pcolr → Orange, ncolr → Brown};
dDiagram[A_, OptionsPattern[]] := Module[{d, sqrs, t1, t2, t3, t4, PS, PT, PB, PA},
    d = Max[Total[#] &/@ Keys[A]];
    sqrs = Select[Keys[A], KeyExistsQ[A, # + {1, 0}] &&
            KeyExistsQ[A, # + {1, 1}] && KeyExistsQ[A, # + {0, 1}] &];
    PS := Show[Table[sqAssoc[sqrSeq[p, A]] @@ p, {p, sqrs}]];
    t1 = Select[Keys[A], #[[1]] + #[[2]] == d - 1 &];
    t2 = Select[Keys[A], -#[[1]] + #[[2]] == d - 1 &];
    t3 = Select[Keys[A], -#[[1]] - #[[2]] == d - 1 &];
    t4 = Select[Keys[A], #[[1]] - #[[2]] == d - 1 &];
    PT := Show[Table[t1Assoc[triseq1[p, A]] @@ p, {p, t1}],
        Table[t2Assoc[triseq2[p, A]] @@ p, {p, t2}], Table[t3Assoc[triseq3[p, A]] @@ p,
          {p, t3}], Table[trAssoc[triseq4[p, A]] @@ p, {p, t4}]];
    PB := ListLinePlot[{{{-d, 0}, {d, 0}}, {{0, -d}, {0, d}, {d, 0}, {0, -d}, {-d, 0}, {0, d}}},
        PlotStyle → Black, Axes → False];
    PA := plotAssoc[A, OptionValue[pcolr], OptionValue[ncolr]];
    Show[PS, PT, PB, PA,  PlotRange → d + 1]
  ];
```

(9.A6) sDiagram: Square diagram, used for Descartes's law, and the descartesSketch function.

```
In[301]:= Options[sDiagram] = {pcolr → Orange, ncolr → Brown};
sDiagram[A_, OptionsPattern[]] := Module[{d, sqrs, PS, PA},
    sqrs = Select[Keys[A], KeyExistsQ[A, # + {1, 0}] &&
            KeyExistsQ[A, # + {1, 1}] && KeyExistsQ[A, # + {0, 1}] &];
    PS := Show[Table[sqAssoc[sqrSeq[p, A]] @@ p, {p, sqrs}]];
    PA := plotAssoc[A, OptionValue[pcolr], OptionValue[ncolr]];
    Show[PS, PA, Axes → True]
  ];
```

(9.A7) descartesSketch:

```
In[304]:= Options[descartesSketch] = {pcolr → Orange, ncolr → Brown};
descartesSketch[A_, d_, OptionsPattern[]] := Module[{SA, T, B, AA},
    T = Association[Table[{x, 1} → 1, {x, -d, d}]];
    B = Association[Table[{x, -1} → 0, {x, -d, d}]];
    AA = Association[Table[{x, 0} → A[{x}], {x, -d, d}]];
    SA = Union[B, AA, T];
    sDiagram[SA, pcolr -> OptionValue[pcolr], ncolr -> OptionValue[ncolr]]
  ]
```

(9.A8) descartesDiagram:

```
In[305]:= descartesDiagram[sa_] := Module[{d, sqrs, PS, PA, ssa, K, A},
    d = Max[Flatten[Keys[sa]]];
    ssa = sa;
    AssociateTo[ssa, Table[{-j} → If[Mod[j, 2] == 0, sa[{j}], 1 - sa[{j}]], {j, d}]];
    descartesSketch[ssa, d, pcolr → Green, ncolr → Red]
  ]
```

(9.A9) s2A: Conversion from Sign sequence to association.

```
In[306]:= s2A[s_] := Module[{n, d, A, K},
    n = Length[s];
    d = Which[ n == 3, 1, n == 6, 2, n == 10 , 3, n == 15, 4,
        n == 21, 5, n == 28, 6, n == 36, 7, n == 45, 8, n == 55, 9, True, Fail];
    K = Select[ddVerts[d], (#[[1]] ≥ 0 && #[[2]] ≥ 0) &];
    A = Association[Table[K[[i]] → s[[i]], {i, n}]]];
```

(9.A10) dvAssoc: Code for dvAssoc described in Chapter 9.2, A must have key $\{i, j\}$ for all non-negative $\{i, j\}$.

```
In[307]:= dvAssoc[A_] := Module[{d, K, V, v, pver},
    K = Keys[A];
    d = Max[Table[Total[p], {p, K}]];
    v = ddVerts[d];
    V = <| |>;
    Do[Which[p[[1]] ≥ 0 && p[[2]] ≥ 0, AppendTo[V, p → A[p]],
        p[[1]] < 0 && p[[2]] >= 0 && Mod[-p[[1]], 2] == 0, AppendTo[V, p → A[Abs[p]]],
        p[[1]] < 0 && p[[2]] >= 0 && Mod[-p[[1]], 2] == 1,
        AppendTo[V, p → (1 - A[Abs[p]])],
        p[[2]] < 0 && p[[1]] >= 0 && Mod[-p[[2]], 2] == 0, AppendTo[V, p → A[Abs[p]]],
        p[[2]] < 0 && p[[1]] >= 0 && Mod[-p[[2]], 2] == 1,
        AppendTo[V, p → (1 - A[Abs[p]])],
        Sign[(-1) ^ p[[1]]] * Sign[(-1) ^ p[[2]]] == 1, AppendTo[V, p → A[Abs[p]]],
        Sign[(-1) ^ p[[1]]] * Sign[(-1) ^ p[[2]]] == -1,
        AppendTo[V, p → (1 - A[Abs[p]])]], {p, v}];
    V];
```

Appendix 1

Functions from Appendix 1: Topics in Numerical Linear Algebra

(A.1) gaussNewton: Method using PseudoInverse to find the least-square solution of three nonlinear equations in two variables. To use this with two equations, use $h = 0$ as the third equation.

```
In[308]:= gaussNewton3[f_, g_, h_, p0_, x_, y_] := Module[{Q, Q0, p1, p2, J0, n},
    n = 30;
    p1 = N[p0];
    Catch[Do[Q = {f, g, h} /. Thread[{x, y} → p1];
        J0 = D[{f, g, h}, {{x, y}}] /. Thread[{x, y} → p1];
        p2 = p1 - PseudoInverse[J0].Q;
        If[Norm[p1 - p2] < 1.*^-7, Throw[p2]];
        p1 = p2, {n}]];
    p2]
```

In[309]:=
```
f = 10 + 15 x + 9 x^2 - 9 y;
g = 13 + 9 x - 21 y + 9 y^2;
h = -26 - 42 x + 30 y + 36 x y;
intersectionMultiplicity[f, g, {-1/3, 2/3}, 1.*^-12]
(* exact solution *)
```

Out[312]= 3

In[313]:=
```
NumberForm[{x, y} /. FindRoot[{f, g}, {x, -.3}, {y, .7}], 10]
```

Out[313]//NumberForm= {-0.3333297065, 0.6666702936}

In[314]:=
```
NumberForm[gaussNewton3[f, g, h, {-.3, .7}, x, y], 10]
```

Out[314]//NumberForm= {-0.3333332694, 0.6666667306}

(A.3) fAssoc: Puts a bivariate polynomial f into association format using CoefficientRules but adding keys for terms with coefficient zero. It returns the association and degree.

In[315]:=
```
fAssoc[f_, x_, y_] := Module[{A, B, d, FA},
    A = Association[CoefficientRules[f, {x, y}]];
    d = Max[Table[Total[p], {p, Keys[A]}]];
    FA = Association[Table[If[MissingQ[A[p]], p → 0, p → A[p]], {p, pExps[d]}]];
    {FA, d}];
```

In[316]:=
```
f = 10 + 15 x + 9 x^2 - 9 y;
FA = fAssoc[f, x, y]
```

Out[317]= {<| {0, 0} → 10, {1, 0} → 15, {0, 1} → -9, {2, 0} → 9, {1, 1} → 0, {0, 2} → 0 |>, 2}

(A.3) shiftFA: Multiplication of the association form of the polynomial by a monomial given as a pair of exponents.

In[318]:=
```
shiftFA[FA_, q_, n_] := Module[{S, K, d},
    K = Keys[FA];
    S = Association[Table[p + q → FA[p], {p, K}]];
    Association[Table[If[MissingQ[S[p]], p → 0, p → S[p]], {p, pExps[n]}]]];
```

In[319]:=
```
f = 10 + 15 x + 9 x^2 - 9 y;
FR = shiftFA[FA[[1]], {1, 2}, FA[[2]] + 3]
FromCoefficientRules[Normal[FR], {x, y}]
Expand[f * x y ^ 2]
```

Out[320]= <| {0, 0} → 0, {1, 0} → 0, {0, 1} → 0, {2, 0} → 0, {1, 1} → 0, {0, 2} → 0, {3, 0} → 0, {2, 1} → 0, {1, 2} → 10, {0, 3} → 0, {4, 0} → 0, {3, 1} → 0, {2, 2} → 15, {1, 3} → -9, {0, 4} → 0, {5, 0} → 0, {4, 1} → 0, {3, 2} → 9, {2, 3} → 0, {1, 4} → 0, {0, 5} → 0 |>

Out[321]= $10 x y^2 + 15 x^2 y^2 + 9 x^3 y^2 - 9 x y^3$

Out[322]= $10 x y^2 + 15 x^2 y^2 + 9 x^3 y^2 - 9 x y^3$

(A.3) sylD1: Sylvester matrix of one bivariate polynomial f with n as the order.

```
In[323]:= sylD1[f_, n_, x_, y_] := Module[{FA, d},
    {FA, d} = fAssoc[f, x, y];
    If[d > n, Print["Must have n≥deg(f)"];
      Abort[]];
    Table[Values[shiftFA[FA, q, n]], {q, pExps[n - d]}]]
```

```
In[324]:= f = 10 + 15 x + 9 x^2 - 9 y;
sylD1[f, 3, x, y] // MatrixForm
```

Out[325]//MatrixForm=

$$\begin{pmatrix} 10 & 15 & -9 & 9 & 0 & 0 & 0 & 0 & 0 & 0 \\ 0 & 10 & 0 & 15 & -9 & 0 & 9 & 0 & 0 & 0 \\ 0 & 0 & 10 & 0 & 15 & -9 & 0 & 9 & 0 & 0 \end{pmatrix}$$

(A.3) sylvesterMatrix: Sylvester matrix of f, g bivariate polynomials with n as the order.

```
In[326]:= sylvesterMatrix[f_, g_, n_, x_, y_] := Join[sylD1[f, n, x, y], sylD1[g, n, x, y]];
```

```
In[327]:= f = 10 + 15 x + 9 x^2 - 9 y;
g = 2 + 3 x + 4 y;
S3 = sylvesterMatrix[f, g, 2, x, y];
S3 // MatrixForm
```

Out[330]//MatrixForm=

$$\begin{pmatrix} 10 & 15 & -9 & 9 & 0 & 0 \\ 2 & 3 & 4 & 0 & 0 & 0 \\ 0 & 2 & 0 & 3 & 4 & 0 \\ 0 & 0 & 2 & 0 & 3 & 4 \end{pmatrix}$$

(A.3) invSyl: Inverse Sylvester matrix converts each row of Sylvester matrix S of order d to a polynomial in x and y. It is a subroutine for dDivide and sqFree.

```
In[331]:= (* using S3 directly above *)
invSyl[row_, d_, x_, y_] :=
    FromCoefficientRules[Table[pExps[d][[i]] → row[[i]], {i, Length[row]}], {x, y}];
```

```
In[332]:= invSyl[S3[[1]], 2, x, y]
```

```
Out[332]= 10 + 15 x + 9 x^2 - 9 y
```

```
In[333]:= invSyl[S3[[3]], 2, x, y]
```

```
Out[333]= 2 x + 3 x^2 + 4 x y
```

(A.3) nDivide: Checks if g is a factor of h and returns quotient if true to tolerance tol.

```
In[334]:= nDivide[h_, g_, x_, y_, tol_] := Module[{d1, d2, P, S, f},
      d1 = tDeg[g, x, y];
      d2 = tDeg[h, x, y];
      If [d1 > d2, Print["Does Not Divide"]; Return[Fail]];
      P = PseudoInverse[N[sylD1[g, d2, x, y]], Tolerance → tol];
      S = Chop[sylD1[h, d2, x, y].P];
      f = invSyl[S[[1]], d2 - d1, x, y];
      If[Norm[Flatten[sylD1[f * g - h, d2, x, y]]] > d2 * tol,
        Print["Does not divide at this tolerance"];
        Return[Fail]];
      f];
```

```
In[335]:= f = 10 + 15 x + 9 x^2 - 9 y;
          g = 2 + 3 x + 4 y;
          h = Expand[f * g]

Out[337]= 20 + 60 x + 63 x^2 + 27 x^3 + 22 y + 33 x y + 36 x^2 y - 36 y^2

In[338]:= nDivide[h, f, x, y, 1.*^-12]

Out[338]= 2. + 3. x + 4. y
```

(A.3) nGCD: Numerical greatest common divisor of *ff*, *hh* using the Kaltofen–Dayton algorithm (step 2 of the Zeng–Dayton algorithm). tol is the tolerance.

```
In[339]:= nGCD[ff_, hh_, x_, y_, tol_] :=
            Module[{S, Sk, U, V, u, v, su, d1, d2, dv, f, h, n, m, k, W, lnull},
             lnull[M_] := Dimensions[M][[1]] - Length[SingularValueList[N[M], Tolerance → tol]];
             d1 = tDeg[ff, x, y];
             d2 = tDeg[hh, x, y];
             If[d1 ≥ d2, f = ff; h = hh; m = d1; n = d2,
              f = hh; h = ff; m = d2; n = d1];
             k = m - 1;
             Sk = {{0}};
             While[k < m + n && lnull[Sk] == 0,
              k++;              (* Main loop *)
              Sk = sylvesterMatrix[f, h, k, x, y]];
             {U, S, V} = SingularValueDecomposition[N[Sk], Tolerance → tol];
             W = Flatten[Take[U, All, -1]];
             dv = k - m;
             If[dv ≤ 0, Return[h]];
             su = Chop[Take[W, Binomial[dv + 2, 2]], tol];
             u = invSyl[su, dv, x, y];
             nDivide[h, u, x, y, tol]]

In[340]:= ff = Expand[(2 + 3 x + 4 y) (4 x^2 - 2 y + 1)]

Out[340]= 2 + 3 x + 8 x^2 + 12 x^3 - 6 x y + 16 x^2 y - 8 y^2

In[341]:= h = 20 + 60 x + 63 x^2 + 27 x^3 + 22 y + 33 x y + 36 x^2 y - 36 y^2;
          nGCD[ff, h, x, y, 1.*^-12]

Out[342]= 45.0777 + 67.6166 x + 90.1554 y

In[343]:= Expand[% / %[[1]] * 2]

Out[343]= 2. + 3. x + 4. y
```

(A.3) sqFree: Square-free decomposition of numerical polynomial *f* to tolerance tol.

```
In[344]:= sqFree[f_, x_, y_, tol_] := Module[{S, Sk, U, V, v, su, dv, h, n, m, k, W},
             h = Expand[{D[f, x], D[f, y]}.RandomReal[{-1, 1}, 2]];
             lnull[M_] :=
              Dimensions[M][[1]] - Length[SingularValueList[N[M], Tolerance → tol]];
             m = tDeg[f, x, y];
             n = tDeg[h, x, y];
             k = m - 1;
```

```
    Sk = {{0}};
    While[k < m + n && lnull[Sk] == 0,
      k++;
      Sk = sylvesterMatrix[f, h, k, x, y]];
    {U, S, V} = SingularValueDecomposition[N[Sk], Tolerance → tol];
    W = Flatten[Take[U, All, -1]];
    dv = k - n;
    su = Chop[Take[W, -Binomial[dv + 2, 2]], tol];
    invSyl[su, dv, x, y]]
```

In[345]:=
```
f3 = Expand[y^3 (x + y)^2]
```

Out[345]= $x^2 y^3 + 2 x y^4 + y^5$

In[346]:=
```
sqFree[f3, x, y, 1.*^-12]
```

Out[346]= $-0.18809\, x\, y - 0.18809\, y^2$

(A.4) difOp: Differential operator where f is the bivariate function, pt is the point, i is the order of x, and j is the order of y. Variables x, y are assumed.

In[347]:=
```
difOp[f_, pt_, i_, j_] :=
    N[(1/Factorial[i]/Factorial[j] D[f, {x, i}, {y, j}]) /. Thread[{x, y} → pt]];
```

In[348]:=
```
f = 10 + 15 x + 9 x^2 - 9 y;
difOp[f, {-1/3, 2/3}, 1, 1]
difOp[x*y*f, {-1/3, 2/3}, 3, 1]
```

Out[349]= 0.

Out[350]= 9.

(A.4) dAss: A recursive subroutine.

In[351]:=
```
dAss[f_, pt_, 0] = <|{0, 0} → 0|>;
dAss[f_, pt_, n_] :=
    dAss[f, pt, n] = Append[dAss[f, pt, n - 1], Table[{i, n - i} → difOp[f, pt, i, n - i], {i, 0, n}]];
```

In[353]:=
```
f = 10 + 15 x + 9 x^2 - 9 y;
dAss[f, {-1/3, 2/3}, 0]
dAss[f, {-1/3, 2/3}, 2]
```

Out[354]=
```
<|{0, 0} → 0|>
```

Out[355]=
```
<|{0, 0} → 0, {0, 1} → -9., {1, 0} → 9., {0, 2} → 0., {1, 1} → 0., {2, 0} → 9.|>
```

(A.4) macaulayD1: A subroutine for a matrix of a single polynomial.

In[356]:=
```
macaulayD1[f_, p_, n_] := Module[{M},
        Table[Values[shiftFA[dAss[f, p, n - Total[q]], q, n]], {q, pExps[n - 1]}]];
```

In[357]:=
```
f = 10 + 15 x + 9 x^2 - 9 y;
macaulayD1[f, {-1/3, 2/3}, 2] // MatrixForm
```

Out[358]//MatrixForm=

$$\begin{pmatrix} 0 & 9. & -9. & 9. & 0. & 0. \\ 0 & 0 & 0 & 9. & -9. & 0 \\ 0 & 0 & 0 & 0 & 9. & -9. \end{pmatrix}$$

(A.4) macaulayMatrix: Macaulay matrix of pair of functions f, g at point p with order n.

```
In[359]:= macaulayMatrix[f_, g_, p_, n_] := Join[macaulayD1[f, p, n], macaulayD1[g, p, n]];
```

```
In[360]:= f = 10 + 15 x + 9 x^2 - 9 y;
g = 13 + 9 x - 21 y + 9 y^2
macaulayMatrix[f, g, {-1/3, 2/3}, 2] // MatrixForm

Out[361]= 13 + 9 x - 21 y + 9 y^2
```

Out[362]//MatrixForm=

$$\begin{pmatrix} 0 & 9. & -9. & 9. & 0. & 0. \\ 0 & 0 & 0 & 9. & -9. & 0 \\ 0 & 0 & 0 & 0 & 9. & -9. \\ 0 & 9. & -9. & 0. & 0. & 9. \\ 0 & 0 & 0 & 9. & -9. & 0 \\ 0 & 0 & 0 & 0 & 9. & -9. \end{pmatrix}$$

(A.4) intersectionMultiplicity: Intersection multiplicity of intersection point p of curves f, g. A default tolerance of 10^{-9} is assumed but may be changed in the code. This version does not, by default, print out intermediate values of null space, the depth, or the breadth, but the option HF → On will give this information.

```
In[363]:= Options[intersectionMultiplicity] = {HF → Off};
intersectionMultiplicity[f_, g_, p_, tol_, OptionsPattern[]] :=
  Module[{M, c, r, m, n, del, maxm, i, L, x, y},
    maxm = 20;
    If[Abs[f /. Thread[{x, y} → p]] > tol, Print["p not zero of f"]; Abort[]];
    If[Abs[g /. Thread[{x, y} → p]] > tol, Print["p not zero of g"]; Abort[]];
    del = 1;
    m = 1;
    L = {1};
    For[i = 1, i < maxm && del > 0, i++,
      M = macaulayMatrix[f, g, p, i];
      c = Dimensions[M][[2]];
      r = Length[Select[SingularValueList[N[M]], # > tol &]];
      n = c - r;
      del = n - m;
      m = n;
      AppendTo[L, m]];
    If[m == maxm, Print["Multiplicity Greater or equal to ", maxm]];
    If[OptionValue[HF] == On, Print["depth = ", i - 2, " HF = ", L]];
    m]
```

```
In[365]:= f = 10 + 15 x + 9 x^2 - 9 y;
g = 3 + 9 x - 21 y + 9 y^2;
intersectionMultiplicity[f, g, {-1/3, 2/3}, 1.*^-12]

Out[367]= 3
```

```
In[368]:= intersectionMultiplicity[f, g, {-1/3, 2/3}, 1.*^-12, HF → On]

depth = 2 HF = {1, 2, 3, 3}

Out[368]= 3
```

(A. 5) p2aMatrix: A subroutine for converting polynomial parametric functions to algebraic curves.

```
In[369]:= p2aMatrix[F_, d_, t_, x_, y_] := Module[{m, n, M, c, ca, ra},
   m = Max[Exponent[F[[1]], t], Exponent[F[[2]], t]];
   n = m * d;
   M = {};
   Do[c = CoefficientRules[Expand[(x ^ p[[1]] y ^ p[[2]]) /. Thread[{x, y} → F]], t];
    ca = Association[c];
    ra = Association[Table[If[MissingQ[ca[{k}]], {k} → 0, {k} → ca[{k}]], {k, 0, n}]];
    AppendTo[M, Values[ra]], {p, pExps[d]}];
   M];

In[370]:= PA = p2aMatrix[{t ^ 2, t ^ 3}, 3, t, x, y]

Out[370]= {{1, 0, 0, 0, 0, 0, 0, 0, 0, 0}, {0, 0, 1, 0, 0, 0, 0, 0, 0, 0},
  {0, 0, 0, 1, 0, 0, 0, 0, 0, 0}, {0, 0, 0, 0, 1, 0, 0, 0, 0, 0},
  {0, 0, 0, 0, 0, 1, 0, 0, 0, 0}, {0, 0, 0, 0, 0, 0, 1, 0, 0, 0}, {0, 0, 0, 0, 0, 0, 1, 0, 0, 0},
  {0, 0, 0, 0, 0, 0, 0, 1, 0, 0}, {0, 0, 0, 0, 0, 0, 0, 0, 1, 0}, {0, 0, 0, 0, 0, 0, 0, 0, 0, 1}}
```

(A.5) p2aEq: A subroutine that converts a sequence of coefficients to an equation. The typical use is to apply it to a row of the left null space of p2aMatrix.

```
In[371]:= p2aEq[u_, x_, y_] := Module[{P, d},
   If[Length[Dimensions[u]] > 1, Print["u must be a vector"]; Abort[]];
   d = Switch[Length[u], 3, 1, 6, 2, 10, 3, 15, 4, 21, 5, 28, 6, 36, 7, 45, 8, 55, 9];
   If[Length[d] > 1, Abort[]];
   P = pExps[d];
   FromCoefficientRules[Table[P[[k]] → u[[k]], {k, (d + 2) (d + 1) / 2}], {x, y}]];

In[372]:= (* uses matrix PA directly above *)
 NS = NullSpace[Transpose[PA]]

Out[372]= {{0, 0, 0, 0, 0, -1, 1, 0, 0, 0}}

In[373]:= p2aEq[NS[[1]], x, y]

Out[373]= x^3 - y^2
```

(A.6) singularFactor: Factors curve f when it contains too many singularities. Its use requires functions from Chapter 5, and it may be dependent on the tolerance tol.

```
In[374]:= singularFactor[f_, x_, y_, tol_] := Module[{h, csp, d, sp, M, M1, cr, c0, rhs, rr, XX, h1},
   csp = Chop[complexSingularPoints[f, x, y]];
   d = tDeg[f, x, y];
   If[d < 3, Print["degree must be at least 3"]; Abort[]];
   If[Length[csp] < ((d - 1) (d - 2) / 2 + 1), Print["Not enough affine singular points"];
    Abort[]];
   csp = RandomSample[csp, (d - 1) (d - 2) / 2 + 1];
   sp = If[d == 3, csp, Join[csp, realRegularPoints[f, x, y, d - 3]]];
   XX = pExps[d - 2];
   M = Table[Which[e == {0, 0}, 1, e[[1]] == 0, p[[2]] ^ e[[2]], e[[2]] == 0,
      p[[1]] ^ e[[1]], True, p[[1]] ^ e[[1]] * p[[2]] ^ e[[2]]], {p, sp}, {e, XX}];
   rr = Table[0, {Length[XX]}];
   rr[[-1]] = 1;
```

```
        M1 = Append[M, RandomReal[{-1, 1}, Length[XX]]];
        rhs = Table[0, {Length[XX]}];
        rhs[[-1]] = 1;
        cr = LeastSquares[M1, rhs];
        c0 = Min[Select[Abs[cr], # > .1 &]];
        h = FromCoefficientRules[Table[XX[[i]] → cr[[i]], {i, Length[XX]}], {x, y}];
        h = ComplexExpand[Re[h]];
        h1 = nGCD[f, h, x, y, tol];
        If[tDeg[h1, x, y] == 0, Print["GCD not found, Returning h"];
          Return[{h, 1}]];
        Chop[{h1, nDivide[f, h1, x, y, tol]}, 10*tol]];
```

In[375]:=
```
sf = Expand[(x^2+y^2-1) (x^2-y)]
```

Out[375]= $-x^2+x^4+y-x^2 y+x^2 y^2-y^3$

In[376]:=
```
singularFactor[sf, x, y, 1.*^-7]
```

Out[376]= $\{-0.808963\,x^2+0.808963\,y,\ 1.23615-1.23615\,x^2-1.23615\,y^2\}$

(A.6) singularFactorInf: Treats the case where some singularities are infinite.

In[377]:=
```
singularFactorInf[f_, x_, y_, tol_] := Module[{d, fh, f1, z, A, h, k},
    d = tDeg[f, x, y];
    A = iTransform[RandomReal[{-1, 1}, 3].{x, y, 1}, x, y];
    f1 = FLT[f, A, x, y];
    {h, k} = singularFactor[f1, x, y, tol];
    Chop[{FLT[h, Inverse[A], x, y], FLT[k, Inverse[A], x, y]}, 10*tol]
  ];
```

In[378]:=
```
singularFactorInf[x^3-x y, x, y, 1.*^-7]
```

Out[378]= $\{5.73976\,x,\ 0.174223\,x^2-0.174223\,y\}$

References

1 Abhyankar, Shreeram S. 1990. *Algebraic Geometry for Scientists and Engineers.* Providence, RI: American Mathematical Society.

2 Antonio, Igor C. "Colored Klein Bottle" from the Wolfram Demonstrations Project—A Wolfram Web Resource. demonstrations.wolfram.com/ColoredKleinBottle.

3 Bochnak, Jacek, Michel Costa and Marie-Francoise Roy. 1998. *Real Algebraic Geometry*. New York, NY: Springer.

4 Brown, Abby. "Conic Sections: Polar Equations" from the Wolfram Demonstrations Project—A Wolfram Web Resource. demonstrations.wolfram.com/ConicSectionsPolarEquations.

5 Burkert, Walter. 1972. *Lore and Science in Ancient Pythagoreanism*. Cambridge, MA: Harvard University Press.

6 Dayton, Barry H., Tien-Yien Li and Zhonggang Zeng. 2011. "Multiple Zeros of Nonlinear Systems." *Mathematics of Computation* 80, no. 276 (February): 2143–2168. doi.org/10.1090/S0025-5718-2011-02462-2.

7 Dickau, Robert. "Hilbert and Moore Fractal Curves" from the Wolfram Demonstrations Project—A Wolfram Web Resource. demonstrations.wolfram.com/HilbertAndMooreFractalCurves.

8 Fischer, Gerd. 2001. *Plane Algebraic Curves*. Translated by Leslie Kay. Providence, RI: American Mathematical Society.

9 Garrity, Thomas, Richard Belshoff, Lynette Boos, Ryan Brown, Carl Lienert, David Murphy, Junalyn Navarra-Madsen, Pedro Poitevin, Shawn Robinson, Brian Snyder and Caryn Werner. 2013. *Algebraic Geometry: A Problem Solving Approach*. Providence, RI: American Mathematical Society/Institute for Advanced Study.

10 Hafner, Izidor and Marko Razpet. "Examples of Zindler Curves" from the Wolfram Demonstrations Project—A Wolfram Web Resource. demonstrations.wolfram.com/ExamplesOfZindlerCurves.

11 Joseph, George Gheverghese. 1991. *The Crest of the Peacock: Non-European Roots of Mathematics*. Princeton, NJ: Princeton University Press.

12 Kozlowski, Andrzej. "Newton Polygon and Branching of Algebraic Curves" from the Wolfram Demonstrations Project—A Wolfram Web Resource. demonstrations.wolfram.com/NewtonPolygonAndBranchingOfAlgebraicCurves.

13 Leykin, Anton, Jan Verschelde and Ailing Zhao. 2006. "Newton's Method with Deflation for Isolated Singularities of Polynomial Systems." *Theoretical Computer Science* 359, no. 1–3 (August): 111–122. doi.org/10.1016/j.tcs.2006.02.018.

14 Macaulay, Francis Sowerby. 1916. *The Algebraic Theory of Modular Systems.* Cambridge, UK: Cambridge University Press.

15 Maes, Chris. "Newton's Method" from the Wolfram Demonstrations Project—A Wolfram Web Resource. demonstrations.wolfram.com/NewtonsMethod.

16 Martos, Eduardo J. Oliver. "Cassini Ovals" from the Wolfram Demonstrations Project—A Wolfram Web Resource. demonstrations.wolfram.com/CassiniOvals.

17 Naeemi, Ali. "Cross-Cap Surface and Variations" from the Wolfram Demonstrations Project—A Wolfram Web Resource. demonstrations.wolfram.com/CrossCapSurfaceAndVariations.

18 Nochella, Jesse. "3D Chebyshev Lissajous Figures" from the Wolfram Demonstrations Project—A Wolfram Web Resource. demonstrations.wolfram.com/3DChebyshevLissajousFigures.

19 Popescu-Pampu, Patrick. 2016. *What is the Genus*? New York, NY: Springer.

20 Sommese, Andrew J. and Charles W. Wampler, II. 2005. *The Numerical Solution of Systems of Polynomials Arising in Engineering and Science*. Hackensack, NJ: World Scientific.

21 Strang, Gilbert. 1976. *Linear Algebra and Its Applications*. Cambridge, MA: Academic Press.

22 Struik, Dirk Jan. 1986. *A Source Book in Mathematics, 1200-1800*. Princeton, NJ: Princeton University Press.

23 Uspensky, James Victor. 1948. *Theory of Equations*. New York, NY: McGraw-Hill.

24 Viro, Oleg. 1979. "Constructing Multicomponent Real Algebraic Surfaces." *Soviet Mathematics - Doklady* 20, no. 3: 991–995.

25 Viro, Oleg. 1979. "Construction of M-Surfaces." *Functional Analysis and Its Applications* 13, no. 3 (July–September): 212–213. doi.org/10.1007/BF01077489.

26 Viro, Oleg. 1980. "Curves of Degree 7, Curves of Degree 8 and Ragsdale Conjecture." *Soviet Mathematics - Doklady* 22, no. 2: 566–570.

27 Viro, Oleg. 2006. "Patchworking Real Algebraic Varieties." arxiv.org/abs/math/0611382.

28 Wolfram, Stephen. 2017. *An Elementary Introduction to the Wolfram Language*. Champaign, IL: Wolfram Media, Inc.

29 Zeng, Zhonggang and Barry H. Dayton. 2004. "The Approximate GCD of Inexact Polynomials," in *Proceedings of the 2004 International Symposium on Symbolic and Algebraic Computing*, Santander, Spain (July): 320–327.

Made in the USA
Monee, IL
07 May 2026

49714897R00217